Praziquantel

Thomas Spangenberg
Editor

Praziquantel

Discovery and Development of an Anthelmintic Drug

Springer

Editor
Thomas Spangenberg
Global Health R&D of Merck Healthcare
Ares Trading S.A., (a subsidiary of Merck KGaA, Darmstadt, Germany)
Eysins, Switzerland

ISBN 978-3-031-97396-3 ISBN 978-3-031-97397-0 (eBook)
https://doi.org/10.1007/978-3-031-97397-0

This Springer imprint is published by the registered company Springer Nature Switzerland AG
The registered company address is: Gewerbestrasse 11, 6330 Cham, Switzerland

To my curious daughters,
Sophia & Louisa

In memory of my brother, Robert

Preface

Discovering and developing new therapies is both scientifically stimulating and highly challenging. It draws from a wide range of disciplines, all working toward the common goal of delivering transformative drugs to patients in need. The journey is not without hurdles; adapting to newly generated evidence during a long and sustained effort where collaboration is central to success.

When dealing with infectious diseases, the challenges and rewards are no different. This book modestly intends to provide the next generation of curious minds with an insightful overview of the research and development (R&D) efforts behind praziquantel, a key anthelmintic drug.

Discovered half a century ago through a collaborative effort, praziquantel's history is deeply intertwined with the evolution of the pharmaceutical industry. To provide context, it is important to first explore how the industry has advanced over the centuries, shaping the processes that lead from initial discovery to patient care. The "bed-to-bench-to-bed" approach exemplifies this evolution, emphasizing a patient-centric mindset that informs R&D efforts and ultimately brings therapies to those in need. With this historical perspective in mind, we turn our attention to schistosomiasis, a disease provoked by infection with a parasitic worm that significantly impacts patients' lives. Understanding the landscape of available treatments and identifying unmet patient needs helps clarify the therapeutic goals driving the discovery and development of new interventions.

It is now time to provide an overview of praziquantel, a small molecule that has evolved into a widely used drug. The trip continues with the unresolved question: how does the drug work at the molecular level? This inquiry has led to a better understanding of praziquantel's anthelmintic spectrum and the potential for drug resistance. However, its antiparasitic properties have primarily been characterized at the phenotypic level—through studies conducted on whole organisms and in animal models.

Regardless of how active a compound is against the parasite; it cannot be considered a drug if it cannot enter the patient's body. Hence, praziquantel must be absorbed, reach the parasites, exert its antiparasitic effect, and then be eliminated. Equally important, a drug must be safe for patients, making its toxicological characterization essential for defining the therapeutic window.

Moving into the realm of drug development, we introduce a unique approach to assess the early efficacy of antischistosomal therapeutics: a controlled human schistosome worm infection. The next critical phase involves human clinical trials, where effectiveness of praziquantel in curing schistosomiasis patients is assessed in real-world settings. Given that pharmaceutical R&D is highly regulated to ensure patient safety and an appropriate benefit-risk ratio, these points are also discussed. Finally, we explore the intellectual property aspects of praziquantel, recognizing the costs associated with innovation.

Throughout this book, the reader will have the opportunity to rediscover an established drug with fresh eyes. To provide scientists with a logical approach to drug discovery and development, the topics are presented sequentially. Some aspects may be anachronistic, as they were not explored during the original development phase, while others may be outdated, reflecting changes in practices or regulations over time.

After decades of effort, the R&D saga of praziquantel is coming to an end and I hope this book will spark your interest, deepen your understanding of your peers' work, and inspire your own discoveries—contributions that could ultimately elevate humanity.

Geneva, Switzerland
January 18, 2025

Thomas Spangenberg

Competing Interests

Open access was funded by Merck KGaA, Darmstadt, Germany.

Acknowledgments

I would like to express my gratitude to Dr. Diego González Cabrera for his unwavering support throughout this project, to Dr. Sabine Bernschneider-Reif for guiding me through Merck's rich corporate history, and to Marc Lemuel Gonzales (@sinaglemuel) for designing the book cover and selected figures. I would like to extend my warmest thanks to the contributors of this book, the colleagues who reviewed the chapters, and my wife, Sarah, for her constants encouragement throughout this endeavor.

Contents

Editor and Contributors

About the Editor

Thomas Spangenberg With Merck since 2015, Thomas is responsible for global health research. He leads drug discovery programs from drug screening to preclinical candidate nomination. Through open innovation, he supports malaria and schistosomiasis assets in clinical development with translational research including the generation of Real-World Data and preclinical evidences.

Before joining Merck, Thomas worked in the discovery team of the non-for-profit R&D organization Medicines for Malaria Venture (Switzerland) and led the open-source Malaria Box and Pathogen Box initiatives designed to catalyze drug discovery research in neglected diseases.

Thomas holds a PhD in organic chemistry from the Universities of Strasbourg (France) and Freiburg im Breisgau (Germany). In 2009, he was appointed as a post-doctoral fellow at Harvard University (USA) where he contributed to the total synthesis of the mycolactones and the development toward the point-of-care diagnosis of Buruli ulcer, a neglected necrotizing skin disease.

He authored and co-authored over 50 peer-reviewed articles and is a co-inventor on several patents.

Contributors

Wilhelmina M. Bagchus Retired from the Merck Institute for Pharmacometrics, Ares Trading S.A., (an affiliate of Merck KGaA, Darmstadt, Germany), Lausanne, Switzerland

Peter Ballard PB DMPK Consulting Ltd., Chinley, UK

Christian Brenneis Chemical and Preclinical Safety, Merck Healthcare KGaA, Darmstadt, Germany

Diego González Cabrera Independent, Gran Canaria, Spain

Nicola Caldwell Wellcome Centre for Anti-Infectives Research, Division of Biological Chemistry and Drug Discovery, School of Life Sciences, University of Dundee, Dundee, UK

Clarissa Prazeres da Costa Institute for Medical Microbiology, Immunology and Hygiene, Center for Global Health, TUM School of Medicine and Health, Technical University Munich, Munich, Germany

Center for Global Health, TUM School of Medicine and Health, Technical University Munich, Munich, Germany

Claudia Demarta-Gatsi Global Health R&D of Merck Healthcare, Ares Trading S.A., (a subsidiary of Merck KGaA, Darmstadt, Germany), Eysins, Switzerland

Ian Gilbert Wellcome Centre for Anti-Infectives Research, Division of Biological Chemistry and Drug Discovery, School of Life Sciences, University of Dundee, Dundee, UK

Arno Hartmann Merck Healthcare KGaA, Darmstadt, Germany

Philip Hewitt Chemical and Preclinical Safety, Merck Healthcare KGaA, Darmstadt, Germany

Marie-Astrid Hoogerwerf Leiden University Center for Infectious Diseases, Leiden University Medical Center, Leiden, The Netherlands

Nicole Hübler Chemical and Preclinical Safety, Merck Healthcare KGaA, Darmstadt, Germany

Jennifer Keiser Department of Medical Parasitology and Infection Biology, Swiss Tropical and Public Health Institute, Allschwil, Switzerland

University of Basel, Basel, Switzerland

David Maillard Central Process Development – Downstream Processing Services, Merck Electronics KGaA, Darmstadt, Germany

Jonathan S. Marchant Department of Cell Biology, Neurobiology and Anatomy, Medical College of Wisconsin, Milwaukee, WI, USA

Mathew Njoroge Holistic Drug Discovery and Development Centre (H3D), University of Cape Town, Rondebosch, South Africa

Sang-Kyu Park Department of Cell Biology, Neurobiology and Anatomy, Medical College of Wisconsin, Milwaukee, WI, USA

Joachim Richter Institute of International Health, Charité, Berlin, Germany

Center for Global Health, Charté Universitätsmedizin, Berlin, Germany

Swiss Tropical and Public Health Institute, Allschwill, Switzerland

Meta Roestenberg Leiden University Center for Infectious Diseases, Leiden University Medical Center, Leiden, The Netherlands

Katja Schüz Merck Healthcare KGaA, Darmstadt, Germany

Paul M. Selzer Eberhard Karls University Tübingen, Interfaculty Institute of Biochemistry, Tübingen, Germany

Thomas Spangenberg Global Health R&D of Merck Healthcare, Ares Trading S.A., (a subsidiary of Merck KGaA, Darmstadt, Germany), Eysins, Switzerland

Daniel J. Sprague Department of Cell Biology, Neurobiology and Anatomy, Medical College of Wisconsin, Milwaukee, WI, USA

Yara Sylla Global Project Leadership Department of Merck Healthcare KGaA, Darmstadt, Germany

Stefan Weigt Chemical and Preclinical Safety, Merck Healthcare KGaA, Darmstadt, Germany

1 A Brief History of the Pharmaceutical Industry

Diego González Cabrera, Paul M. Selzer, and Thomas Spangenberg

Abstract

The evolution of the pharmaceutical industry has been shaped by groundbreaking discoveries and significant challenges. Before 1900, medicines were primarily derived from natural remedies, and there was limited understanding of infectious diseases. The development of antitoxins and vaccines laid the foundation for the growth of the pharmaceutical sector. This was followed by the introduction of synthetic dyes and their by-products, which played a vital role in creating early chemotherapeutic agents used during wartime to treat infections. These advancements set the stage for the transformative post-World War II era, characterised by rapid progress in research, technology, and the implementation of stringent regulations to ensure drug safety. In the 1960s and 1970s, significant advancements in drug discovery, analysis, and production resulted in the development of numerous new medications, including antiparasitic drugs essencial for both animal and global health. Together, these milestones highlight the pharmaceutical industry's enduring impact on public well-being and modern drug development.

D. G. Cabrera (✉)
Independent, Gran Canaria, Spain

P. M. Selzer
Eberhard Karls University Tübingen, Interfaculty Institute of Biochemistry, Tübingen, Germany

T. Spangenberg
Global Health R&D of Merck Healthcare, Ares Trading S.A., (a subsidiary of Merck KGaA, Darmstadt, Germany), Eysins, Switzerland

T. Spangenberg (ed.), *Praziquantel*,
https://doi.org/10.1007/978-3-031-97397-0_1

Keywords

History · Pharmaceutical industry · Infectious diseases · Chemotherapeutics · Vaccine · Antitoxins · Vitamins · Antibiotics · Regulation · Drug discovery and development

Medicines Before 1900

For centuries, medical treatments were often concocted through a blend of empiricism and prayer. Mummy powder was used extensively as a remedy for bruising, and the corpses of healthy and young humans were considered as elixir of life [1] (Fig. 1.1).

Other practices, such as bloodletting and administering potent purgatives and cathartics, were common based on the belief that removing harmful "bodily fluids" would restore health. In response to these practices and fueled by the technological advancements of the Industrial Revolution, chemists began to extract and concentrate plant-based remedies from traditional medicine. They also started modifying dyes and by-products from coal-tar—a thick, dark liquid containing compounds like benzene and naphthalene—leading to the creation of synthetic dyes. Some of these dyes were also investigated for their potential therapeutic effects [2].

Therefore, the modern pharmaceutical industry has its roots in two primary sources. On one hand, apothecaries transitioned into the wholesale production of drugs derived from natural products, such as morphine, quinine, and strychnine. For

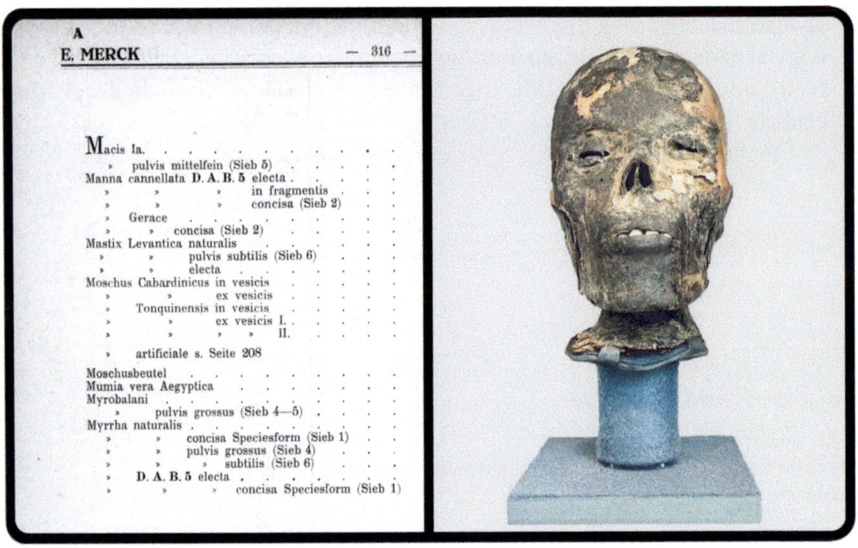

Fig. 1.1 "*Mumia vera Aegyptica*" was for sale by E. Merck, Darmstadt, Germany, as indicated in the price list until 1923 (Merck Corporate History)

Fig. 1.2 Natural product extracts (Merck Corporate History)

example, Merck (Darmstadt, Germany), founded in 1668, began large-scale drug production in the 1820s. On the other hand, dye and chemical companies, such as Bayer (Barmen, Germany) and Hoechst (Höchst, Germany), both founded in 1863, which established research laboratories and began identifying medical uses for their products. Scientific breakthroughs, technological advances, and legislative initiatives throughout the nineteenth century further accelerated the growth of the pharmaceutical industry [3] (Fig. 1.2).

Awareness of Infectious Diseases

As early as 430 BC, it was widely known that survivors of smallpox gained immunity to the disease. Before the discovery of vaccination, the most effective method to combat smallpox was inoculation (from the Latin word *inoculare*, meaning "to graft"), also known as variolation. This procedure involved taking fresh matter from a ripe pustule of an infected person with a "mild" case of the disease and introducing it subcutaneously into the arms or legs of a nonimmune individual. There were, however, concerns that 2–3% of recipients might die from the disease, spread it to others, or contract other illnesses such as tuberculosis and syphilis through the procedure. Inoculation was likely practiced in Africa, India, and China long before the eighteenth century, when Lady Mary Wortley Montague introduced it to Europe and, despite its risks, it quickly gained popularity.

The practice of using smallpox and cowpox to prevent disease was widely common among country physicians in the dairy regions of eighteenth-century England. A major breakthrough came in 1796, when Edward Jenner inoculated an 8-year-old boy with material from a dairymaid's fresh cowpox lesions. The boy developed only a mild reaction, and when Jenner later inoculated him with matter from a fresh smallpox lesion, the boy did not develop the disease. Jenner named this new procedure vaccination, from the Latin word for cowpox, *vaccinia*. His work marked one of the first scientific efforts to initiate the body's immune response against harmful microorganisms. Vaccination rapidly spread in England and, by 1800, it had reached most European countries and the United States of America (USA). Over time, it replaced variolation, which was officially prohibited in England in 1840 [4].

The handwashing theory, introduced by Ignaz Semmelweis in 1847 to prevent sepsis, initially faced opposition from doctors who doubted that unwashed hands could transmit disease from deceased or sick patients to healthy individuals. Less than two decades later, Joseph Lister's pioneering work on antisepsis laid essential groundwork for the emerging germ theory of disease. Subsequent breakthroughs by Louis Pasteur and Robert Koch solidified microorganisms as significant contributors to diseases. The discovery of *Escherichia coli* in 1885 offered researchers a safe and cultivable bacterium for laboratory studies. Paired with advancements in growth media, sterile techniques, and bacterial isolation methods, significantly enhanced research capabilities. Additionally, the development of new instruments by lens makers and microscopists, in collaboration with laboratory scientists, further propelled scientific progress [5]. Pasteur's development of the attenuated rabies vaccine in 1885 was also a significant milestone in the fight against infectious diseases. This breakthrough paved the way for a rapid succession of vaccine developments targeting various pathogens in the subsequent decades. Around the same time, in 1890, Emil von Behring and Shibasaburo Kitasato isolated the first antitoxins - antibodies effective against tetanus and, shortly thereafter, diphtheria. These antitoxins proved to be effective cures when injected into infected individuals [6]. Together with vaccines, they laid the foundation for a modern pharmaceutical industry, leading to the decline of older medical practices. Despite these advancements, rural physicians, especially in the USA, continued to recommend bleeding and purgatives as cures well into the early twentieth century [5].

Dyes as Therapeutics

Along with advancements in biology, chemistry began to play a transformative role in medicine in 1856. This pivotal moment occurred when William Perkin, while attempting to synthesize the antimalarial natural product quinine, accidentally discovered mauveine, the first synthetic coal tar dye. The discovery spurred widespread interest in coal tar derivatives, leading to the development of numerous synthetic dyes - some of which were later found to have therapeutic properties [7]. One such compound, hexamethyl pararosaniline chloride (crystal violet) became a widely used topical antiseptic and continues to be used as a stain for visualizing cells and tissues under a microscope, playing a key role in the Gram staining procedure. Phenazone, produced from quinoline derivatives in 1883, was one of the first antipyretic and anti-inflammatory drugs to be sold in a ready-dosage, prepackaged form. Antifebrin, an antifever drug derived from aniline, and phenacetin, introduced in 1887 as a pain reliever, were also notable developments, though phenacetin was withdrawn in the 1970s due to its carcinogenic effects. The nineteenth century

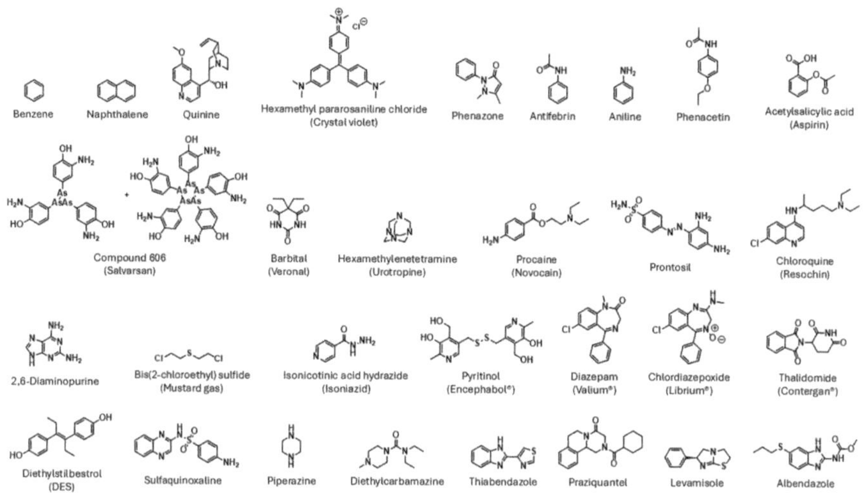

Fig. 1.3 Chemical structures of selected chemicals and drugs

concluded with a landmark achievement in pharmaceutical chemistry: the development of acetylsalicylic acid (aspirin) by Felix Hoffman at Bayer [8] (Fig. 1.3).

Legislative reforms played a pivotal role in the expansion of the pharmaceutical industry. These included stricter quality and safety controls, the regulation of drug labeling, and the establishment of clear distinctions between prescription and over-the-counter medications. New regulatory bodies were created in both Europe and the USA to manage the growing volume of vaccines and antitoxins being developed. Additionally, regulatory measures were implemented to establish stricter qualifying examinations and create a registry of chemists and druggists. Many of the more advanced pharmaceutical companies originated in Germany, then the global leader in this domain, where scientists from major firms such as Bayer, Hoechst and Merck, often working in collaboration with academia, conducted extensive fundamental research [8] (Fig. 1.4).

First Chemotherapeutics

Paul Ehrlich, recognized as one of the foremost scientists of his time, introduced the innovative "magic bullet" theory in 1906 [9]. This concept proposed that synthetic chemicals could be designed to selectively target and neutralize disease-causing microorganism - such as parasites and bacteria - without harming the host, laying the foundation for modern targeted therapies. Ehrlich's work was supported by empirical data from trials involving vaccines, antitoxins, and the study of chemical dye structures. His pivotal testing of synthetic dyes for trypanocidal activity in mice, aimed at facilitating the importation of European horses and cattle into Germany's

Fig. 1.4 Various products on Pyoktanin® (crystal violet) from E. Merck, Darmstadt, Germany (Merck Corporate History)

African colonies, marked the beginning of screening synthetic compounds for therapeutic use in animal health. This approach culminated in 1910 with the introduction of compound 606, an arsenic-based drug marketed by Hoechst as Salvarsan, which became the first systematically developed and effective treatment for syphilis [3].

Pharmaceuticals During War Times

Although patents and key production methods for veronal—a potent hypnotic commonly used to alleviate battlefield injuries—had already been established in the USA, the outbreak of World War I in 1914 created an urgent need for self-sufficiency in medical supplies. With access to German pharmaceuticals disrupted, the USA and several European nations were compelled to intensify the production of critical drugs and replicate German manufacturing processes for aspirin and Salvarsan. To address shortages, committees were established to identify medicines that could be sourced from domestic or alternative suppliers or required full-scale domestic production. For instance, within a year, Burroughs Wellcome in the United Kingdom (UK) developed a substitute for Salvarsan and began manufacturing versions of aspirin and urotropine, an essential drug for treating typhoid. Additionally, many university laboratories contributed to the production of key drugs, including Novocain and other anesthetics [10].

After successfully replacing key German therapeutics and introducing new techniques for drug development, pharmaceutical firms in the USA, Europe, and Japan experienced rapid growth. During this period, significant investments were made in research, development, and marketing. Companies expanded their in-house research and development capabilities while continuing to collaborate with academic researchers. Positioned at the crossroads of scientific innovation, medical need, and expanding healthcare markets, the pharmaceutical sector was poised for continued growth. However, the 1918 influenza pandemic revealed the limitations of medical science in combating certain infectious diseases [6].

During the 1920s and 1930s, significant advancements in scientific research and pharmaceutical technology led to groundbreaking discoveries. This period saw the emergence of synthetic vitamins, sulfonamides, antibiotics, hormones, psychotropics, antihistamines, and new vaccines, significantly expanding our understanding of body chemistry and therapeutic interventions. One of the most pivotal achievements was the isolation and development of insulin, which became widely available due to advancements in bulk production and purification techniques [11]. Around the same time, hormones such as estrogen and testosterone were also synthesized and made accessible to patients. Despite advances in medical treatments, there remained a significant need for more effective therapies for bacterial infections. In 1935, Gerhard Domagk discovered the antibacterial properties of Prontosil, an orange-red azo dye later discovered to contain a sulfonamide group. This discovery marked the introduction of one of the first synthetic antibacterial drugs, effective against certain gram-positive bacterial infections, such as those caused by streptococci. Although thousands of sulfonamide derivatives were synthesized in subsequent years, only 15 proved clinically effective [12] (Fig. 1.5).

In 1928, Alexander Fleming's observation of a large blue-green colony of mold killing bacteria on an agar plate led to the discovery of penicillin. Although Fleming

Fig. 1.5 Dr. Theodor Moll's laboratory in the Department of Bacteriology where penicillin research was conducted during the war (1930)

published his findings in 1929, it was not until the late 1940s that penicillin saw extensive use in the medical community [2]. Around the same time, significant progress was made in virology, with biochemist Wendell Stanley successfully crystallizing the tobacco mosaic virus in 1935. However, antiviral research did not achieve the same level of success as antibiotic development. This period also saw growing recognition of the role of vitamins in health, with studies on vitamins A, D, and K gaining momentum. This led to a significant increase in vitamin consumption, fueled by widespread belief in their extraordinary benefits, creating a lucrative opportunity for the pharmaceutical industry [12] (Fig. 1.6).

Despite these advancements, the pharmaceutical industry was subject to very few regulations during the 1930s. Most medicines were sold without a prescription, with nearly half compounded locally by pharmacists. Vitamins and therapeutics faced minimal oversight, leading to tragedies such as the elixir of sulfanilamide incident, in which a toxic antifreeze substance was mixed with sulfanilamide, resulting in the deaths of 100 people, including children. In response, new legislation, such as the 1938 Food, Drug, and Cosmetic Act, required safety testing for new drugs and, in many cases, mandated prescriptions in the USA for their dispensing. However, it was not until the 1960s, with the Kefauver-Harris Amendments, that manufacturers were required to substantiate the claims about the effectiveness of their products [3].

In 1939, Rene Dubos, working at the Rockefeller Institute for Medical Research, achieved a significant milestone by isolating an antibiotic called tyrothricin from a soil bacterium, *Bacillus brevis*. While penicillin had already been discovered, Dubos' work represented one of the earliest successes in identifying antibiotics through systematic screening of soil bacteria. This substance exhibited the ability to kill Gram-positive bacteria and is considered the first intentionally discovered antibiotic. This discovery marked the culmination of extensive global efforts involving the screening of thousands of soil samples to identify antimicrobial agents [12]. A surge of antibiotic discoveries followed, including streptomycin (MSD), chlortetracycline (Lederle), chloramphenicol (Parke-Davis), erythromycin (Abbott and Eli Lilly), and tetracycline (Pfizer). The commercial success of these antibiotics brought substantial profits to pharmaceutical companies, enabling them to invest in large-scale research facilities in pursuit of further breakthroughs [13]. In addition to antibiotics, the 1930s also saw the development of chloroquine, a widely used antimalarial drug originally synthesized by Bayer under the name Resochin. Chloroquine's effectiveness against malaria became widely recognized, during World War II. However, it was not until 1946 that the United States Food and Drug Administration (FDA) formally approved its use [12].

World War II spurred major advancements in the mass production and refinement of pharmaceutical instruments and processes, revolutionizing research and development. Wartime efforts significantly accelerated the large-scale manufacturing of penicillin, antimalarials, cortisone, and other antibacterials, marking the true onset of the antibiotic era. Chemists Ernst Boris Chain, Norman G. Heatley, Edward P. Abraham, and others from the Dunn School played a pivotal role in supplying Howard Walter Florey with sufficient penicillin for the clinical trials conducted at

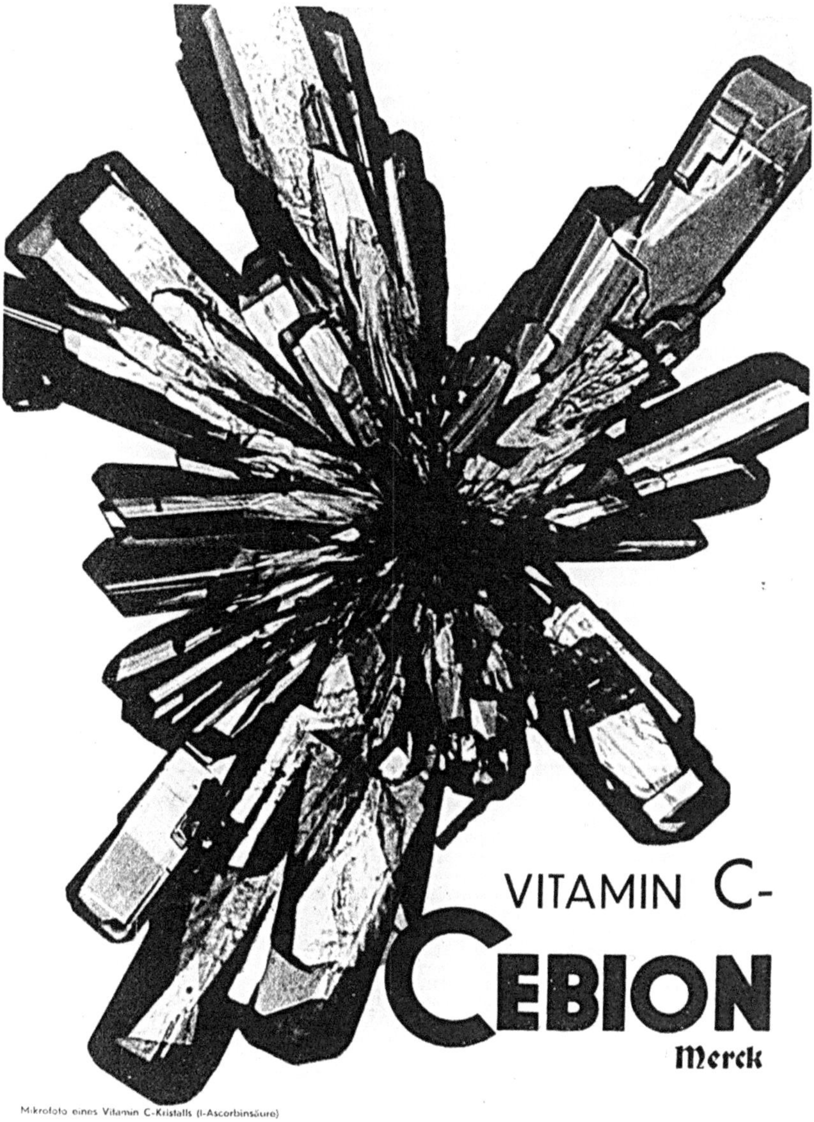

Fig. 1.6 Advertisement on its Vitamin C product Cebion® on the 1930s, E. Merck, Darmstadt, Germany

the Radcliffe Infirmary in Oxford in 1941. The initial improvement of the first patient - suffering from *Staphylococcus aureus* and *Streptococcus pyogenes* infections - highlighted penicillin's potential. However, due to the limited availability, the patient eventually succumbed after experiencing a relapse. This prompted urgent efforts to increase penicillin production. By 1943, several US pharmaceutical companies were mass-producing purified penicillin to treat war-related injuries and diseases, including syphilis and gonorrhea. Production rates soared, reaching over 600 billion units per month by 1944, and continued to rise during 1945, significantly aiding wartime medical care [14].

World War II also prompted the need to define ethical boundaries for medical experiments on human subjects, leading to the establishment of the Nuremberg Code in 1947. This document mandated voluntary consent, scientific qualification for researchers, the minimization of unnecessary suffering, and the protection of subjects from harm, setting ethical standards for human experimentation. Simultaneously, significant therapeutic breakthroughs were made for a range of diseases, including the synthesis of new antibiotics, cancer treatments, and antiviral agents. Notable advances included the synthesis of 2,6-diaminopurine by Albert Hitchings and Gertrude Elion, which contributed to the development of chemotherapy agents and the development of mustard gas derivatives, such as mechlorethamine, for cancer treatment. Progress in experimental genetics, biology, and virology were exemplified by the work of John Franklin Enders, Thomas H. Weller, and Frederick Chapman Robbins, who in 1949 developed a technique to grow poliovirus in human tissue cultures - an achievement that revolutionized virology research. Concurrently, analytical innovations such as partition chromatography, infrared and mass spectrometry, and nuclear magnetic resonance (NMR) transformed pharmaceutical science by enabling precise drug analysis, structural identification, and the design of novel therapeutic agents, heralding a new era of progress [14].

Post-World War II Era

After World War II, an unprecedented baby boom fueled mass consumerism, with a focus on housing, appliances, automobiles, and luxury goods. Declassified scientific advances found new civilian applications in technologies such as silicone-based products, microwave ovens, and radar, while other wartime innovations, such as plastics, nylon stockings, long-playing vinyl records, and computing devices, became more widely adopted. Massive government spending on science and technology, particularly in the USA, propelled innovation across various sectors, including healthcare, defense, and consumer goods. In the pharmaceutical industry, the mechanization of production enabled large-scale drug manufacturing, while emerging media like radio and television facilitated mass-marketing. The quest for improved antibiotics, which began in the 1940s, continued into the postwar years, leading to significant breakthroughs in controlling bacterial infections. These advances included both new natural products and synthetically modified compounds such as nystatin, erythromycin, novobiocin, and kanamycin, which broadened the therapeutic arsenal against a wide range of bacterial diseases [15].

Substantial funding for science, including medical research, flowed into the university sector, nurturing a highly trained workforce essential in transforming the industry. Mechanized mass-production techniques honed during World War II expanded to meet the soaring demand for therapeutics. Increased private funding, driven by philanthropic institutions and public contributions, further supported scientific research aimed at curing diseases. The development of the polio vaccine by Jonas Salk in 1954, predominantly funded by the public, exemplified this trend. The 1950s also witnessed the discovery of the first effective tuberculosis, isoniazid, identified after the screening 5000 compounds. Isoniazid proved particularly effective when used as part of a combination therapy with streptomycin and *para*-aminosalicylic acid (PAS) [16].

This decade saw a surge in new instrumentation and analytical techniques that transformed medical research and pharmaceutical development. Innovations such as gas–liquid partition chromatography, silicon transistors (which enhanced analytical instruments and early computers), electromagnetic microbalances, and medical research reactors for cheaper production of medical isotopes greatly expanded scientific capabilities. Techniques such as automated bacterial colony counting, polyacrylamide gel electrophoresis (PAGE) for nucleic acid and protein analysis, and X-ray diffraction also emerged during this period [15]. Using X-ray diffraction, Rosalind E. Franklin produced images essential to James Watson and Francis Crick's structural elucidation of desoxyribonucleic acid (DNA). Additionally, discoveries of molecules such as polynucleotide phosphorylase, ribosomes, DNA polymerase, and transfer ribonucleic acid (RNA) paved the way for Crick's theory of the one-way transfer of genetic information from DNA to proteins. In another breakthrough Frederick Sanger determine the complete amino acid sequence of bovine insulin's two polypeptide chains, the B chain in 1951 and the A chain in 1952. Revealing a 30-amino acid B chain and a 21-amino acid A chain connected by disulfide bonds. This achievement marked a significant advancement in understanding protein structure and function [11].

Golden Age of the 1960s

Building on the technological foundations laid during and immediately after World War II, the 1960s experienced a surge in innovation that transformed both industry and daily life. This period also saw major advances in drug discovery, analytical techniques and pharmaceutical manufacturing, resulting in the development of numerous new medications for contraception, mental health, blood pressure regulation, and disease prevention. A notable example is pyritinol, developed by E. Merck, Darmstadt, Germany, in 1963 by linking two pyridoxine molecules with a disulfide bond. Originally used as a psychostimulant with minimal adverse effects, pyritinol's applications later expanded to the treatment of several cognitive disorders across various countries (Fig. 1.7).

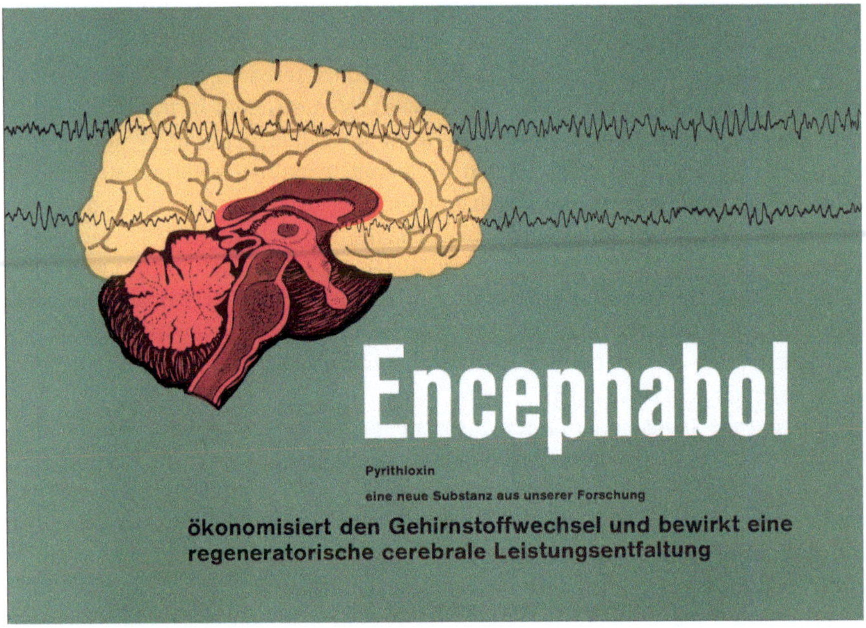

Fig. 1.7 Advertisement for Encephabol® (Pyritinol)

The introduction of the oral contraceptive pill revolutionized birth control practices. At the same time, central nervous system (CNS) drugs from the benzodiazepine family, such as Valium and Librium, quickly became the most prescribed medications in history, thanks to mass production and widespread marketing. Major advancements in medical science also transformed healthcare practices. Organ transplantation became increasingly feasible with the development of the first immune-suppressant drugs, which enabled routine procedures. In 1964, the introduction of intravenous nourishment techniques marked a major improvement in patient care. That same year, the development of a new measles vaccine represented a major advance in preventive medicine, followed by the discovery of a live-virus mumps vaccine 3 years later. Scientists successfully deciphered the genetic code, uncovering the complex relationship between mutations in DNA and the resulting changes in protein structure and function. Additionally, the introduction of automated systems for analyzing peptides and nucleic acids accelerated research into cellular and genetic mechanisms [17]. Innovations in chromatography led to the development and refinement of techniques such as high-performance liquid chromatography (HPLC), capillary gas chromatography (GC), and gas chromatography/mass spectrometry (GC/MS). Meanwhile, thin-layer chromatography (TLC) became widely adopted. These techniques, combined with the transformative impact of proton NMR in molecular structural analysis, significantly improved the precision and speed of chemical characterization. Advances in tissue culture techniques strengthen in vitro testing, allowing for more detailed and reliable assessments of compounds. Additionally, powerful computer systems became integral to laboratory workflows, particularly in molecular analysis, greatly improving efficiency and accuracy in research and experimentation [17].

This decade marked a pivotal shift in drug development practices, driven by a stronger emphasis on protecting human participants in clinical trials. The World Medical Association's 1964 Declaration of Helsinki introduced ethical principles for medical research, notably requiring informed consent for all trial subjects. This landmark declaration prompted pharmaceutical companies to allocate greater resources to preclinical and clinical testing to align with these new standards [3]. At the same time, public demand for protections against unsafe or fraudulent medications spurred increased government involvement in drug development, manufacturing, and marketing.

Thalidomide, introduced in the late 1950s and early 1960s as a sedative and treatment for pregnancy-related nausea, was widely marketed across Europe, Canada, South America, and parts of Asia. In 1961, it was linked to severe birth defects, particularly limb deformities in approximately 10,000 children worldwide. This tragedy underscored the urgent need for rigorous clinical testing and stringent regulatory oversight of new drugs. In Germany, where thalidomide use was widespread, the 1961 drug safety law was reinforced, and by 1964, new regulations required the registration of all new medications and their classification as prescription drugs. In the USA, the drug was never approved thanks to FDA researcher Frances Kelsey, who withheld authorization due to safety concerns. Public outrage over the European cases prompted the US Congress to pass the Kefauver-Harris Amendment in 1962, which significantly strengthen drug safety requeriments[18]. In response to increasing concerns about drug safety, many countries introduced legislation to reinforce drug regulation including clearer distinctions between over the counter (OTC) and prescription medications. As a result, pharmaceutical companies increasingly focused on developing prescription drugs, which offered greater profitability. This shift, combined with advances in research and technology, led to greater specialization and innovation in drug development, particularly in areas such as antibiotics, psychotropics, and chronic disease treatments [2].

Modern Times

In the 1970s, the pharmaceutical industry began expanding significantly, laying the groundwork for the modern mega-industry that exits today. Although drug discovery and testing methodologies were still evolving, it was in the subsequent decades that transformative advances such as biotechnology and the emergence of a more targeted approach known as rational drug design began to revolutionize the process. Rather than relying solely on trial-and -error screening, rational drug design introduced a methodical strategy for creating molecules with desired biological functions. This approach relies on the ability to predict how a molecule's structure affects its behavior using computational models and biological assays. Through iterative design and testing, researchers can modify lead compounds to

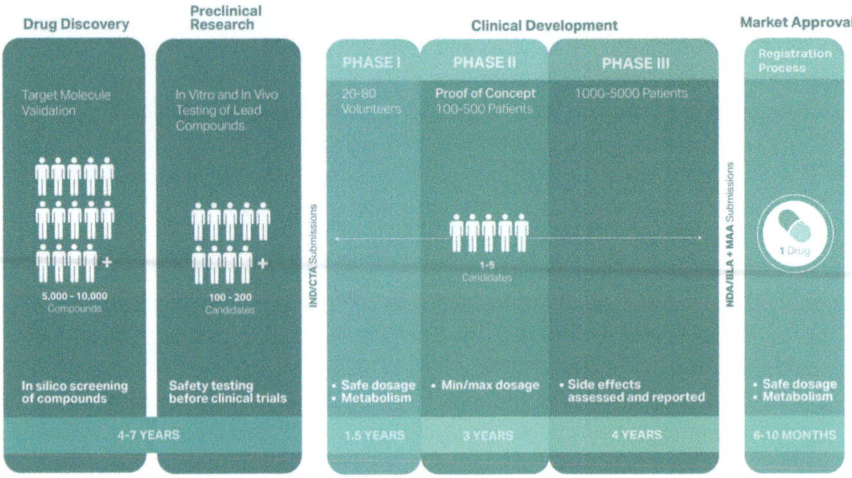

Fig. 1.8 Process of drug discovery, development, and registration with timeline

enhance selectivity, potency, and safety. As a result, rational drug design has become a foundational tool in modern drug discovery and development process, which typically follows a structured, multi-step process (Fig. 1.8):

Drug discovery The drug discovery process begins with basic research aimed at understanding disease mechanisms and identifying potential therapeutic targets, such as proteins, enzymes, or receptors. This stage involves various approaches, including cellular assays, genomic profiling, and proteomic analyses. Scientists then search for small molecules, typically organic compounds with molecular weights below 900 Da (Chap. 4) that can modulate these targets. During the discovery phase, thousands to millions of small molecules—sometimes even billions using computer-aided design—are tested in various biological assays. Promising candidates are then evaluated in animal models of human diseases. In parallel, absorption, distribution, metabolism, and elimination (ADME) studies are conducted to assess the pharmacokinetic (PK) properties of the compounds (Chap. 7).

Preclinical development Preclinical development focuses on understanding the mechanism of action, assessing toxicity, and identifying the most suitable drug formulation. This stage involves both in vitro and in vivo studies to evaluate efficacy and safety. As preparations for clinical trials move forward, an application is submitted to regulatory agencies. This comprehensive document includes data from animal efficacy studies, toxicity assessments conducted under Good Laboratory Practice (GLP) standards (Chap. 8), manufacturing details (Chap. 4), and proposed clinical trial protocols. Upon approval by regulatory authorities, clinical trials can begin (Chap. 11).

Clinical development This stage involves the systematic evaluation of the drug candidate in humans to determine safety, efficacy, and optimal dosing. Clinical development typically progresses through the following phases (Chap. 10):

- *Phase I*: The safety and tolerability of the drug candidate are evaluated in a small group, typically ranging from 20 to 80 healthy volunteers. The primarily goals are to assess safety, pharmacokinetics, and potential for drug accumulation. The data collected help establish a safe starting dose and informs dosing adjustments for later phases.
- *Phase II*: Involving 100 to 500 patients, often across multiple sites. Trials are usually randomized and controlled, aiming to establish whether the drug produces the intended therapeutic effect.
- *Phase III*: Conducted in a large and diverse population, typically from 1000 to 5000 participants, often across multiple sites. These trials provide comprehensive data on efficacy and safety under real-world conditions. Phase III trials are resource-intensive and serve as the primary basis for regulatory approval.

Market approval Following successful clinical trials, a New Drug Application (NDA) or Biologics License Application (BLA) is submitted to regulatory authorities containing comprehensive data from both preclinical and clinical studies for evaluation (Chap. 11). If the review confirms that the drug meets the required standards of safety, efficacy, and quality, marketing authorization is granted. As part of this approval, labeling requirements are established to define the drug's approved indications, dosage instructions, potential side effects, and necessary warnings. Once authorized, the drug can be marketed and prescribed to patients [19]. The drug discovery and development process is characterized by its lengthy duration, high costs, and substantial attrition rate, with many candidates failing at different stages. It encompasses a wide range of scientific disciplines, including biochemistry, pharmacology, and clinical research, and requires years of dedicated effort and significant financial investment. Additionally, it entails the acquisition and utilization of vast amounts of data, often generated through high-throughput technologies, to refine and optimize drug candidates.

Continuing with the 1970s, significant progress was made in combating infectious diseases, yet cancer treatment continue to pose major challenges. In response, additional funding was directed toward cancer research, particularly in basic science and prevention strategies. By 1979, scientists had developed techniques to use DNA from malignant cells to induce tumor formation in cultured mouse cells, providing a powerful model for cancer investigation. Despite these advances, surgery remained the primary treatment for well-defined tumors in accessible areas, while chemotherapy and radiation were employed to shrink tumors, alleviate symptoms, and prolong survival [20]. A major public health achievement of the decade was the eradication of smallpox. In 1974, the World Health Organization (WHO) launched the Expanded Program on Immunization (EPI), to protect children against preventable diseases such as polio, measles, diphtheria, whooping cough, tetanus, and tuberculosis. That same year, the WHO published the first Model List of Essential Medicines, marking a critical step toward equitable access to essential treatments worldwide. Regulatory practices also evolved in response to safety concerns. Following the discovery that diethylstilbestrol (DES) caused birth defects and reproductive issues in offspring of pregnant women, regulatory agencies such as the FDA introduced stringent clinical guidelines. These included recommendations to exclude women of childbearing potential from early-phase trials, except in cases of life-threatening illnesses. This marked a significant shift toward standardized government-regulated clinical testing. As a result of these stricter regulations, a delay emerged between drug approvals in other countries and in the USA. This delayed prompted USA-based firms to diversify their portfolios into fragrances, cosmetics, and other consumer products to mitigate potential revenue losses. At the same time, technological innovation accelerated the pace of drug development. Computers became essential tools for managing compound libraries and performing complex calculations. Analytical instrumentation advanced rapidly with high-performance liquid chromatography-mass spectrometry (HPLC/MS) becoming a cornerstone to compound analysis. In medical imaging, the late 1970s saw breakthroughs in magnetic resonance imaging (MRI), pioneered by Raymond Damadian and colleagues building on Paul Lauterbur's earlier work from 1973. Concurrently, the development of positron emission tomography (PET), provided a new, non-invasive way to visualize brain function, revolutionizing diagnostic capabilities [20].

In the mid-1980s, major pharmaceutical companies began forming strategic alliances with smaller biotechnology firms. This trend was further accelerated by a wave of mergers and acquisitions that reshaped the industry and integrated biopharmaceutical innovations into larger corporate frameworks. Today, the pharmaceutical sector operates within a highly regulated framework, adhering to rigorous international standards governing the development, manufacturing and approval of both chemical and biological drugs for human and veterinary use.

Antiparasitic Drugs in Animal Health

As discussed earlier, infectious diseases were a major focus of medical research in the first half of the twentieth century. In this section, we will specifically delve into parasitic diseases. One key element that enables the rapid assessment of substances on parasites is phenotypic assays, which involve whole parasitic organisms and allow for proof-of-principle studies at an early stage of drug discovery. This prompted the opportunity of repurposing or repositioning old drugs for new indication, or compounds developed to treat a disease other than the one for which they were originally intended. Repurposing holds significant promise for providing innovative treatments for rare and neglected diseases. It offers a cost-effective and time-efficient strategy by circumventing certain early-stage hurdles, as the compounds have already undergone extensive safety and efficacy testing. Furthermore, this approach has been particularly successful in discovering anti-infectives for both human and veterinary medicine. The methodologies used in veterinary drug discovery parallel those in the human pharmaceutical sector. However, phenotypic screening still dominates the veterinarian drug discovery phase, and molecular target-based approaches are scarce (Chap. 6). Moreover, unlike human health, where drug candidates are typically assessed in model organisms, veterinary research extends further by developing a clinical profile specifically for target animals.

In the 1950s, several drugs initially developed for human use were repurposed for parasite control in veterinary practice. This was essential not only for the welfare of animals but also to mitigate the risk of parasite transmission to humans. For example, sulfaquinoxaline, an antibiotic from the sulfa drug class introduced in the 1940s for human bacterial infections, became widely used in the 1950s as an effective treatment for coccidiosis, a parasitic disease affecting birds and cattle. Piperazine became widely used in the 1950s as an anthelmintic for treating various roundworm infections in animals, such as ascarids, pinworms, threadworms, and hookworms. Similarly, diethylcarbamazine, originally used in humans to treat filariasis, was later adopted in veterinary medicine to manage filarial infections in animals. Therefore, it is important to recognize that these drugs were initially developed for human applications, not necessarily for parasitic conditions, before being adapted for veterinary use [22–24].

Drug repurposing is more complex than it might initially appear, particularly regarding the adaptation of human pharmaceuticals for veterinary use. In many instances, especially with antiparasitics, the trend has historically—and continues to—move in the opposite direction, with drugs originally developed for veterinary health being repurposed for human applications. This shift in the general flow of antiparasitic drugs from human to veterinary use began in the late 1950s with the adoption of thiabendazole for human use. Initially developed to treat veterinary parasites, thiabendazole found a new purpose in humans, particularly for treating various gastrointestinal infections caused by parasitic worms. Its efficacy extended across a spectrum of helminths, including roundworms, hookworms, whipworms, and pinworms. This pattern was consolidated in the 1980s, when several drugs originally developed for veterinary use were repurposed as human medicines. Notable examples include praziquantel, ivermectin, levamisole, and albendazole, all of which are listed in the WHO Model List of Essential Medicines [21, 24]. In addition, many compounds marketed in the animal health sector originate from collaborations with crop science companies. This is especially true for ectoparasiticides (drugs or agents specifically designed to eliminate parasites that live on the surface of their host, such as fleas, ticks, and lice), as crop science companies typically have large programs on insect pests. Prominent examples include fipronil and the neonicotinoids, such as imidacloprid, nitenpyram, and acetamiprid, along with the more recent isoxazolines, including afoxolaner, fluralaner, sarolaner, and lotilaner. Isoxazolines, which are highly potent ectoparasiticides, were first introduced to the animal health market in 2014 and have attracted interest for their potential applications in human pharmaceuticals. This class of compounds has been extensively studied and marketed by the animal health industry [26]. Currently, there is

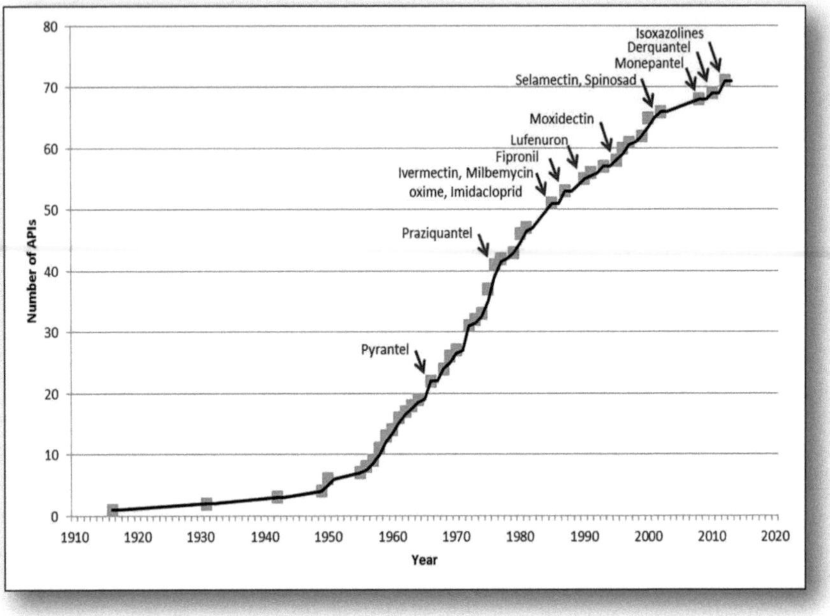

Fig. 1.9 The introduction of new antiparasitic during the twentieth century and beyond (API: active pharmaceutical ingredient)

significant interest from biotech companies, such as Tarsus Pharmaceuticals, in exploring the potential applications of these compounds for human health [27] (Fig. 1.9).

Acknowledgments We would like to thank Dr. Sabine Bernschneider-Reif for reviewing the manuscript for accuracy regarding Merck-related content and for providing the historical visual material from Merck Corporate History.

Competing Interests Diego González Cabrera has received consultant honoraria from Merck Healthcare KGaA. Paul M. Selzer and Thomas Spangenberg have no conflicts of interest to declare that are relevant to the content of this chapter.

References

1. Sugg R (2008) Corpse medicine: mummies, cannibals, and vampires. Lancet 371(9630):2078–2079. https://doi.org/10.1016/s0140-6736(08)60907-1
2. Drews J (2000) Drug discovery: a historical perspective. Science 287(5460):1960–1964. https://doi.org/10.1126/science.287.5460.1960

3. Wang ML (2009) The modern pharmaceutical industry: history, current position and challenges. In: Global health partnerships. Palgrave Macmillan, London. https://doi.org/10.1057/9780230582873_2
4. Riedel S (2005) Edward Jenner and the history of smallpox and vaccination. Proc Bayl Univ Med Cent 18(1):21–25. https://doi.org/10.1080/08998280.2005.11928028
5. Chast F (2008) A history of drug discovery. In: Wermuth CG (ed) The practice of medicinal chemistry, 3rd edn. Elsevier, Oxford, pp 3–62
6. Lesney MS (2000) The pharmaceutical century: ten decades of drug discovery. In: Patents & potions, 1st edn. American Chemical Society, Washington, DC
7. Wainwright M (2008) Dyes in the development of drugs and pharmaceuticals. Dyes Pigments 76(3):582–589. https://doi.org/10.1016/j.dyepig.2007.01.015
8. Daemmrich A, Bowden ME (2005) A rising drug industry. In: Emergence of pharmaceutical science and industry: 1870–1930, C & EN, 83(25):28–42
9. Ehrlich P (1906) Über den jetzigen Stand der Chemotherapie. Berl Klin Wochenschr 43:169–172. https://doi.org/10.1002/cber.19090420105
10. Corley T (2003) The pharmaceutical industry: a guide to historical records. In: Richmond L, Stevenson J, Turton A (eds) The British pharmaceutical industry since 1851, 1st edn. Routledge, London, pp 1–33
11. Vecchio I, Tornalli C, Bragazzi NL, Martini M (2018) The discovery of insulin: an important milestone in the history of medicine. Front Endocrinol 9(613):1–8. https://doi.org/10.3389/fendo.2018.00613
12. Pizzi RA (2000) The pharmaceutical century: ten decades of drug discovery. In: Salving with science, 1st edn. American Chemical Society, Washington, DC
13. Grabowski H (2011) The evolution of the pharmaceutical industry over the past 50 years: a personal reflection. Int J Econ Bus 18(2):161–176. https://doi.org/10.1080/13571516.2011.584421
14. Miller JB (2000) The pharmaceutical century: ten decades of drug discovery. In: Antibiotics and isotopes, 1st edn. American Chemical Society, Washington, DC
15. Tweedy BD, Lesney MS (2000) The pharmaceutical century ten decades of drug discovery. In: Prescriptions and polio, 1st edn. American Chemical Society, Washington, DC
16. Hirsch PM (1975) Organizational effectiveness and the institutional environment. Adm Sci Q 20(3):327–344. https://doi.org/10.2307/2391994
17. Frey R, Lesney MS (2000) The pharmaceutical century ten decades of drug discovery. In: Anodynes and estrogens, 1st edn. American Chemical Society, Washington, DC
18. Vargesson N, Stephens T (2021) Thalidomide: history, withdrawal, renaissance, and safety concerns. Expert Opin Drug Saf 20(12):1455–1457. https://doi.org/10.1080/14740338.2021.1991307
19. Singh N, Vayer P, Shivalika T, Poyet J-L, Tsaoiun K, Villoutreix O (2023) Drug discovery and development: introduction to the general public and patient groups. Front Drug Discov 3:1–11. https://doi.org/10.3389/fddsv.2023.1201419
20. Lesney MS, Frey R (2000) The pharmaceutical century ten decades of drug discovery. In: Chemistry, cancer, and ecology, 1st edn. American Chemical Society, Washington, DC
21. World Health Organization (2023) WHO Model List of Essential Medicines – 23rd list. https://www.who.int/publications/i/item/WHO-MHP-HPS-EML-2023.02. Accessed 12 Nov 2024
22. Noack S, Harrington J, Carithers DS, Kaminsky R, Selzer PM (2021) Heartworm disease – overview, intervention, and industry perspective. Int J Parasitol Drugs Drug Resist 16:65–89. https://doi.org/10.1016/j.ijpddr.2021.03.004
23. Noack S, Chapman HD, Selzer PM (2019) Anticoccidial drugs of the livestock industry. Parasitol Res 118(7):2009–2026. https://doi.org/10.1007/s00436-019-06343-5
24. Geary TG, Woods DJ, Williams T, Nwaka S (2009) Target identification and mechanism-based screening for anthelmintics: application of veterinary antiparasitic research programs to search for new antiparasitic drugs for human indications. In: Selzer PM (ed) Antiparasitic and antibacterial drug discovery: from molecular targets to drug candidates, 1st edn. WILEY-VCH Verlag GmbH & Co. KGaA, Weinheim, pp 1–15

25. Weber T, Selzer PM (2016) Isoxazolines: a novel chemotype highly effective on ectoparasites. Chem Med Chem 11(3):270–276. https://doi.org/10.1002/cmdc.201500516
26. Selzer PM, Epe C (2021) Antiparasitics in animal health: *quo vadis*? Trends Parasitol 37(1):77–89. https://doi.org/10.1016/j.pt.2020.09.004
27. Tarsus Pharmaceuticals Inc (2024). https://tarsusrx.com/. Accessed 12 Nov 2024
28. The Selection and use of Essential Medicines (2025) WHO Model List of Essential Medicines for Children, 10th list

The Parasite's Life Cycle and the Patient

2

Joachim Richter and Clarissa Prazeres da Costa

Abstract

In this chapter we explore the intricate relationship between schistosomes and their human hosts, delving into the biology, pathogenesis, clinical manifestations, and diagnostics of schistosomiasis. Prevalent for millennia, the disease remains a significant global health challenge, particularly in sub-Saharan Africa. The life cycle of schistosomes involves freshwater snails, with transmission to the definitive host, such as humans, linked to water exposure. Acute and chronic stages manifest with diverse symptoms, ranging from fever and hematuria to severe organ damage, including fibrosis and bladder cancer. Advanced diagnostics, such as molecular and serological methods, complement traditional microscopy but are less accessible in endemic regions. Clinical interventions focus primarily on praziquantel, with significant reversibility of symptoms even in late stages. The chapter underscores schistosomiasis as a "One Health" issue, requiring integrated approaches targeting humans, snails, and environmental water sources. It highlights the disease's socioeconomic impact and its interplay with other infections, including HIV. Comprehensive elimination strategies encom-

J. Richter
Institute of International Health, Charité, Berlin, Germany

Center for Global Health, Charté Universitätsmedizin, Berlin, Germany

Swiss Tropical and Public Health Institute, Allschwill, Switzerland

C. P. da Costa (✉)
Institute for Medical Microbiology, Immunology and Hygiene, Center for Global Health, TUM School of Medicine and Health, Technical University Munich, Munich, Germany

Center for Global Health, TUM School of Medicine and Health, Technical University Munich, Munich, Germany

German Center for Infection Research (DZIF), partner site Munich, Munich, Germany
e-mail: clarissa.dacosta@tum.de

T. Spangenberg (ed.), *Praziquantel*,
https://doi.org/10.1007/978-3-031-97397-0_2

passing diagnostics, treatment, and environmental control are emphasized to combat this ancient yet persistent parasitic challenge.

Keywords

Schistosomiasis · Pathology · Neglected Tropical Disease · One Health · Diagnostic methods

Discovery of Schistosomiasis

Schistosomiasis is an ancient disease likely originating in water birds, with its initial spread believed to have started around the great African lakes, extending from the Red Sea down to Mozambique in southeastern Africa. Evidence of the disease dates back over 6000 years. Terminally spined eggs, a distinctive feature of *Schistosoma* species, have been discovered in skeletal remains at Tell Zeidan, a settlement in Syria. Ancient Egyptian papyri from as early as 1500 BC described a disease with symptoms such as discharge from the penis, which may indicate schistosomiasis, though it could also be other conditions like gonorrhea (Fig. 2.1). The term "a-a-a disease", frequently mentioned in ancient medical papyri, is most likely linked to schistosomiasis, suggesting the widespread prevalence of the disease. To prevent the infection, ancient Egyptians were advised to avoid contaminated water, and those with regular water contact, such as fishermen and farmers, were recommended to wear linen penile coverings.

Description of schistosomiasis-related symptoms such as hematuria (blood in urine) or bladder discomfort can also be found in Assyrian (ancient civilization that inhabited areas corresponding to present-day Iraq, northeastern Syria, southeastern Turkey, and northwestern Iran) medical texts where it was termed "musu." Much later, Napoleon's soldiers stationed in Egypt from 1799 to 1801 suffered severely from hematuria from unknown cause. The identification of worms from the genus *Schistosoma* as the causative agents of a disease long known in endemic regions like Egypt is relatively recent. It dates back to 1851 when Theodore Bilharz and his colleague Carl Theodor Ernst, discovered the adult worm during an autopsy of a

Fig. 2.1 The Egyptian hieroglyphics from the Ebers Papyrus; written in 1500 B.C. representing the widespread "a-a-a disease" as a phallus dripping fluid which is thought to be blood

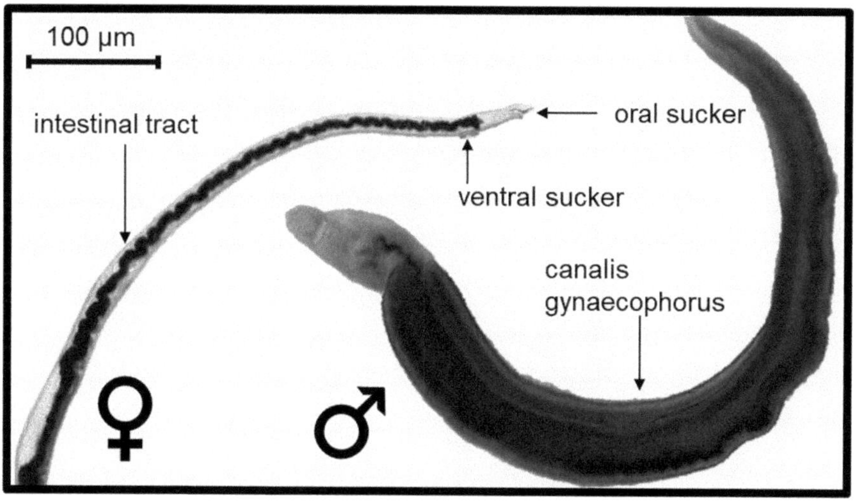

Fig. 2.2 Adult male and female *Schistosoma mansoni* worms. (Light microscopy photography; Source: C Prazeres da Costa/TUM©)

deceased soldier at the Kasr-El-Eini hospital in Cairo, Egypt. Bilharz, the chief of the surgical department at the time, was a renowned parasitologist who also discovered the tapeworm *Hymenolepis nana* and the flatworm *Heterophyes heterophyes*. The newly discovered parasite was firstly termed *Distomum haematobium* before being renamed *Bilharzia haematobium* in 1859 in honor of the discoverer, which is still occasionally used. The name *Schistosoma* eventually became widely accepted. It is derived from the Greek words "skhistos," meaning split or cleaved, and "soma," meaning body, referring to the distinctive flattened, elongated shape of the adult worms, which appear as if their bodies are split or cleaved (Fig. 2.2). From a historical perspective, it is noteworthy that as early as 1918, tartar emetic (antimony potassium tartrate) was administered intravenously to treat schistosomiasis, despite sometimes causing severe side effects such as vomiting and dehydration. Nonetheless, improved formulations of antimonials were used until the 1970s, when the advent of praziquantel (PZQ) completely replaced them.

Nomenclature and Life Cycle of the Parasite

Schistosomes are separately sexed flatworms classified as trematodes of the family *Schistosomatidae*. Most of schistosomiasis cases are attributed to *Schistosoma haematobium*, *S. japonicum*, or *S. mansoni* (Chap. 3). The African continent harbors both *S. haematobium* and *S. mansoni*, with approximately 90% of all cases worldwide concentrated in sub-Saharan Africa. The life cycle of schistosomes is complex, involving humans and other mammals as definitive hosts [1] and freshwater snails as intermediate hosts (Fig. 2.3).

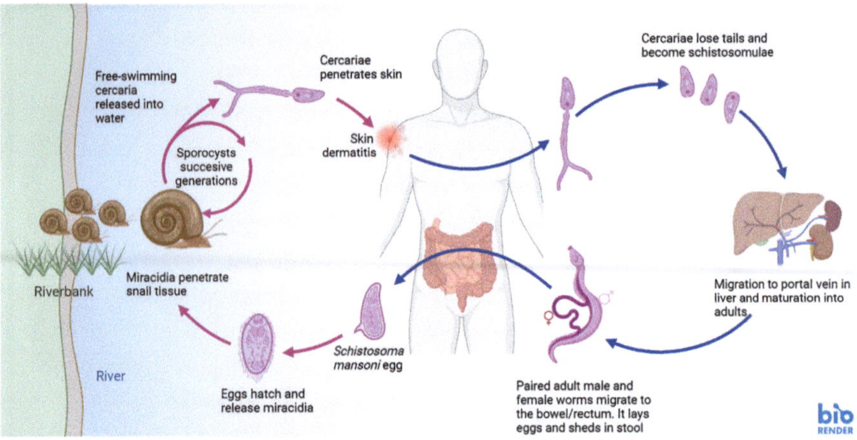

Fig. 2.3 *S. mansoni* life cycle. Humans (in the middle) are the definitive host with adult worms, which live in the venous system of the liver and intestines and freshwater snails (on the left, *Biomphalaria glabrata*) are the intermediate host of the parasite (Made with Biorender®)

Miracidia

Miracidia are larvae with numerous cilia on their surface, enabling them to swim rapidly (2 mm/s) and cover considerable distances (several meters). They hatch from eggs upon contact with freshwater (Fig. 2.3). Partially attracted by genus-specific chemotactic molecules of freshwater snails, miracidia must penetrate a suitable intermediate host within about 1 day, depending on the water temperature, to be able to survive. Penetration occurs through the snail's foot, leading to the transformation of the miracidia into tubular mother sporocysts. These mother sporocysts then generate multiple generations of secondary sporocysts, which ultimately develop into numerous cercariae. In optimal intermediate hosts, a single miracidium can give rise to thousands of cercariae. This developmental process is highly temperature-dependent, taking several weeks for African schistosomes (*S. mansoni*, *S. haematobium*) and several months for Southeast Asian schistosomes (*S. japonicum*).

Cercariae

In human schistosomes, the emigration of mature cercariae (400–600 μm in length) from the snail is triggered by bright sunlight. In contrast, zoonotic schistosomes that infect nocturnal rodents initiate their cercarial emigration in response to darkness. This adaptation increases the cercariae's chance of finding a suitable host despite their short survival period of up to 2 days. Each day, a single snail can release up to several hundred cercariae. These larvae, consisting of a body (or head) and a long, forked tail (Fig. 2.3), can be transported over long distances by water currents and, upon contact, penetrate the intact skin of a suitable mammalian host, such as humans or cattle.

Schistosomula

After penetrating the skin, cercariae shed their tails and outermost layer (glycocalyx), transforming into schistosomula. Within a few hours, their characteristic double-membrane skin (tegument) forms. The schistosomula reach the lungs via the bloodstream after a few days, from where they migrate to the liver through the hepatic artery. During this migratory and developmental stage, they are insensitive to PZQ, the only broadly available drug against schistosomiasis emphasising the need for new therapeutic alternatives [2]. They then undergo further metamorphosis into young or juvenile worms. In the intrahepatic portal vein branches, the schistosomula develop into male and female worms, with juvenile females maturing fully after mating with a male. Mature worms then migrate to small mesenteric veins (in cases of intestinal schistosomiasis) or to the venous plexuses of the lesser pelvis and urinary bladder through blood vessel connections (anastomoses), where they lay eggs. This developmental process within the host takes about 40 days, and adult worms typically live for 3–5 years, though in rare cases, they may survive up to 40 years.

Adult Worms

Adult worms typically reside in male-female pairs (blood flukes) within the lumen of blood vessels. Depending on the species, they can be found in the intestinal or mesenteric veins (*S. mansoni*, *S. intercalatum*, *S. guineensis*, *S. japonicum*, and *S. mekongi*) or in the urogenital veins and other veins of the pelvic region (*S. haematobium*). Female worms release eggs into the vascular lumen, where they become lodged mainly in small venules at the end of the vascular system. This triggers an inflammatory response (granuloma), which helps the eggs pass through the tissue into the lumen of the urinary bladder, ureters, or intestines, where they are eventually excreted in urine or stool.

Adult worms range in length from 6 to 28 mm, with *S. mansoni* being the shortest and *S. japonicum* the longest. The more robustly developed male displays a whitish coloration and has a dorsoventrally flattened body, meaning it is flat from both the upper (dorsal) and lower (ventral) surfaces. The male forms a groove (canalis gynecophorus) by curling its ventral body margins inward, encasing the thinner, longer, and more rounded female (Fig. 2.2). Both sexes have an oral and ventral sucker, the latter used for attachment to the inner surface of blood vessels (endothelium). Schistosomes feed on the host's blood, and digestion of red blood cells (erythrocytes) in their intestinal tract gives them a characteristic blackish coloration (Fig. 2.2). This is particularly noticeable in females, which consume blood at a rate about eight times higher than males (approximately 3×10^5 erythrocytes per hour).

The blackish pigment, known as hemozoin, is a heme-based waste product formed during the digestion of erythrocytes.

The large oval eggs of schistosomes, produced by paired adult male and female worms, possess a lateral or terminal spine. The position and expression of this spine, along with variations in egg size and shape, enable species differentiation. For instance, Asian schistosomes produce smaller, roundish eggs with a markedly reduced spine compared to other species. The daily egg production of a pair of worms in human infections is 250–400 for *S. mansoni*, 200–1000 for *S. haematobium*, and 1500–3500 for *S. japonicum*. Eggs contain highly motile miracidia and are preferentially deposited in the smallest venules of the urogenital or intestinal system, often forming chains. When they become trapped in tissues or are carried by the venous bloodstream to the liver and/or other organs, they cause disease. This is due to decaying eggs and dying miracidia, which triggers a strong inflammatory reaction that leads to tissue scarring. Epidemiologically, only those eggs that successfully traverse tissue intact and reach freshwater contribute to the transmission. Miracidia are highly sensitive and can quickly die from desiccation or exposure to water temperatures above 35 °C. They remain inside the eggs until they come into contact with freshwater, where hatching is triggered by the low osmotic pressure, often causing the eggshell to rupture within 10 min. In *S. haematobium*, the excretion of eggs in urine follows a circadian rhythm that aligns with the ecological patterns of the snails. This synchronization ensures an optimal probability of infection of the snails by the miracidia, with maximum egg excretion occurring around midday. Recent advances in sequencing have revealed hybridization between schistosome species, though the implications of such events for disease severity and epidemiology remain to be fully understood [3].

Pathogenesis and Clinical Manifestations

Parasitic adult worms do not multiply in their hosts. The number of worm pairs present is determined by multiple factors, including the specific local conditions, behavior, and the intensity of exposure (extent of infection and reinfection with cercariae upon contact with water) as well as the host immune responses.

Acute Schistosomiasis

Acute schistosomiasis, also known as Katayama Syndrome, is a febrile illness that generally manifests several weeks after infection, usually between 4 and 6 weeks, though it can range from 1 to 12 weeks. It is most common and severe in infections caused by *S. japonicum* and *S. mansoni* and results from an acute systemic hypersensitivity reaction. This reaction occurs as the body's immune system responds to the immature schistosomes (schistosomula) traveling through the bloodstream from the lungs to establish themselves in the veins around the intestines or bladder.

The symptoms can range from mild to severe and typically include fever, fatigue, headache, neck and limb pain, muscle aches, dry cough, upper abdominal pain, nausea, and sometimes bloody diarrhea. Varying degrees of liver and spleen enlargement may also be observed. In some cases, individuals may develop an itchy rash known as "swimmer's itch," caused by the influx of eosinophilic granulocytes (immune cells) into the skin in response to the invasion of schistosomula. This response is typically less pronounced in infants who still benefit from immunological protection passed from their mothers. Katayama fever affects only a small proportion of infected individuals and is more likely to occur in nonimmune individuals with a first-time or minor previous infection history, particularly following exposure to a large number of cercariae (known as slug fever). This acute febrile stage of the disease primarily manifests in travelers and immigrants from nonendemic areas, while it is less frequently seen in the native population of endemic areas. Recovery from Katayama fever is typically spontaneous and occurs within 2–10 weeks. The disease is usually self-limiting and resolves on its own during this time [4]. However, in cases of massive exposure or initial infections in nonimmune individuals, it can progress to a life-threatening condition. Fatalities attributed to cardiac and cerebral complications have been documented in cases of severe infections, especially with *S. japonicum* and *S. mansoni*.

Chronic Schistosomiasis

The damage during the chronic stage is determined by eggs retained in the body and the immune response triggered against them. About half of the eggs are excreted through the intestinal and/or bladder mucosa, while a variable proportion is trapped in tissues. These retained eggs remain near the adult worms in and around the small veins of the intestine and urogenital tract, but they could also be carried by the bloodstream to other organs, particularly the liver and the lungs, where they become lodged in the presinusoidal portal veins or in the pulmonary arterioles. Since the immune system cannot degrade large eggs, they are encapsulated within an aggregation of cells known as a granuloma, triggering an immune response in the host that remains active over an extended period (chronically activated). Initially, inflammatory granulomas, characterized by a high presence of eosinophils—white blood cells crucial for combating multicellular parasites and certain infections—form around newly deposited schistosome eggs in various organs. Over time, these granulomas diminish in size and experience shifts in cellular composition due to immunomodulation [5]. Eosinophils are gradually supplanted by tissue macrophages and T-helper cells, leading to significant fibrosis, particularly in the outer regions of the granuloma (Fig. 2.4). These pathological changes underlie the eventual onset of disease symptoms, although severe organ damage only manifests in a subset of patients. The variation in disease severity is influenced not only by the number of eggs but also by genetic factors. As a result, individuals within a population exposed to identical conditions may exhibit significant disparities in symptom expression, with notable variations even among family members. Early-stage fibrotic changes

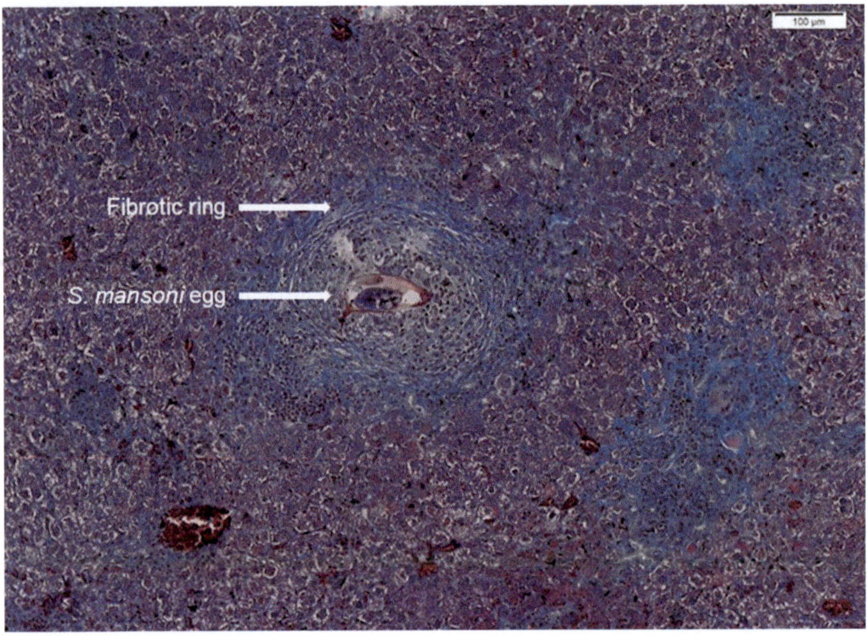

Fig. 2.4 Liver granuloma with a *Schistosoma mansoni* egg containing a micaridium surrounded by a variety of immune cells such as macrophages, eosinophils, dendritic cells, and T and B cells and a fibrotic ring (Massons staining) (Photo by C Prazeres da Costa/TUM©)

can often reverse with treatment or spontaneously but become increasingly irreversible in later stages of the disease.

Urogenital Schistosomiasis

In endemic areas, a significant number of infected individuals develop chronic disease, which can progress even without reinfection or superinfection. Because adult worms typically reside in the mesenteric venules of the bowel and rectum, as well as in the venous plexus of the bladder, for extended periods of time, large quantities of eggs can accumulate in localized areas, leading to the formation of focal lesions. The urinary bladder (causing bladder schistosomiasis) and the distal portion of the ureter are particularly affected by these accumulations. For example, eggs from *S. haematobium* are especially prone to becoming lodged as they traverse the layers of the urinary system. Calcification of the dead eggs can lead to the formation of large, shell-like deposits in the bladder wall. Granulomas that develop around eggs remaining in the bladder and ureter walls often grow to a visible size, presenting as nodules up to 1–2 mm in diameter (pseudotubercules). This granulomatous inflammation can result in ulcerations, which may cause localized mucosal bleeding (hemorrhage), inflammatory masses, and abnormal tissue growths (polyps) on the bladder wall. The progressive replacement of the chronically inflamed wall layers by fibrotic connective tissue leads to thickening of the bladder wall, which is frequently accompanied by scarring and shrinkage. As fibrosis advances, the mucosa develops brownish-yellow patches with a leathery or sandy, rough surface. These "sandy patches," typically located in the interior region of the bladder (trigonum), are characteristic of urogenital schistosomiasis and can also occur throughout the genitourinary system. During this process, the trigonum and bladder neck may be pulled forward, potentially causing a blockage at the bladder neck or where one or both ureters enter the bladder (unilateral or bilateral obstructive uropathy). This condition can result in an obstruction in the upper urinary tract, impacting the kidneys and upper ureters, leading to ureteral dilation and kidney swelling (hydronephrosis). If the ureteral wall weakens and dilates, it can allow urine to reflux back into the kidney. This reflux increases the risk of recurrent kidney infections, exacerbating kidney disease and potentially culminating in life-threatening kidney failure.

Symptoms of urogenital schistosomiasis generally emerge around 10–12 weeks after infection, though in some instances, symptoms may not manifest for several years. Symptoms typically begin with painful urination (dysuria) and increased frequency of urination (micturition), often accompanied by a sensation of incomplete bladder emptying. Hematuria usually follows, often occurring at the end of urination (terminal hematuria) and can even be present in infancy. In highly endemic areas, hematuria is so prevalent among adolescent males that it is likened to menstruation as a sign of sexual maturation. However, it often appears intermittently, with microhematuria persistently detectable between episodes. In more severe infections, individuals often experience an increased urge to urinate (pollakisuria), along with recurrent inability to control urination (enuresis), or incontinence. Pain is typically felt in the urethra but may also be present above the pubic bone (suprapubic) and in the area between the anus and external genitalia (perineal). The intensity of these symptoms can vary widely, ranging from pronounced to entirely absent.

Bladder cancer is one of the most prevalent malignancies in countries where schistosomiasis is endemic. For instance, in Egypt, it accounted for approximately a quarter of all cancers before national treatment campaigns were deployed. Bladder carcinomas can develop due to chronic infection with *S. haematobium*, hence its classification as a carcinogenic agent (Group 1) [6]. Clinically, the symptoms resemble those of uncomplicated bladder schistosomiasis, with urination difficulties and hematuria being suggestive symptoms. Schistosomiasis-associated bladder carcinomas predominantly affect men aged between 35 and 50 years, although isolated cases have been reported in children.

Eggs and egg granulomas often lead to subsequent confluent sclerosis and calcifications in other pelvic organs, particularly in the genital organs, a condition known as genital schistosomiasis, which has historically been neglected. In adolescent and young women with female genital schistosomiasis (FGS), granulomas often develop in the tissues of the vulva, vagina, and cervix uteri. These are visible macroscopically as ulcerations, sandy patches, and papillomata. Such manifestations can sometimes be mistaken for sexually transmitted diseases (STDs), potentially leading to stigmatization, particularly if no eggs are detected in the urine. FGS can manifest as early as childhood. Over time, the internal genital organs experience progressive fibrosis, inflammatory scarring, and narrowing (stricture) of one or both fallopian tubes. These developments can lead to the formation of granulomatous masses in the pelvic region and may also contribute to the onset of vesicovaginal fistulas. Vesicovaginal fistulas manifest as urinary incontinence and leakage of urine from the vagina, often accompanied by poor healing outcomes. Damage to the genital mucosa caused by schistosomiasis may increase the risk of STDs, including an increased risk of human immunodeficiency virus (HIV) infection. Coinfections involving genital schistosomiasis and human papillomavirus (HPV), as well as other STDs, are also frequently observed. Perineal pain and discomfort during sexual intercourse could be indicative of genital schistosomiasis. In severe cases, tubal occlusion from schistosomiasis-related salpingitis may lead to acute life-threatening intraperitoneal hemorrhage during a ruptured ectopic pregnancy. Occasionally, intraperitoneal adhesions caused by schistosomiasis are mistaken for other conditions such as chronic pelvic inflammatory disease or cancerous growths or tumors. Schistosomiasis affecting the internal genital organs can lead to sterility and may also complicate pregnancy by exacerbating conditions like ureteral obstruction or iron deficiency (anemia). While eggs can be detected in the placenta, they do not pass to the unborn child. Male genital schistosomiasis (MGS) typically presents with altered consistency or a bloody appearance of ejaculate, accompanied by perineal or testicular pain. Additional symptoms may include scrotal swelling (hydrocele) or fibrosis of the testes. The seminal vesicles and prostate are frequently affected in MGS. This condition can result from infections caused by various schistosome species, but it is predominantly associated with *S. haematobium* and *S. intercalatum*. In a significant number of patients treated with medication, the associated changes may regress notably. This regression extends to complications within the urinary tract, such as bladder obstruction due to schistosomiasis.

Intestinal Schistosomiasis

Typically, some of the eggs remain lodged in the intestinal wall, resulting in chronic granulomatous inflammation. While eggs are commonly found in the small intestine wall, significant lesions predominantly occur in the colon. These lesions are typically focal, with the distal colon and rectosigmoid region often showing more severe involvement. Depending on the stage and extent of the lesions, hyperemia of the mucosa, raised nodules, typical sandy patches (areas of high egg concentrations), ulcerations, and hemorrhages may be observed.

Mild chronic infections are usually asymptomatic, with the majority of people in endemic areas remaining largely symptom-free. In more severe infections, abdominal pain and bloody diarrhea may occur. More commonly, blood and mucus are intermittently excreted in the stool without diarrhea. The presence of numerous polyps (polyposis) in the colon, as seen in *S. mansoni* infection, manifests similarly with bloody diarrhea and a sensation of incomplete bowel evacuation (tenesmus). This condition is associated with significant loss of enteric blood and protein, which can lead to anemia and edema. Large granulomatous lesions in the colon and rectosigmoid, which can be mistaken for tumors or inflammatory bowel disease, may cause bleeding, dysfunction in bowel movement, ileus, intussusception, or rectal prolapse. Serious pathological changes with significant clinical manifestations can also occur with *S. mekongi* and *S. japonicum* infections. In addition, infections with *S. intercalatum* and *S. guineensis* primarily result in egg deposition within the rectal tract.

Hepatosplenic Schistosomiasis

The morbidity and mortality of intestinal schistosomiasis are primarily determined by hepatosplenic schistosomiasis rather than by the intestinal manifestations. The granulomatous-fibrotic reaction around the eggs in the liver causes progressive portal fibrosis. As the small portal vein branches become obstructed and scarred, the larger branches and the main portal vein trunk are also affected. This fibrosis originates from the portal branches, leading to scarring or occlusion, which increases pressure in the portal vascular network and results in presinusoidal portal hypertension, eventually causing the accumulation of intra-abdominal fluid (ascites) (see Fig. 2.5). Despite these changes, liver function typically remains preserved, although cholestasis parameters are often mildly to moderately elevated due to scarring and strictures of small intrahepatic bile duct branches. Symptoms of hepatic insufficiency such as jaundice, hepatic skin manifestations, coagulopathy, and encephalopathy are rare. Hepatic encephalopathy and hepatic coma usually occur only when other factors such as variceal hemorrhage, viral hepatitis, alcohol abuse, secondary infection, shunt surgery, or malnutrition are present. Typical symptoms of hepatosplenic schistosomiasis include increasing general weakness, abdominal discomfort, weight loss, muscle atrophy, and anemia. Amenorrhea, infertility, loss of libido, and impotence may also develop. Some patients remain largely symptom-free until the disease is well advanced, and complications of portal hypertension manifest. These complications include hypertensive gastropathy and esophageal or gastric varices, which carry a high risk of massive bleeding. Additionally, collateral vessels

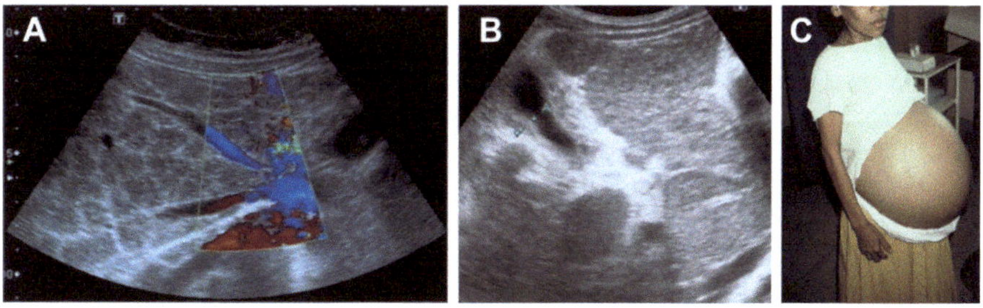

Fig. 2.5 (**a**) Network fibrosis (liver sonography) (Photo by J Richter©). (**b**) Gallbladder fibrosis (*S. mansoni*) (Photo by J Richter©). (**c**) Patient from the Philippines with hepatosplenic schistosomiasis (*S. japonicum*) and decompensated portal hypertension with massive ascites (Photo by C Prazeres da Costa/TUM©)

may form and become visible on the skin of the abdomen (caput medusae). Symptoms and signs of advanced disease, such as liver fibrosis and portal hypertension, typically appear after several years of infection. Disease progression is generally faster with *S. japonicum* infection (after 3–5 years on average) than with *S. mansoni* (approximately 5–15 years).

In patients with *S. mansoni* infection and portal fibrosis, the gallbladder is often affected by progressive loss of contractility (schistosomiasis cholecystopathy). In some cases, gallbladder fibrosis may develop before hepatosplenic infestation becomes evident. Conversely, in Asian forms of schistosomiasis, gallbladder changes are atypical and suggest a different etiology.

Hepatosplenic schistosomiasis is often exacerbated by malnutrition, infectious factors such as viral hepatitis, or toxic factors such as aflatoxins, alcohol abuse, and khat consumption. When hepatosplenic schistosomiasis occurs in childhood, it can be associated with developmental delays, dwarfism, and delayed sexual development.

Cardiopulmonary Schistosomiasis

Clinically apparent lung involvement usually occurs in cases of long-standing hepatosplenic schistosomiasis, in which the worm eggs are transported by collateral veins beyond the liver, potentially affecting other organs. Pulmonary hypertension may develop due to massive egg embolization and progressive vasculitis, an autoimmune disease that causes inflammation of blood vessels. With increasing pressure in the pulmonary circulation, right heart strain develops, leading to the dilatation of the pulmonary artery, hypertrophy of the right ventricle, and eventually right heart failure (pulmonale). Frequently observed drumstick fingers (clubbing), which are not typical of right ventricular failure alone, are attributed to damage to arteriovenous anastomoses (connections between small arteries and veins) in the pulmonary vascular architecture caused by granulomas and fibrosis.

Ectopic and Neuro-Schistosomiasis

As previously noted, eggs and even adult worms can bypass the liver through collateral veins, thereby potentially involving multiple organs in the body. Clinically significant pathological changes have been reported, although rarely, in various ectopic locations such as the skin, stomach, pancreas, genital organs, pericardium, myocardium, thyroid gland, or eyes. For instance, in *S. haematobium* infection, chronic or acute appendicitis is the most frequently observed ectopic manifestation. Ectopic lesions are most notable in the central nervous system (CNS). Egg emboli and granulomas can result in cerebral schistosomiasis, reported in 1–4% of hospitalized patients with *S. japonicum* infection. The predominant symptom is epileptic seizures. Less common presentations include focal motor and sensory deficits, cranial nerve disturbances, signs of increased intracranial pressure, or stroke-like manifestations (apoplectiform). In *S. mansoni* and *S. haematobium* infections, CNS involvement often manifests as acute or chronic spinal cord inflammation (myelitis), most commonly as transverse myelitis with segmental sensory and motor deficits. Symptoms typically start with pain in the lumbar region and lower extremities, progressing to sensorimotor deficits, including bladder and rectal paralysis. If the cervical cord is affected, paralysis of all limbs (quadriplegia) may develop within a few days.

Other Consequences of Schistosomiasis and Coinfections

Children with severe *S. mansoni* or *S. japonicum* infections experience growth disorders including dwarfism. The extent to which these are related to the generally poor nutritional conditions and frequent occurrence of other worm infections in the mostly poor rural populations, and what part schistosomiasis plays in their development, is difficult to clarify. Schistosomiasis also appears to have an influence on the endocrine system, although this has not yet been adequately studied. Beyond growth retardation, severely infected children commonly show delayed sexual development. Furthermore, schistosomiasis during pregnancy has an impact on birth weight, can cause maternal anemia, and can have a lasting effects on allergy and vaccine responses [7].

In bladder schistosomiasis, bacterial kidney infections (pyelonephritis) are more common in advanced cases with obstructive disease. Additionally, bacterial superinfections can occur from scratch excretions in the presence of cercarial dermatitis or other pyodermas, potentially leading to staphylococcal abscesses. In advanced hepatosplenic schistosomiasis, immunodeficiency increases the risk of secondary bacterial infections. Furthermore, the interaction between genital schistosomiasis and viral infections caused by HIV, human T-cell leukemia/lymphoma virus (HTLV), hepatitis viruses, human papillomavirus, or agents of other STDs has recently been investigated and, in fact, the risk of HIV infection increases almost four-fold with coexisting genital lesions. Traditional methods of treating schistosomiasis-related morbidities can also lead to parenteral transmission of pathogens, for example, when papillomatous changes of the external genitalia are ablated with nonsterile razor blades. The role of genital and placental schistosomiasis in mother-to-child transmission of other infectious agents is unclear.

Diagnosis

The current state-of-the-art diagnostics for schistosomiasis have evolved to include both traditional and modern molecular techniques. Microscopy, using the Kato-Katz method, however, remains the gold standard for detecting *Schistosoma* eggs in stool or urine, particularly in endemic areas. However, this method has limitations in sensitivity, especially in cases of low-intensity infections. Here, serological tests that detect antibodies or circulating antigens offer noninvasive alternatives as they are more sensitive but may not distinguish between past and current infections.

Molecular diagnostics, particularly polymerase chain reaction (PCR)-based assays, provide high sensitivity and specificity by detecting *Schistosoma* DNA in blood, stool, or urine. These assays are increasingly used for early detection and in low-transmission settings, though their use is limited by the need for specialized equipment and relatively high costs. Loop-mediated isothermal amplification (LAMP) and point-of-care (PoC) rapid diagnostic tests (RDTs) are currently being developed to improve accessibility and cost-effectiveness in endemic regions. Overall, while traditional methods remain in use widely, there is a growing need to shift toward more sensitive and noninvasive molecular tools for more accurate diagnosis and monitoring in times of large-scale elimination efforts.

Parasitological Examinations

Urine Examination

Excretion of *S. haematobium* eggs begins 30–40 days after initial infection, occasionally after several months. If egg excretion is high, detection is possible by examining the last portion (10–50 mL) of spontaneous urine. Egg excretion is highest in the late morning hours (midday urine) and after physical activity (exercise urine). The sediment obtained by simple sedimentation, filtration, or centrifugation should be assessed microscopically, and eggs quantified. Since egg excretion can vary, it may be necessary to repeat the examination several times.

Stool Examination

Microscopic examination of a simple thin stool smear typically detects schistosome eggs only in cases with high egg excretion. The Kato-Katz thick smear method is more reliable for field conditions and basic laboratories [8]. Because a single negative test cannot rule out infection, at least three independent stool specimens should be examined to confirm the absence of active infection or to evaluate treatment efficacy. However, the sensitivity of these tests ranges roughly from 30% to 70%.

The level of infection is expressed as eggs per 10 mL of urine or eggs per gram of stool (EPG). These methods have low sensitivity and the observation of 1 egg in a slide corresponds to detection of 20–40 EPG, or 5000–10,000 eggs per diurnal fecal portion of 250 g (in *S. mansoni* infections, 1–99 EPG are considered low-intensity infections). Egg excretion can exhibit significant day-to-day, intraspecimen, and diurnal variation.

Immune and Molecular Diagnostics

Several methods are available for detecting antibodies—proteins produced by B cells of the immune system in response to exposure to antigens. Antigens from adult worms or eggs of *S. mansoni* are mostly used because this is the most accessible parasite species for life cycle maintenance in a laboratory setting (Chap. 6); cercariae antigens can also be used for serodiagnosis of fresh infections. Under technical conditions, *Schistosoma*-specific antibodies can be detected in more than 95% of infected individuals using enzyme-linked immunosorbent assay (ELISA), immunoblotting, or immunofluorescence testing. The sensitivity of antibody detection is independent of the antigen source and remains consistent across different phases of the disease, with antibodies often detectable long after successful therapy. Antibody detection methods are crucial in travel medicine for preliminary diagnosis of acute or low-grade infections that may not be identified through conventional testing methods. They can provide the only diagnostic information during the 1–3-month prepatency period of initial infections, when eggs are not yet shed. Furthermore, antibody tests are suitable for epidemiological investigations, such as determining the infestation levels of a population before administering mass therapy.

Direct detection of circulating adult worm antigens, such as circulating anodic antigen (CAA) [9] and circulating cathodic antigen (CCA), in urine and serum is becoming increasingly important. These antigens correlate with the intensity of infestation or the extent of egg excretion (Chap. 9). Rapid tests have been developed for antigen detection, which can be used as PoC diagnostics using urine samples; the POC-CCA test is now commercially available. However, it should be noted that the sensitivity of these tests is significantly lower for mild infections. Several studies show that while sensitivity is high for infections with *S. mansoni*, it is much lower for *S. haematobium* and other species. Additionally, specificity may be relatively low, as indicated by initial large field studies.

Polymerase chain reaction (PCR) methods can detect schistosome-specific DNA segments in stool, urine, vaginal fluid, or rectal swabs. Serum PCR can be utilized in cases of suspected Katayama syndrome and during the prepatency period, when microscopic and serological tests may still be negative. Although these tests generally offer very high sensitivity and specificity and can differentiate between different schistosome species, they are technically more complex and expensive, making them available only in a few specialized laboratories. DNA extraction methods are not yet fully developed. PCR is also not suitable for monitoring the success of treatment, as a positive result does not necessarily indicate the presence of living schistosomes; DNA from dead worms and eggs can be detected long after treatment. More advanced antigen and PCR testing methods are being developed, including those using lateral-flow or LAMP technologies. Currently, no commercial tests are available for these advanced methods. Nevertheless, in an era of post-mass drug administration, such highly sensitive methods clearly play an important role [10].

Other Laboratory Tests

In acute schistosomiasis, there is typically significant blood eosinophilia (10–70%), often leading to an increased total leukocyte count. The erythrocyte sedimentation rate (ESR) is considerably accelerated, and transaminases, alkaline phosphatase, and lactate dehydrogenase levels are usually moderately elevated. In the chronic stage, eosinophilia is less pronounced and only partially present, with hematologic and biochemical laboratory parameters generally unremarkable in most infected individuals. In advanced stages of the disease, anemia is present in both urogenital and intestinal schistosomiasis, the ESR is moderately accelerated, and concentrations of immunoglobulin G (IgG) and immunoglobulin E (IgE) in serum are markedly elevated. In hepatosplenic disease, leukopenia and thrombocytopenia are common, while coagulopathy, hypoalbuminemia, and hypergammaglobulinemia, along with elevated bilirubin levels or serum transaminases, suggest hepatic comorbidity. Liver congestion parameters are typically normal or mildly elevated. In bladder schistosomiasis, proteinuria is often present even in the asymptomatic stage, indicating glomerulopathy. A rapid rise in retention levels is a prognostically unfavorable sign in both obstructive uropathy and hepatosplenic schistosomiasis.

In endemic regions, using ready-reagent test strips to detect microhematuria is a straightforward and cost-effective approach for a preliminary diagnosis of *S. haematobium* infection. Test strip examination for fecal occult blood varies in sensitivity based on the endemicity of the area. Detection of eosinophil cationic protein (ECP) in urine, secreted by activated eosinophil granulocytes, indicates an inflammatory response, although this method remains experimental.

Imaging and Endoscopic Examinations

X-rays can be used to detect transient and self-resolving lung infiltrates (Löffler infiltrates) in the early stages of infection, but their main application is in assessing cardiac size as an indicator of pulmonary hypertension. Occasionally, urography is used to diagnose ureteral strictures after injecting a contrast medium containing iodine intravenously, which is subsequently excreted through the kidneys. Ultrasound is an essential noninvasive tool for diagnosing and monitoring bladder schistosomiasis, assessing liver–spleen enlargements, and evaluating liver fibrosis extent [11]. It can efficiently detect upper urinary tract obstruction, hydroureter, and hydronephrosis without radiation exposure. Bladder masses should be evaluated for malignancy, particularly if they remain irregular and do not shrink after treatment. Ultrasound is also effective in detecting prostate fibrosis, scrotal masses, and hydrocele, though it does not typically identify other genital lesions (Fig. 2.5). Additionally, it can reveal nonspecific bowel wall thickening and colon polyps.

In liver–spleen involvement, ultrasound shows clear, echo-rich areas along the portal vein, indicating portal fibrosis [12]. In *S. japonicum* infected patients, it shows a distinctive network-like liver pattern (tortoise back) (Fig. 2.5a). Ultrasound can also assess the severity of portal hypertension, evaluate the risk of esophageal variceal bleeding as well as estimate the extent of fibrosis in other organs such as the gallbladder (Fig. 2.5b). Duplex ultrasound, which combines Doppler with traditional ultrasound imaging, is capable of detecting portal vein thrombosis and evaluating blood flow both before and after certain shunt procedures [13]. However, Doppler ultrasound may underestimate portal hypertension due to increased splenic blood flow. For detecting CNS lesions, computed tomography (CT) or magnetic resonance imaging (MRI) is typically utilized. All endoscopic procedures, such as cystoscopy, colposcopy, esophagogastroduodenoscopy, and colonoscopy, are invasive, time-consuming, and less suitable for field conditions. These procedures are primarily performed in tertiary healthcare centers in endemic regions or for travelers from high-income countries suspected of having the disease. However, they enable prompt intervention, such as the removal of bladder or colon polyps. In women, speculum examinations can identify genital lesions and facilitate biopsies. For suspected vesicovaginal fistula, dye can be introduced into the bladder to check for leakage through the vagina. Colposcopy aids in identifying suspicious genital lesions, particularly in cases where cancer is suspected. Esophagogastroduodenoscopy (EGD) is employed to investigate suspected varices and allows for therapeutic procedures like band ligation or sclerotherapy.

Prognosis

With a highly effective antiparasitic therapy for schistosomiasis, the prognosis is excellent, provided there is no significant organ damage. Even advanced changes can regress substantially. In patients with pronounced bladder wall changes and obstructive uropathy, complete regression can often be observed after drug therapy alone. Failure to regress a bladder polyp and hematuria must raise suspicion of bladder cancer. Similarly, in intestinal *S. mansoni* infection, extensive or complete regression of polyposis of the colon, even if marked, is still possible [14]. With early treatment of hepatosplenic schistosomiasis, there is significant regression of changes in the liver and spleen, even if reinfection continues. As few as two PZQ treatment courses taken at least 6 weeks apart can reduce the risk of developing progressive hepatic fibrosis even if reexposure persists. Patients with advanced stages of chronic schistosomiasis should also never be deprived of antiparasitic therapy, even if egg excretion is difficult to detect [15]. Surgical measures in cases of obstructive uropathy and decompensated portal hypertension should be guided by preemptive and informative schistosomiasis diagnostics since significant clinical improvement can already occur after conservative treatment. Prior to splenectomy, clinicians should assess malaria exposure and ensure administration of preventive vaccinations to lower the risk of overwhelming postsplenectomy infection syndrome (OPSI).

Summary

Schistosomes are parasites that have coevolved with mammals, including humans, over a long period. As a result, they have developed various strategies to persist within their hosts for extended durations while minimizing harm. These strategies include partially evading immune system recognition and employing the immune system to their advantage by enhancing the development of cells that suppress antiparasitic responses. As a result, clinical disease, such as fibrosis, typically develops slowly though it can sometimes lead to severe morbidity. This gradual progression is largely driven by the host's immune responses rather than by direct effects of the parasite, such as toxins. From a diagnostic perspective, detecting schistosomes remains challenging, especially with ongoing global efforts to eliminate the disease in endemic countries [16]. These regions are mainly located in sub-Saharan Africa, where many countries cannot afford the highly sensitive diagnostics required [17]. Furthermore, due to this gap and the fact that schistosomes need to be considered a One Health disease, thriving both in the environment (freshwater) [18] and the animal kingdom (snails), elimination strategies need to be multifold to ensure persistent success [19, 20].

Competing Interests This study was funded by DFG (Deutsche Forschungsgemeinschaft) CO 1469/14–1, 15–1, 16–1, and DZIF (German Centre for Infection Research) TTU Grant Number 03.818.

References

1. Anisuzzaman et al (2021) Host-specific serum factors control the development and survival of *Schistosoma mansoni*. Front Immunol 12:635622
2. Vejzagic N et al (2021) Identification of hit compounds with anti-schistosomal activity on in vitro generated juvenile worms in cell-free medium. PLoS Negl Trop Dis 15(5):e0009432
3. Mone H et al (2015) Introgressive hybridizations of Schistosoma haematobium by *Schistosoma bovis* at the origin of the first case report of schistosomiasis in Corsica (France, Europe). Parasitol Res 114(11):4127–4133
4. Langenberg MCC et al (2019) Katayama syndrome without *Schistosoma mansoni* eggs. Ann Intern Med 170(10):732–733
5. McManus DP et al (2020) Schistosomiasis-from immunopathology to vaccines. Semin Immunopathol 42(3):355–371
6. IARC Working Group on the Evaluation of Carcinogenic Risks to Humans (2012) Biological agents. IARC Monogr Eval Carcinog Risks Hum 100(Pt B):1–441
7. Lacorcia M, Prazeres da Costa CU (2018) Maternal schistosomiasis: immunomodulatory effects with lasting impact on allergy and vaccine responses. Front Immunol 9:2960
8. Katz N, Chaves A, Pellegrino J (1972) A simple device for quantitative stool thick-smear technique in Schistosomiasis mansoni. Rev Inst Med Trop Sao Paulo 14(6):397–400
9. Corstjens P et al (2020) Circulating anodic antigen (CAA): a highly sensitive diagnostic biomarker to detect active *Schistosoma* infections-improvement and use during SCORE. Am J Trop Med Hyg 103(1_Suppl):50–57
10. Mutapi F et al (2017) Human schistosomiasis in the post mass drug administration era. Lancet Infect Dis 17(2):e42–e48
11. el Scheich T et al (2014) The WHO ultrasonography protocol for assessing hepatic morbidity due to *Schistosoma mansoni*. Acceptance and evolution over 12 years. Parasitol Res 113(11):3915–3925
12. Richter J et al (2015) Severe liver fibrosis caused by *Schistosoma mansoni*: management and treatment with a transjugular intrahepatic portosystemic shunt. Lancet Infect Dis 15(6):731–737
13. Group, N.W (1996) Ultrasound in Schistosomiasis. A practical guide to the standardized use of ultrasonography for the assessment of schistosomiasis-related morbidity. Second International Workshop
14. Richter J (2003) The impact of chemotherapy on morbidity due to schistosomiasis. Acta Trop 86(2–3):161–183

15. Bergquist R, Utzinger J, Keiser J (2017) Controlling schistosomiasis with praziquantel: how much longer without a viable alternative? Infect Dis Poverty 6(1):74
16. Colley DG, Andros TS, Campbell CH Jr (2017) Schistosomiasis is more prevalent than previously thought: what does it mean for public health goals, policies, strategies, guidelines and intervention programs? Infect Dis Poverty 6(1):63
17. Faust CL et al (2020) Schistosomiasis control: leave no age group behind. Trends Parasitol 36(7):582–591
18. Steinmann P et al (2006) Schistosomiasis and water resources development: systematic review, meta-analysis, and estimates of people at risk. Lancet Infect Dis 6(7):411–425
19. Pointier JP, Jourdane J (2000) Biological control of the snail hosts of schistosomiasis in areas of low transmission: the example of the Caribbean area. Acta Trop 77(1):53–60
20. World Health Organization. Ending the neglect to attain the sustainable development goals: a road map for neglected tropical diseases 2021–2030; 2020 [cited 14 November 2024]. Available from: www.who.int/publications/i/item/9789240010352

The Medical Need for Patients with Schistosomiasis

3

Nicola Caldwell and Ian Gilbert

Abstract

Schistosomiasis affects around 140 million people worldwide, with the highest impact on people living in low- or middle-income countries. Although the number of deaths due to schistosomiasis seems relatively low, there is a long-term health burden on patients with chronic or repeat infections. Many historical anthelmintic drugs approved for the treatment of schistosomiasis are no longer in use due to a lack of efficacy or intolerable side effects. Praziquantel remains the only available treatment, but it has some considerable disadvantages—suboptimal efficacy against juvenile worms and a potential risk of developing resistance. Despite an obvious need for alternative antischistosomals, no new treatment options have emerged in the last 40 years and there are very few drug discovery programs working in this area. To tackle schistosomiasis, there is a need to establish a community-agreed target product profile to outline desirable criteria for developing new treatments. Recognizing the clinical impact of schistosomiasis and identifying the key requirements will help inform drug discovery efforts.

Keywords

Schistosomiasis · Disease burden · Anthelmintics · Target product profile · Drug discovery

N. Caldwell (✉) · I. Gilbert
Wellcome Centre for Anti-Infectives Research, Division of Biological Chemistry and Drug Discovery, School of Life Sciences, University of Dundee, Dundee, UK
e-mail: n.z.caldwell@dundee.ac.uk

T. Spangenberg (ed.), *Praziquantel*,
https://doi.org/10.1007/978-3-031-97397-0_3

Epidemiology

Epidemiology is the study of the determinants, occurrence, and distribution of disease in a defined population. Epidemiological analysis is fundamental to understanding how particular diseases affect global health and aids in developing strategies for prevention, control, and management. Some key terminology associated with epidemiology is used throughout this chapter to describe the global impact of schistosomiasis [1]. For example, "prevalence" is the total number of cases of a particular disease in a specified population over a designated time period. "Morbidity" refers to the condition of having a particular disease or the rate of disease within a population, whereas "mortality" refers to the number or rate of deaths. "Risk factors" are particular aspects associated with lifestyle, environmental exposure, or genetic characteristics that may increase susceptibility of developing a disease.

Prevalence

Schistosomiasis, an infectious disease caused by *Schistosoma* species, is the second most pathogenic human parasitic infection after malaria [2]. It is endemic in 78 countries worldwide, with around 800 million people at risk of infection. The majority of people (>90%) at risk of developing infection live in sub-Saharan Africa (Fig. 3.1) [3]. Since this disease adversely affects people in low- and middle-income countries (LMIC), the World Health Organization (WHO) has designated it as a neglected tropical disease (NTD) and set targets for its elimination as a public health

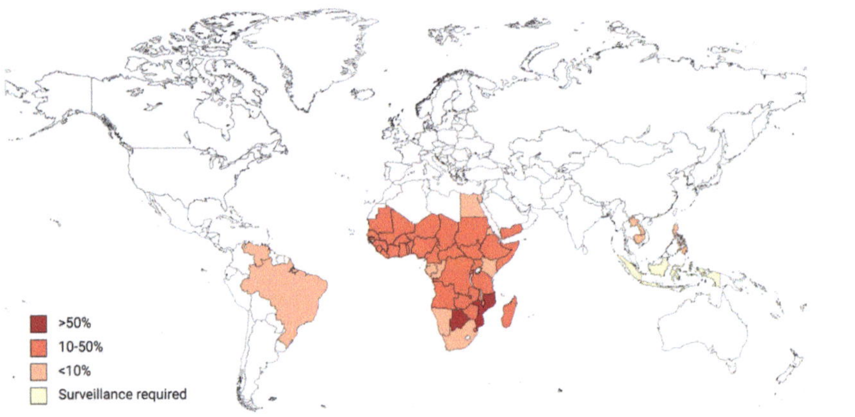

Fig. 3.1 Global prevalence of schistosomiasis in 2022 showing percentage of population requiring preventative chemotherapy (Map created with mapchart.net using data from World Health Organization Preventative Chemotherapy and Transmission Control (WHO PCT) Databank 2023 [6])

Table 3.1 Ranking of schistosomiasis prevalence compared to the "Big Three" infectious diseases

Disease	Prevalence (millions)
Tuberculosis	1912
Malaria	174
Schistosomiasis	151
HIV/AIDS	40

problem by 2030 [4]. To achieve this goal, the WHO has implemented a preventative treatment program relying on mass drug administration (MDA) with praziquantel (PZQ) in areas endemic for *Schistosoma mansoni* (*S. mansoni*) [5]. In 2022, it was estimated that >260 million people required annual preventative chemotherapy with PZQ [6].

Compared to tuberculosis, malaria, and HIV/AIDS, which are considered the "Big Three" infectious diseases [7], schistosomiasis ranks highly in prevalence (Table 3.1). Data from the most recent Global Burden of Disease Study estimated around 151 million cases in 2021. Only malaria and tuberculosis had a higher number of cases in this timeframe. Compared to the total number of cases in 2011 (197 million) [2], this represents a 23% decrease in prevalence of schistosomiasis over a 10-year period, likely as a result of preventative chemotherapy programs with PZQ. However, the concomitant number of deaths due to schistosomiasis in the same period only fell by 16% (15,500 in 2011 compared with 13,000 in 2021), suggesting that additional or replacement treatments are needed to tackle this disease [8].

Well-defined target product profiles (TPP) have been published, which outline specific use-cases and desirable criteria for developing new antimalarials [9]. Similarly, a suite of in vitro assays is available to profile early antituberculosis hits as well as established animal models to support lead optimization of advanced compounds toward preclinical candidate selection [10]. These tools are used to direct drug discovery efforts in these disease areas. Despite the comparable prevalence of schistosomiasis, no TPP or compound progression criteria has yet been established that delineate the path toward preclinical candidate selection for new treatments. A community-agreed TPP would help to support future efforts to tackle the high global burden of schistosomiasis.

Geographical Distribution

Three main *Schistosoma* species are responsible for the majority of human infection—*S. mansoni*, *S. japonicum*, and *S. haematobium*. Each of these species has a different geographical distribution relating to the habitats of freshwater snail intermediate hosts in those locations (Table 3.2). Both *S. mansoni* and *S. haematobium* are endemic in Africa, which has the highest prevalence of schistosomiasis. Mixed infections of both species have been reported in Senegal [11]. *S. mansoni* is also endemic in regions of South America (Brazil, Suriname, and Venezuela) whereas

Table 3.2 Geographical distribution of major *Schistosoma* species that cause human infection [12, 13]

Disease	*Schistosoma* species	Snail species	Geographical distribution
Intestinal schistosomiasis	*S. mansoni*	*Biomphalaria* spp.	Sub-Saharan Africa, Sudan, Egypt, Brazil, Suriname, Venezuela, Caribbean islands
	S. japonicum	*Oncomelania* spp.	Indonesia, China, Philippines
	S. mekongi	*Neotricula aperta*	Cambodia, Laos
	S. intercalatum	*Bulinus* spp.	Central and West Africa
	S. guineensis	*Bulinus forskali*	Central and West Africa
Urogenital schistosomiasis	*S. haematobium*	*Bulinus* spp.	Sub-Saharan Africa, Northwest Africa, Middle East, Corsica

S. haematobium is endemic in the Middle East, with more recent transmissions identified in Corsica [12]. *S. japonicum* is mainly found in China, Indonesia, and the Philippines. Other minor species of *Schistosoma* (*S. mekongi*, *S. intercalatum*, and *S. guineesis*) are less prevalent and are mainly distributed in localized areas (Table 3.2) [13].

In addition to human-infective *Schistosoma* species, several species (*S. bovis*, *S. curassoni*, and *S. mattheei*) occur in livestock. These species are mainly located in Africa and hybrid variants with *S. haematobium* have been described in co-endemic areas. Hybrid variants occur as a result of humans or animals becoming infected with more than one *Schistosoma* species. Subsequent cross-species pairing of adult worms inside the mammalian host can result in the production of eggs that share phenotypes of both species. Genetic sequencing of eggs retrieved from infected patients has confirmed this hybrid forms [14]. Such interspecies hybrids have been identified in infected children in Malawi and Mali [15, 16]. The prevalence and impact of human infection by these hybrids is not well understood, yet they contribute to transmission.

Climate Change and Trends

Climate change has an impact on the geographical distribution of schistosomiasis; however, these effects can be difficult to predict on a global scale through computational modeling. Schistosome life cycle development within the intermediate snail host is particularly sensitive to temperature, with an optimal range between 15 and 31 °C. Outside of this range, snail numbers remain constant, but parasite survival and fecundity decline [17]. Therefore, global shifts in temperature and weather patterns can result in new geographical areas becoming suitable habitats for schistosome development within the intermediate host. An example of this is the outbreak on the island of Corsica (France), a region that was previously suboptimal for schistosome survival due to the cold temperatures of inland waters in winter. As a result of rising temperatures, *S. haematobium* parasites originating from Senegal were transmitted by travelers and eggs excreted into freshwater sources were able to infect and grow within susceptible snail hosts [18].

Conversely, increasing temperatures in tropical regions can also affect schistosome development within the snail and may negatively impact parasite survival. It is believed that this may reduce the prevalence of schistosomiasis in regions of sub-Saharan Africa, where infection is endemic. Other climate-related changes such as altered rainfall patterns and rising water levels may expand suitable snail habitats or facilitate their movement to other regions [17].

Finally, it is important to recognize that the consequences of climate change on human health may exacerbate the severity of schistosomiasis and other NTDs, leading to increased morbidity and mortality. Flood, famine, drought, and malnourishment, for example, could indirectly affect disease prevalence and progression [17].

Clinical Impact

Schistosoma parasites have a complicated life cycle, involving asexual replication in the freshwater snail host prior to human infection (Chap. 2). Snails release tens of thousands of free-swimming cercariae into water sources, which can infect humans through penetration of the skin. During penetration, cercariae lose their tails to become schistosomula, which enter the bloodstream and migrate through the veins to the liver via the lungs. In the liver, schistosomula develop into juveniles and then adult worms over a period of 7 weeks. Adult worm pairs subsequently migrate to different tissues or organs in the body depending on the species. Of the three species that cause the majority of human infections, *S. mansoni* and *S. japonicum* migrate to the mesenteric venous system (intestinal schistosomiasis), and *S. haematobium* migrate to the venous plexus of the bladder and other pelvic organs (urogenital schistosomiasis). Sexually mature male-female pairs can reside within these areas of the body for years where they mate and females release eggs into the systemic circulation. Some of these eggs are excreted in urine or feces and can hatch, releasing miracidia into freshwater sources that can infect snails, completing the parasite life cycle, and leading to transmission of infection (Fig. 3.2). Eggs that are not excreted can become trapped in surrounding tissues, triggering an immuno-inflammatory response. The resulting granulomas and fibrosis caused by immune cells are responsible for the development of clinical symptoms of schistosomiasis.

Intestinal and Urogenital Schistosomiasis

Due to the migration of worms in the body and subsequent deposition of eggs in different tissues, the symptoms of schistosomiasis are wide-ranging. Common symptoms of intestinal schistosomiasis, caused mainly by *S. mansoni* and *S. japonicum*, include abdominal pain, inappetence, and diarrhea. Chronic infection and recurrent exposure can also result in gastrointestinal bleeding from ulcerations. In addition, if eggs become trapped in the liver, inflammation and liver fibrosis can

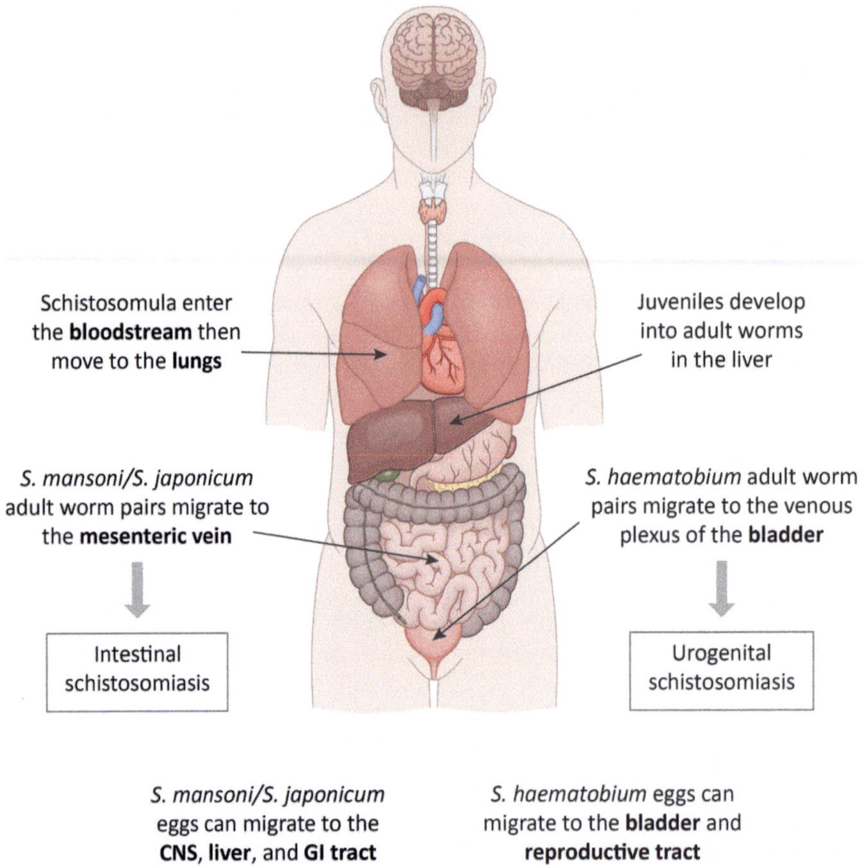

Fig. 3.2 Schistosomiasis life cycle highlighting infected organs in humans (Diagram created using BioRender®)

occur [19]. Schistosomiasis is a common cause of severe liver disease. The most serious complication of advanced hepatic fibrosis is bleeding from the upper gastrointestinal tract, which can be fatal. Liver enlargement (hepatomegaly) is also commonly detected by ultrasound in up to 80% of infected children although this is less frequently observed in adults [20]. As this is a secondary effect of infection, it is not known whether hepatomegaly resolves after treatment of schistosomiasis [21].

Granuloma formation in the central nervous system (CNS) is more common with *S. japonicum* as the eggs are smaller and can therefore migrate more easily to the brain. In contrast, *S. mansoni* eggs are more likely to accumulate in the spinal cord [22]. Symptoms are wide-ranging and include intracranial hypotension (causing headache, nausea, and vomiting), focal neurological deficits, and seizures. Since most of these symptoms manifest in the absence of systemic disease, it can be challenging to diagnose schistosomiasis as parasites are rarely detected in urine or stool in these cases [19]. Infection in children can lead to long-term cognitive impairment and learning disorders.

Urogenital schistosomiasis is caused by *S. haematobium* infection in the bladder and reproductive tract [19]. Again, inflammation and subsequent granuloma formation is responsible for the related symptoms. Chronic fibrosis in the urinary tract, particularly in the bladder outlet, can cause obstructions. Blockages in this region of the body can ultimately lead to kidney failure [20]. The most common clinical presentation of infection in the bladder is hematuria (blood in the urine). There is also an increased risk of squamous-cell carcinoma of the bladder as the granulomatous lesions can intensify the exposure of the bladder epithelium to mutagenic substrates [20]. Consequently, *S. haematobium* is classified as a carcinogen.

In urogenital schistosomiasis, inflammation and subsequent fibrosis can also develop in the seminal vesicles and prostate in males and the uterine cervix, vagina, and vulva in females. Fibrotic areas typically present as characteristic "sandy patches" in these regions [23]. Female genital schistosomiasis (FGS) has a significant impact on women's reproductive health and is associated with increased risk of contracting HIV [24]. Common symptoms include pain, incontinence, infertility, and increased risk of abortion [25]. Male genital schistosomiasis (MGS) commonly presents with pain, hematospermia, hematuria, and genital disfigurement [26]. Chronic infection may be associated with infertility and sexual dysfunction.

In general, symptoms of schistosomiasis are nonspecific and can easily be confused with other medical conditions [19]. Urogenital schistosomiasis is particularly challenging to diagnose as there may not be detectable eggs in the urine and the symptoms are also frequently associated with sexually transmitted infections (STIs). The related stigma impacts on the number of patients seeking medical care. There is also a lack of awareness of FGS, and its consequent effect on women's health among healthcare professionals and people living in endemic communities [27]. As a result, FGS is severely underdiagnosed [28].

There is limited evidence of the effectiveness of PZQ for treatment of FGS [28]. Ideally, new treatments for schistosomiasis should have efficacy against intestinal and urogenital forms of the disease caused by *S. mansoni*, *S. japonicum*, and *S. haematobium*. However, parasite-killing alone may not be sufficient for treatment of FGS and MGS. Current dosing regimens with PZQ do not appear to resolve lesions and additional symptomatic treatment, for example with anti-inflammatories, may be required. Ultimately, a separate use-case scenario and TPP may be necessary for treatment of FGS and MGS. In addition, public health education and screening campaigns that accompany the implementation of new treatments will also be needed to increase awareness of FGS/MGS among healthcare professionals (HCPs) and people living in endemic countries [27].

Table 3.3 Estimated mortality and morbidity figures for schistosomiasis globally in 2021 [2]

Measure	Number[a]
Deaths	12,857
Disability-adjusted life years (DALYs)	1,746,333
Years of life lost (YLLs)	500,461
Years lived with a disability (YLDs)	1,245,872

[a]Data from Global Burden of Disease Study 2021 (GBD 2021) [2]

Mortality and Morbidity

Schistosomiasis is reported to directly cause around 13,000 deaths annually worldwide (Table 3.3) although this figure is likely to be higher due to both underdiagnosis of the disease and underreporting of death rates [29]. In addition to deaths resulting directly from schistosomiasis, it is likely to be a contributing factor to mortality of other conditions. An analysis of death certificates in Brazil between 1999 and 2018 identified that schistosomiasis was an associated cause of death in 4168 cases, in addition to being directly responsible for 10,251 deaths. The disease was predominantly associated with causes of death due to disorders of the digestive system, especially liver diseases [30].

Another metric used to assess the global health impact of schistosomiasis is disability-adjusted life years (DALYs). One DALY represents the loss of the equivalent of 1 year of healthy life and is calculated as to the sum of years of life lost (YLL) and years lived with a disability (YLD) as a result of a particular disease or health condition. In 2021, DALYs due to schistosomiasis were estimated to be around 1.7 million [2], reflecting the long-term health burden. Although mortality rates are not as high as other infectious diseases such as malaria, morbidity caused by granuloma formation and fibrosis severely impacts the health of chronic sufferers.

Demographics

DALYs can be analyzed by grouping into different demographics (Fig. 3.3) to explore the health burden on particular populations. The sociodemographic index (SDI) is a composite measure that takes into account income, education, and fertility rate and can be used to represent social and economic development in countries across the world. People living in countries with a low SDI (Fig. 3.3a) account for over half (51%) of the total number of DALYs caused by schistosomiasis. Similarly, a significant proportion of DALYs can be attributed to people living in countries with low-middle (30%) and middle (18%) SDI. Conversely, high-middle (1%) and high (0%) SDI areas have only a low percentage of the total number of DALYs. These data correlate with the high prevalence of disease in sub-Saharan Africa and reflect the disproportionate impact of schistosomiasis on LMICs.

Breakdown of DALYs by age group (Fig. 3.3b) shows a higher impact on young adolescents, presumably due to repeated exposure to infection throughout

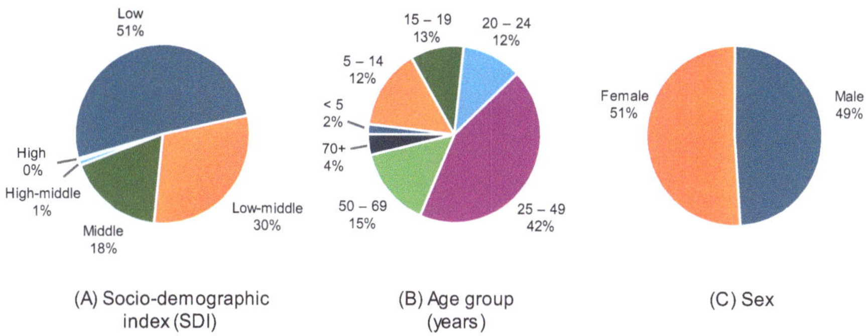

Fig. 3.3 Disability-adjusted life years (DALYs) in 2021 categorized by (**a**) sociodemographic index, (**b**) age group, and (**c**) sex. DALYs for each demography are expressed as a percentage of the total number of DALYs (Data from Global Burden of Disease Study 2021 (GBD 2021) [2])

childhood [25]. Worms can survive for a number of years within the human host (average 3–10 years) [25], so the health burden is also high in adults, particularly if they have frequent contact with contaminated water and are at risk of repeated infection. There does not appear to be a significant difference in DALYs according to sex (Fig. 3.3c).

These statistics highlight the continued need for schistosomiasis control and treatment programs in LMICs. New therapies must be suitable for treating individuals of any age including infants, school-aged children, and adults. In 2022, the WHO reported that 93 million people globally received preventative chemotherapy with PZQ. When categorized by age group, this represents 51.9% treatment coverage for school-aged children but only 17.6% treatment coverage for adults [6]. Therefore, there is a concern that there are gaps in treatment coverage and disease monitoring for at-risk adults in endemic areas [31]. In addition, pregnant women are often excluded from MDA programs, so new treatments must be assessed for their safety in these patient populations. Implementation of new treatments and ensuring access for all at-risk groups is needed to ensure adequate treatment coverage and reduce transmission.

Schistosomiasis Treatments

Historical Treatments

Schistosomiasis treatment has been reported since the beginning of the twentieth century (Fig. 3.4), with the use of antimonials that were commonly used at the time to treat leishmaniasis [32]. In 1918, it was discovered that antimony potassium tartrate could be used to treat hematuria presumably caused by *S. haematobium* [33]. Antimonials prevented the release of eggs in schistosomes and also inhibited carbohydrate metabolism, leading to a loss of muscle tone and subsequent detachment of male worms from the blood vessels [34]. Although egg-induced morbidity was not

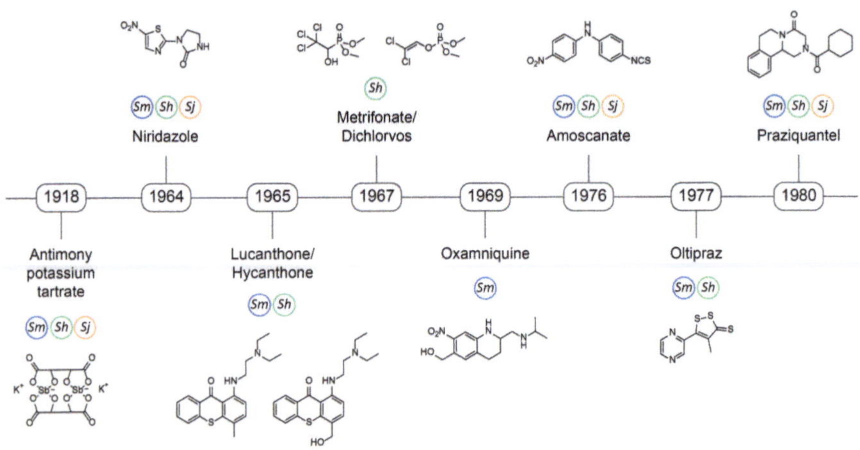

Fig. 3.4 Timeline of approved schistosomiasis drugs from past to present showing estimated year drug was first used [32]. *Sm* = *S. mansoni*, *Sh* = *S. haematobium*, *Sj* = *S. japonicum*; indicating which species compounds are active against

well-understood at the time, it was noted that treatment of schistosomiasis with antimonials correlated with faster healing time of urinary fistulae that were surgically operated on in chronic patients [35]. Antimonials were mainly restricted to inpatient settings and were administered intravenously. Increasing doses up to a maximum of 2 g were given on alternating days [33]. However, the accumulation of antimony in the body led to extensive side effects including nausea, vomiting, diarrhea, myalgia, dry cough, and skin rashes. More serious adverse events such as severe pain, portal hypertension, and cardiac dysrhythmia were often observed and were fatal in some cases. Due to the associated toxicity, antimonials suffered from poor patient compliance but, despite this, remained the only available schistosomiasis treatment for over 40 years.

In the 1960s and 1970s, an endeavor to identify novel, oral treatments led to the identification of several alternatives to antimonials. Metrifonate was originally identified as a treatment for urogenital schistosomiasis in 1967 due to its selective effects on *S. haematobium* worms. Metrifonate is a prodrug that releases dichlorvos, an organophosphate acetylcholinesterase inhibitor that is believed to cause paralysis in adult *S. haematobium* worms. It is not well understood why these effects do not extend to other *Schistosoma* species [32]. Due to the relatively mild side effects (mainly gastrointestinal distress) observed in humans, metrifonate was the predominant treatment for urogenital schistosomiasis until it was superseded by PZQ. The recommended dosing regimen for metrifonate was 7.5–10 mg/kg orally, which was repeated twice after intervals of 2 weeks [34]. In addition to their use as antischistosomals in humans, metrifonate and dichlorvos were widely used as insecticides in household and agricultural settings but have since been banned or restricted in many countries due to concerns regarding their toxicity and environmental impact [36].

As a result of these restrictions and the superior efficacy observed with PZQ, metrifonate was discontinued in 1999 [37].

Niridazole was first introduced as an alternative to antimonials in 1964 and was used to treat schistosomiasis through the 1960s and 1970s. It is believed that nitroreductive metabolism followed by covalent binding of metabolites to schistosome macromolecules in adult worms is responsible for the parasite killing ability of this compound [38]. The major advantage of niridazole compared to antimonials was that it could be administered orally and the recommended treatment regimen was 25 mg/kg per day for 7 days or 35 mg/kg per day for 5 days, preferably split over two or three doses per day [34]. It demonstrated moderate efficacy against *S. mansoni* and *S. japonicum* but was particularly effective against *S. haematobium* [39]. However, niridazole caused serious adverse events, including neurological side effects, which meant that it was contraindicated in patients with a history of mental illness or epilepsy [32]. Many other side effects were noted and niridazole was also found to be carcinogenic and mutagenic [40]. Due to the high toxicity, its use in MDA was precluded.

Lucanthone was used clinically to treat schistosomiasis caused by *S. mansoni* and *S. haematobium* prior to evidence that the biologically active component was actually its hydroxylated metabolite, hycanthone. From 1965 onward, hycanthone replaced lucanthone as the preferred therapy due to its superior activity when administered as a single low dose (3–3.5 mg/kg) intramuscular injection [32]. Hycanthone is hypothesized to bind irreversibly to acetylcholine receptors, leading to paralysis in the worm digestive tract, preventing them from feeding [34]. Some liver toxicity was observed but could be managed by lowering the dose. However, it was subsequently discovered that hycanthone was both mutagenic and carcinogenic. It can bind to DNA by intercalating between base pairs, distorting the helical structure, and causing mutations [41]. For this reason, its use was phased out by 1972.

Exploration of analogues of hycanthone led to the discovery of oxamniquine, a compound that also possesses superior activity as the hydroxylated metabolite. Oxamniquine acts as a prodrug, a compound that is in itself inactive but is metabolized within the body to release an active metabolite. In the case of oxamniquine, it is sulfonated by a sulfotransferase enzyme inside the parasite [42]. However, oxamniquine only displays activity against *S. mansoni* worms, which possess this sulfotransferase enzyme target. Once formed, the resulting sulfate ester is unstable and decomposes to form an electrophilic species that can alkylate DNA and disrupt metabolic functions, leading to death of the parasite. As the mechanism of activation is parasite-specific, oxamniquine displays relatively minor side effects on the human host and has been reported to be well-tolerated [43]. This drug has been widely used in Brazil where *S. mansoni* is endemic. Its oral formulation, low cost, and nonserious side effects mean that it is amenable to MDA and is most commonly given as a single 15–20 mg/kg dose [32]. It is still in clinical use today but PZQ tends to be favored due to its efficacy against additional *Schistosoma* species.

Amoscanate is another example of a drug that has been used extensively in a specific location. Primarily used in China (6–7 mg/kg per day for 3 days) to treat infections caused by *S. japonicum*, it is a broad spectrum antiparasitic with activity against the three main *Schistosoma* species [32]. Amoscanate is believed to operate via inhibition of cyclic adenosine monophosphate (cAMP)-phosphodiesterase [44] that disturbs regulation of metabolism, resulting in worms being swept into the liver and eventually killed. The isocyanate moiety is essential for activity [32]. Ultimately, the use of amoscanate was discontinued in favor of PZQ due to observed liver toxicity and CNS disturbances [34].

Finally, oltipraz was another clinically active drug that preceded PZQ and showed efficacy against *S. mansoni* and *S. haematobium*. Oltipraz decreases glutathione levels in schistosomes, causing oxidative stress and affecting parasite survival [45]. It has also been shown to inhibit parasite thioredoxin glutathione reductase, which may account for this activity [46]. Oltipraz was administered orally at doses of 20–35 mg/kg per day. Similar to previous treatments, oltipraz was withdrawn due to toxicity—primarily phototoxicity that caused paresthesia (abnormal sensation or feelings in the skin) and acute pain in fingertips [32].

Of all the examples described above, the majority of these treatments have fallen out of clinical use due to concerns about toxicity, or a lack of efficacy against the three main species responsible for human infection. Research during the period of the 1960s–1980s culminated in the introduction of PZQ, which had an enhanced safety profile compared to its progenitors and was active against *S. mansoni*, *S. japonicum*, and *S. haematobium*. To date, PZQ remains the sole preferred treatment for intestinal and urogenital schistosomiasis.

Praziquantel

Revolution of schistosomiasis treatment came with the introduction of PZQ. Originally synthesized as part of a series of tranquilizer compounds later tested for anthelmintic activity, PZQ was first approved as a veterinary treatment before being approved for use in humans in 1980. It is administered as a racemic mixture, of which the (*R*)-enantiomer is the most active (Chap. 4 & 6) [47]. Although the mechanism of action is not fully understood, PZQ is believed to operate via modulation of voltage-operated calcium channels [48]. Engagement of a transient receptor potential melastatin ion channel ($TRPM_{PZQ}$) is consistent with the in vitro effects observed in schistosomes [49] (Chap. 5); however, activity at additional targets (potentially GPCRs) is likely to contribute to activity in humans through immunoregulatory pathways [48]. Treating adult worms with PZQ causes rapid muscle contraction and damage to the tegument (outer surface membrane) of adult worms, resulting in paralysis as well as exposing schistosome antigens to attacks by the host immune system (Chap. 6) [50].

Compared to previous treatments, PZQ had a much-improved safety profile and was well-tolerated in patients with only minor side effects such as abdominal discomfort, headache, and nausea (Chap. 8 & 10). Such side effects may not be directly drug-related and could occur due to the clearance of paralyzed worms from the body. Other benefits included its oral formulation, low cost, and activity against all three main *Schistosoma* species. For these reasons, PZQ remains the recommended treatment for schistosomiasis for more than 40 years and has been used extensively in MDA programs in endemic areas.

The WHO has several recommendations for preventative chemotherapy with PZQ depending upon the prevalence of infection in endemic communities. Prevalence is defined as the percentage of the population that is infected with any species of *Schistosoma*. There is no difference in WHO recommendations for infection caused by different species. In countries with prevalence ≥10%, annual preventative chemotherapy with a single dose (usually at 40–60 mg/kg) is recommended to control morbidity [5]. The aim is to achieve ≥75% treatment coverage in all age groups to target elimination of schistosomiasis as a public health problem by 2030 [4]. In countries where this level of treatment coverage is met and there is a lack of response, the WHO recommends a biannual dose of PZQ instead of annual preventative chemotherapy [5].

In lower prevalence settings (infection <10%), if there has been a regular program of preventative chemotherapy, the WHO suggests continuing this at the same or reduced frequency to prevent transmission of infection. If there is no existing MDA program, a test-and-treat approach is recommended instead [5].

Both prevalence and cure rate in these cases are determined by counting the number of eggs in feces or urine using Kato-Katz or urine filtration techniques and lab-based methods for preparing stool samples for examination by microscopy. By this measure, mean cure rates of 77% have been reported for both *S. mansoni* and *S. haematobium* using the WHO-recommended treatment program. Cure rates up to 95% have been reported in the case of *S. japonicum* [51]. However, it is thought that diagnostics based on egg counting may be overestimating the cure rate of PZQ. More sensitive methods that measure worm-derived antigens suggest that parasites can remain in humans even after multiple treatments with PZQ (Chap. 9) [52]. Despite the advancement of more sensitive diagnostics, these are not widely used in the field due to poor reproducibility, or the need for expensive equipment and trained personnel [53]. Therefore, there is a lack of sufficient clinical data to support replacing conventional methods. As a result, there is no clear consensus on the most appropriate way to assess drug efficacy, which will be a challenge for the implementation of new treatments.

Preventative chemotherapy programs have been successful at reducing prevalence and morbidity however, have thus far failed to eradicate infection, suggesting that PZQ alone may not be sufficient to completely interrupt transmission. Additional or replacement treatments may be needed to completely eliminate schistosomiasis as a public health problem.

A major drawback with this drug is that it is only effective against adult worms. The failure to act upon juvenile worms means that a single treatment is unlikely to completely clear all parasites and prevent reinfection. Furthermore, suboptimal pharmacokinetic properties (extensive first-pass metabolism and low oral bioavailability) are likely to impact on its ability to reach all parasites in the body and completely clear infection.

The lack of efficacy against all life cycle stages of the parasite, combined with PZQ's large-scale use in preventative chemotherapy programs, raises concerns about the development of drug resistance. As the sole treatment relied upon for treating schistosomiasis, resistance toward PZQ could have devastating consequences for global health. Alternative therapies will be urgently needed in this event.

Drug Resistance

Determining the existence and/or extent of resistance in the field is complicated by a lack of robust methods of detection [54]. Reported cases of reduced susceptibility to PZQ treatment have not revealed any convincing evidence of resistance and can be explained by epidemiological factors such as high initial worm burden and rapid reinfection [55]. Despite the fact that widespread resistance is not presently apparent, the risk that it may emerge remains a concern.

The low clinical efficacy of PZQ could be due to multiple factors. Resistance is a possible explanation; however, there is currently little evidence to support this. The schistosomula and juvenile worms are found in a wide variety of tissue types. It is possible that there is insufficient tissue distribution of PZQ to reach all the parasites. A related hypothesis is that PZQ is insufficiently potent to kill all the parasites in an infection. Another possibility is the presence of persister parasites which have reduced susceptibility to PZQ.

Target Product Profiles

Suggested Criteria for Next Generation of Anthelmintics/Antischistosomal Drugs

There is a need for new compounds for the treatment of schistosomiasis. Unfortunately, there is little systematic discovery being undertaken. However, to enable this discovery, there is need for a target product profile (TPP).

The drug TPP is a strategic tool used in drug discovery and development to outline the desired characteristics of a potential new drug. It serves as a guide for the development of the drug by defining its intended route of administration, therapeutic uses, safety parameters, and efficacy goals. The TPP helps align the expectations of stakeholders, including researchers, regulators, public health authorities, and commercial teams, and provides a foundation for decision-making throughout the drug development process.

The TPP typically includes information such as the intended patient population, the disease or condition the drug is meant to treat, the desired efficacy outcomes, safety and tolerability considerations, and potential commercial aspects. It is a critical document that helps ensure that the development process is focused and efficient.

In 2022, a workshop was organized to discuss what is needed in terms of a drug for schistosomiasis and hence a TPP. It became apparent that this is a complex issue. A summary of the discussions has been published [26]. There are a number of possible TPPs for schistosomiasis that could be developed: test and treat, MDA, and female/male genital schistosomiasis. There is also the question as to whether a new drug would be given as a single agent or in combination with another agent, most likely PZQ.

For treatment, it is necessary to find compounds that target all the different human life cycle stages of the parasite: the schistosomula, juvenile worms, and adult worms. A drug should be cidal (parasite killing) to clear the parasites. While adult worms are found in either the mesenteric veins or in the blood vessels around the bladder, the juvenile worms and schistosomula are much more widely distributed around the body. Hence, a compound should have a widespread tissue distribution (high volume of distribution). Further, a compound that has a long half-life may reduce the number of doses to be given. Additionally, a long half-life compound should provide some chemoprotective effect following cure, interrupting transmission, which is going to be important for elimination. A compound with improved efficacy to PZQ that could be used with a test-and-treat protocol is urgently needed. To facilitate such a protocol, there is a need for simple, cheap, and sensitive diagnostics to enable better identification of patients and to make sure that they are cured following treatment; thus, more sensitive and accurate approaches than counting eggs in feces are needed.

MDA is currently carried out with PZQ which although it has limited efficacy, is very cheap. A drug for MDA would need to be affordable and suitable for single-dose administration, while possessing the properties described above. This represents a 'high bar'.

Little is known about FGS and MGS. FGS is characterized by granulomas formed around eggs deposited in tissues in the genital area. Over time, this gives rise to fibrosis. In terms of treatment, there is a question over whether killing the eggs is desirable: killing the eggs may give rise to a significant immunological response. However, it has been suggested that some kind of anti-inflammatory should be given together with the antischistosomal compound. Unfortunately, little is known about FGS and MGS and there is a requirement for more research and understanding of the disease biology, prevalence, and the development of animal models of infection. Also, a simple method for diagnosis is required, as the current methods are highly invasive.

Table 3.4 indicates some aspects required for a TPP and some ideal properties. It may be acceptable to be less stringent with some of the criteria. This is not yet complete, and outstanding questions are highlighted.

Table 3.4 Ideal criteria for new treatments and outstanding questions

	Treatment	MDA	FGS	Questions
Target species	Ideally against all species known to infect humans. A minimum against *S. mansoni* and *S. haematobium* may be acceptable			
Target population	Ideally for all people at risk, including pregnant and lactating women and infants <2 years			
Treatment regimen	Ideally single dose, but room for latitude	Single dose	Ideally single dose, but likely to require multiple doses	
Formulation	Oral	Oral	Oral	Is there any possibility of long-lasting injectables?
Efficacy	High efficacy (>95%?)	High efficacy (>90%?)	High efficacy (>90%?)	Need to define how efficacy is measured, with more sensitive easy to use diagnostics
Safety	High level of safety. No monitoring required. Suitable for pregnant and lactating women, women of child-bearing age and infants <2 years			
Drug–drug interactions	No significant drug–drug interactions with commonly used anti-infectives (e.g., for malaria, TB, etc.)			
Shelf life	>2 years in zone IVb[a]			

[a]Hot/higher humidity climate zone

Schistosomiasis Drug Discovery

Challenges

Very little systematic drug discovery has been conducted for schistosomiasis, despite the clinical need. As a consequence, there are a number of challenges that need to be solved. As described above, it is going to be important to develop high-throughput assays that target the schistosomula, the juvenile worms, and the adult worms. This requires multiple different phenotypic assays. There are robust assays for screening against schistosomula at a reasonable throughput. These are obtained by stimulating snails to release the cercariae, followed by transforming these into schistosomula (Chap. 6). This requires maintaining a large snail colony for supply of parasites. Bickle and coworkers [56] developed a high-content screen that measures both motility and morphology of the parasite. However, the life cycle cannot be maintained in vitro, leading to a very limited supply of juvenile and adult worms for screening. Most laboratories obtain juvenile and adult worms from mice or hamsters. Even so, all methods only supply relatively few adult parasites, which limits the amount of screening that can be carried out on juvenile and adult worms. Most scientists screening adult worms rely on a WHO protocol, which involves observing the adult worms microscopically. Worms are assigned a motility score from 0 (no movement) to 4 (normal movement) and this is used to determine compound activity. This method has very low throughput and low reproducibility.

There is also a challenge with the animal models that are used to assess compound efficacy in vivo. The most commonly used is the mouse model of *S. mansoni.* The *S. haematobium* mouse model does not replicate the human disease; the parasites are not found around the bladder. Some initial work has been done on FGS models, but better ones are still needed.

Most screening is done with *S. mansoni*. However, it is important to make sure that compounds are screened against other species, including *S. haematobium*, *S. mansoni*, and *S. japonicum*. In addition, there are a number of hybrid species between human and bovine pathogens; it is not clear how important these are clinically. It will be important to screen these as well, at least until we have a better understanding of their clinical importance.

A common approach to drug screening for schistosomiasis is to test compounds phenotypically in whole-organism assays as described above (Chap. 6). Target-based screening, testing compounds biochemically against a particular enzyme or protein, could offer a higher throughput approach to identifying active hits (Chap. 5). However, there are very few fully validated drug targets against schistosomes. Target-based projects, using validated targets, would greatly facilitate drug discovery due to the challenges of phenotypic screening and maintenance of the whole life cycle. Full target validation is through demonstration of clinical efficacy. However, initial target validation is probably best achieved by establishing compound–target pairs for phenotypically active compounds. This approach requires target deconvolution, which is complex in this parasite. Further, this validation is required in schistosomula, juvenile worms, and adult worms. Genetic validation of at target is more challenging in these parasites, as they are multicellular organisms. Different proteins may be expressed in different cell types at different times of the parasite life cycle.

Drug Discovery Pathway

Despite all these challenges, some drug discovery programs have been taking place. As far as the authors are aware, the most advanced of these is a compound series developed by a collaboration between not-for-profit small molecule drug discovery organization Salvensis and Quentin Bickle's laboratory at the London School of Hygiene and Tropical Medicine. This series is now in late preclinical development. However, little systematic drug discovery is taking place, despite the huge clinical need. A big challenge is the relatively low throughput of phenotypic screening.

The limited throughput of assays has been a constant challenge. Different workflows have been developed. Bickle and Hoffmann teams have developed a workflow in which the primary screening is against the schistosomula, which is a relatively high-throughput assay. Hits are subsequently screened against adult and juvenile worms. In our experience, many of the hits that are obtained against the schistosomula have low selectivity windows compared to human cell lines, such as HepG2 (second-generation hepatocellular carcinoma). It is also important to take into account drug metabolism and pharmacokinetic (DMPK) properties and safety aspects during the discovery phase, working to compound progression criteria.

Other groups have started screening with adult worms. However, this has a very low throughput and is probably best suited for highly curated compounds, which have a high chance of having antiparasitic activity; this method does not have sufficient throughput for diversity screening.

Compound Properties

Drug discovery programs follow Target Candidate Profiles and Compound Progression Criteria, which define the biological activity, physicochemical, pharmacokinetic, and safety properties that are desired for a molecule. Some of these properties depend on the route and frequency of compound administration, while others are dependent on the pathogen and disease biology.

In the case of schistosomiasis, the likely TPP would be an oral compound, with a large tissue distribution, a long half-life, and ideally single dose. The compound properties required for this profile are well understood, albeit challenging to obtain. To obtain good oral bioavailability would require good permeability (typically obeying the Lipinski rule of 5) and high metabolic stability. A large volume of distribution is needed to achieve the desired tissue exposure and half-life, which also demands high metabolic stability and the presence of either a basic center and/or a higher LogD than is normally found in oral drugs. The risk of the basic center and the elevated LogD is that these increase the possibility of toxicity and developability issues [57]. Another potential route of administration could be a long lasting injectable, which would require very different physicochemical properties.

A recent publication [57] analyzed compounds reported to have antischistosomal activity and found that the compounds tended to have high lipophilicity (cLogD >3). The authors point out the need to look for compounds with better developability; Highly lipophilic compounds often have problems during development; they tend to be more toxic and have poor solubility and metabolic instability. For example, currently developed antischistosomal compounds have cLogD $\leq$2.5.

What is not really understood in the case of schistosomiasis is whether there are any barriers to compound entry into the parasite and whether there are issues with compound efflux. If either penetration into the parasite is limited or efflux occurs, it is unclear which compound properties could circumvent these issues, and indeed these properties may vary between the different life cycle stages. In addition, schistosomes are multicellular organisms, so additional factors may affect compound access and distribution.

Summary

Schistosomiasis remains a major health issue across many LMICs, particularly in sub-Saharan Africa. The disease is complicated due to a complex multistage life cycle within the human host. PZQ is the only drug registered for treatment of schistosomiasis and despite several disadvantages, has a high safety window and low cost. While it has been very valuable for treatment and MDA over the last 40 years, it is clearly not sufficient to completely eradicate infection. New, and additional treatments are required to achieve this. There are several different requirements for new drugs, and it is likely that more than one new agent will be needed to fulfil these criteria: (1) treatment, including efficacy against the adult, juvenile, and schistosomula stages; (2) treatment of urinogenital schistosomiasis; (3) mass drug administration. There is still much unknown about the disease and a deeper understanding of its biology is essential to develop novel antischistosomals.

Competing Interests The authors have no conflicts of interest to declare that are relevant to the content of this chapter.

References

1. Global Burden of Disease (GBD) data and tools guide (2021) Institute for Health Metrics and Evolution, Seattle. https://www.healthdata.org/research-analysis/about-gbd/gbd-data-and-tools-guide. Accessed 8 Aug 2024
2. Global Burden of Disease Collaborative Network (2021) Global Burden of Disease Study 2021. Institute for Health Metrics and Evolution, Seattle. https://vizhub.healthdata.org/gbd-results/. Accessed 2 Dec 2024
3. Steinmann P, Keiser J, Bos R, Tanner M, Utzinger J (2006) Schistosomiasis and water resources development: systematic review, meta-analysis, and estimates of people at risk. Lancet Infect Dis 6(7):411–425. https://doi.org/10.1016/S1473-3099(06)70521-7
4. World Health Organization (2020) World Health Organization. Ending the neglect to attain the sustainable development goals: a road map for neglected tropical diseases 2021–2030. www.who.int/publications/i/item/9789240010352. Accessed 14 Nov 2024
5. World Health Organization (2022) WHO guideline on control and elimination of human schistosomiasis. https://www.who.int/publications/i/item/9789240041608. Accessed 14 Nov 2024
6. World Health Organization (2012) PCT Databank – Schistosomiasis. https://www.who.int/teams/control-of-neglected-tropical-diseases/data-platforms/pct-databank/schistosomiasis. Accessed 22 Feb 2024
7. Makam P, Matsa R (2021) "Big three" infectious diseases: tuberculosis, malaria and HIV/AIDS. Curr Top Med Chem 21(31):2779–2799. https://doi.org/10.2174/1568026621666210916170417
8. Chitsulo L, Engels D, Montresor A, Savioli L (2000) The global status of schistosomiasis and its control. Acta Trop 77(1):41–51. https://doi.org/10.1016/s0001-706x(00)00122-4
9. Burrows JN, Duparc S, Gutteridge WE, Hooft van Huijsduijnen R, Kaszubska W, Macintyre F, Mazzuri S, Möhrle JJ, Wells TNC (2017) New developments in anti-malarial target candidate and product profiles. Malar J 16:26. https://doi.org/10.1186/s12936-016-1675-x

10. Dartois V, Barry CE (2013) A medicinal chemists' guide to the unique difficulties of lead optimization for tuberculosis. Bioorg Med Chem Lett 23(17):4741–4750. https://doi.org/10.1016/j.bmcl.2013.07.006
11. Meurs L, Mbow M, Vereecken K, Menten J, Mboup S, Polman K (2012) Epidemiology of mixed *Schistosoma mansoni* and *Schistosoma haematobium* infections in northern Senegal. Int J Parasitol 42(3):305–311. https://doi.org/10.1016/j.ijpara.2012.02.002
12. U. S. Centers for Disease Control and Prevention (2024) DPDx – Laboratory Identification of Parasites of Public Health Concern – Schistosomiasis. https://www.cdc.gov/dpdx/schistosomiasis. Accessed 22 Feb 2024
13. McManus DP, Dunne DW, Sacko M, Utzinger J, Vennervald BJ, Zhou XN (2018) Schistosomiasis. Nat Rev Dis Primers 4. https://doi.org/10.1038/s41572-018-0013-8
14. Stothard JR, Kayuni SA, Al-Harbi MH, Musaya J, Webster BL (2020) Future schistosome hybridizations: will all *Schistosoma haematobium* hybrids please stand-up! PLoS Negl Trop Dis 14:e0008201. https://doi.org/10.1371/journal.pntd.0008201
15. Webster BL, Alharbi MH, Kayuni S, Makaula P, Halstead F, Christiansen R, Juziwelo L, Stanton MC, LaCourse EJ, Rollinson D, Kalua K, Stothard JR (2019) Schistosome interactions within the *Schistosoma haematobium* group, Malawi. Emerg Infect Dis 25(6):1245–1247. https://doi.org/10.3201/eid2506.190020
16. Privat A, Jérôme B, Bakary S, Laurent D, Assitan D, Niaré DS, Ahristode A, Hassim G, Manon B, Sarah D, Moudachirou I, Thomas S, Abdoulaye D (2023) Genetic profiles of *Schistosoma haematobium* parasites from Malian Transmission Hotspot Areas. Parasit Vectors 16. https://doi.org/10.1186/s13071-023-05860-8
17. Adekiya TA, Aruleba RT, Oyinloye BE, Okosun KO, Kappo AP (2020) The effect of climate change and the snail-schistosome cycle in transmission and bio-control of schistosomiasis in sub-Saharan Africa. Int J Environ Res Public Health 17. https://doi.org/10.3390/ijerph17010181
18. De Leo GA, Stensgaard AS, Sokolow SH, N'Goran EK, Chamberlin AJ, Yang GJ, Utzinger J, De Leo GA (2020) Schistosomiasis and climate change. BMJ. https://doi.org/10.1136/bmj.m4324
19. Carbonell C, Rodríguez-Alonso B, López-Bernús A, Almeida H, Galindo-Pérez I, Velasco-Tirado V, Marcos M, Pardo-Lledías J, Belhassen-García M (2021) Clinical spectrum of schistosomiasis: an update. J Clin Med 10. https://doi.org/10.3390/jcm10235521
20. Gryseels B, Polman K, Clerinx J, Kestens L (2006) Human schistosomiasis. Lancet 368(9541):1106–1118. https://doi.org/10.1016/S0140-6736(06)69440-3
21. Dunn MA, Kamel R (1981) Hepatic schistosomiasis. Hepatology 1(6):653–661. https://doi.org/10.1002/hep.1840010615
22. Barsoum RS, Esmat G, El-Baz T (2013) Human schistosomiasis: clinical perspective: review. J Adv Res 4(5):433–444. https://doi.org/10.1016/j.jare.2013.01.005
23. Barsoum RS (2013) Urinary schistosomiasis: review. J Adv Res 4(5):453–459. https://doi.org/10.1016/j.jare.2012.08.004
24. Patel P, Rose CE, Kjetland EF, Downs JA, Mbabazi PS, Sabin K, Chege W, Watts DH, Secor WE (2021) Association of schistosomiasis and HIV infections: a systematic review and meta-analysis. Int J Infect Dis 102:544–553. https://doi.org/10.1016/j.ijid.2020.10.088
25. Colley DG, Bustinduy AL, Secor WE, King CH (2014) Human schistosomiasis. Lancet 383(9936):2253–2264. https://doi.org/10.1016/S0140-6736(13)61949-2
26. Caldwell N, Afshar R, Baragaña B, Bustinduy AL, Caffrey CR, Collins JJ, Fusco D, Garba A, Gardner M, Gomes M, Hoffmann KF, Hsieh M, Lo NC, McNamara CW, Nono JK, Padalino G, Read KD, Roestenberg M, Spangenberg T, Specht S, Gilbert IH (2022) Perspective on schistosomiasis drug discovery: highlights from a schistosomiasis drug discovery workshop at Wellcome collection, London, September 2022. ACS Infect Dis 9(5):1046–1055. https://doi.org/10.1021/acsinfecdis.3c00081

27. Mazigo HD, Samson A, Lambert VJ, Kosia AL, Ngoma DD, Murphy R, Matungwa DJ (2024) We know about Schistosomiasis but We Know Nothing about FGS: a qualitative assessment of knowledge gaps about female genital schistosomiasis among communities living in *Schistosoma haematobium* endemic districts of Zanzibar and Northwestern Tanzania. PLoS Negl Trop Dis 15:e0009789. https://doi.org/10.1371/journal.pntd.0009789
28. Lamberti O, Bozzani F, Kiyoshi K, Bustinduy AL (2024) Time to bring female genital schistosomiasis out of neglect. Br Med Bull 149(1):45–49. https://doi.org/10.1093/bmb/ldad034
29. World Health Organization (2023) Schistosomiasis. www.who.int/news-room/fact-sheets/detail/schistosomiasis. Accessed 26 Fed 2024
30. Silva da Paz W, dos Santos RE, Brito Leal I, Menezes Barbosa Y, Conceição GM, de Araújo K, Ribeiro de Jesus A, Freire de Souza CD, dos Santos AD, Bezerra-Santos M (2021) Basic and associated causes of schistosomiasis-related mortality in Brazil: a population-based study and a 20-year time series of a disease still neglected. J Glob Health 11. https://doi.org/10.7189/jogh.11.04061
31. Faust CL, Osakunor DNM, Downs JA, Kayuni S, Stothard JR, Lamberton PHL, Reinhard-Rupp J, Rollinson D (2020) Schistosomiasis control: leave no age group behind. Trends Parasitol 36(7):582–591. https://doi.org/10.1016/j.pt.2020.04.012
32. Cioli D, Pica-Mattoccia L, Archer S (1995) Antischistosomal drugs: past, present… and future? Pharmacol Ther 68(1):35–85. https://doi.org/10.1016/0163-7258(95)00026-7
33. Christopherson JB (1918) The successful use of antimony in bilharziasis: administered as intravenous injections of antimonium tartaratum (tartar emetic). Lancet 192(4958):325–327. https://doi.org/10.1016/S0140-6736(01)02807-0
34. Shekhar KC (1991) Schistosomiasis drug therapy and treatment considerations. Drugs 42(3):379–405. https://doi.org/10.2165/00003495-199142030-00004
35. Lasbrey FO, Coleman RB (1921) Notes on one thousand cases of bilharziasis treated by antimony tartrate. Br Med J 1(3139):299–301. https://doi.org/10.1136/bmj.1.3139.299
36. Okoroiwu HU, Iwara IA (2018) Dichlorvos toxicity: a public health perspective. Interdiscip Toxicol 11(2):129–137. https://doi.org/10.2478/intox-2018-0009
37. Feldmeier H, Chitsulo L (1999) Therapeutic and operational profiles of metrifonate and praziquantel in *Schistosoma haematobium* infection. Arzneimittelforschung 49(7):557–565. https://doi.org/10.1055/s-0031-1300462
38. Tracy JW, Catto BA, Webster LT (1983) Reductive metabolism of niridazole by adult *Schistosoma mansoni*. Mol Pharmacol 24(2):291–299
39. Fontanilles F (1969) Risks versus benefits in antischistosomal therapy. Ann N Y Acad Sci 160(2):811–820. https://doi.org/10.1111/j.1749-6632.1969.tb15902.x
40. Lago EM, Xavier RP, Teixeira TR, Silva LM, da Silva Filho AA, De Moraes J (2018) Antischistosomal agents: state of art and perspectives. Future Med Chem 10(1):89–120. https://doi.org/10.4155/fmc-2017-0112
41. Hartman PE, Levine K, Hartman Z, Berger H (1971) Hycanthone: a frameshift mutagen. Science 172(3981):1058–1060. https://doi.org/10.1126/science.172.3987.105
42. Valentim CLL, Cioli D, Chevalier FD, Cao X, Taylor AB, Holloway SP, Pica-Mattoccia L, Guidi A, Basso A, Tsai IJ, Berriman M, Carvalho-Queiroz C, Almeida M, Aguilar H, Frantz DE, Hart PJ, LoVerde PT, Anderson TJC (2013) Genetic and molecular basis of drug resistance and species-specific drug action in schistosome parasites. Science 342(6164):1385–1389. https://doi.org/10.1126/science.1243106
43. Ribeiro B, da Silva V, Leiva Campos BRK, Ferreira de Oliveira J, Decout J-L, do Carmo Alves de Lima M (2017) Medicinal chemistry of antischistosomal drugs: praziquantel and oxamniquine. Bioorg Med Chem 25(13):3259–3277. https://doi.org/10.1016/j.bmc.2017.04.031
44. Walter RD, Albiez EJ (1984) Interaction of amoscanate with the cyclic AMP-phosphodiesterases from Schistosoma mansoni, Onchocerca volvulus and bovine heart. Tropenmed Parasitol 35(2):78–80

45. Huang H-H, Rigouin C, Williams DL (2012) The redox biology of schistosome parasites and applications for drug development. Curr Pharm Des 18(24):3595–3611
46. Kuntz AN, Davioud-Charvet E, Sayed AA, Califf LL, Dessolin J, Arnér ESJ, Williams DL (2007) Thioredoxin glutathione reductase from *Schistosoma mansoni*: an essential parasite enzyme and a key drug target. PLoS Med 4:e206. https://doi.org/10.1371/journal.pmed.0040206
47. Meister I, Ingram-Sieber K, Cowan N, Todd M, Robertson MN, Meli C, Patra M, Gasser G, Keiser J (2014) Activity of praziquantel enantiomers and main metabolites against *Schistosoma mansoni*. Antimicrob Agents Chemother 58(9):5466–5472. https://doi.org/10.1128/AAC.02741-14
48. Araujo Nogueira A, Sampaio Lira MG, Lima Licá IC, Carvalho Gomes Frazão GC, Ferreira dos Santos VA, Correia Mendes Filho AC, Mendes Rodrigues JG, Silva Miranda G, Cardoso Carvalho R, Fernandes Nascimento FR (2022) Praziquantel: an update on the mechanism of its action against schistosomiasis and new therapeutic perspectives. Mol Biochem Parasitol 252. https://doi.org/10.1016/j.molbiopara.2022.111531
49. Park S-K, Friedrich L, Yahya NA, Rohr CM, Chulkov EG, Maillard D, Rippmann F, Spangenberg T, Marchant JS (2021) Mechanism of praziquantel action at a parasitic flatworm ion channel. Sci Transl Med 13:eabj5832. https://doi.org/10.1126/scitranslmed.abj5832
50. Park S-K, Marchant JS (2020) The journey to discovering a flatworm target of praziquantel: a long TRP. Trends Parasitol 36(2):182–194. https://doi.org/10.1016/j.pt.2019.11.002
51. Zwang J, Olliaro PL (2014) Clinical efficacy and tolerability of praziquantel for intestinal and urinary schistosomiasis – a meta-analysis of comparative and non-comparative clinical trials. PLoS Negl Trop Dis 8:e3286. https://doi.org/10.1371/journal.pntd.0003286
52. Hoekstra PT, Casacuberta-Partal M, van Lieshout L, Corstjens PLAM, Tsonaka R, Assaré RK, Silué KD, N'goran EK, N'gbesso YK, Brienen EAT, Roestenberg M, Knopp S, Utzinger J, Coulibaly JT, van Dam GJ (2022) Limited efficacy of repeated praziquantel treatment in *Schistosoma mansoni* infections as revealed by highly accurate diagnostics, vol 16. PLoS Negl Trop Dis, PCR and UCP-LF CAA (RePST Trial), p e0011008. https://doi.org/10.1371/journal.pntd.0011008
53. Ally O, Kanoi BN, Ochola L, Nyanjom SG, Shiluli C, Misinzo G, Gitaka J (2024) Schistosomiasis diagnosis: challenges and opportunities for elimination. PLoS Negl Trop Dis 18:e0012282. https://doi.org/10.1371/journal.pntd.0012282
54. Summers S, Bhattacharyya T, Allan F, Stothard JR, Edielu A, Webster BL, Miles MA, Bustinduy AL (2022) A review of the genetic determinants of praziquantel resistance in *Schistosoma mansoni*: is praziquantel and intestinal schistosomiasis a perfect match? Front Trop Dis 3. https://doi.org/10.3389/fitd.2022.933097
55. Gryseels B, Mbaye A, de Vlas SJ, Stelma FF, Guissé F, van Lieshout L, Faye D, Diop M, Ly A, Tchuem-Tchuenté LA, Engels D, Polman K (2001) Are poor responses to praziquantel for the treatment of *Schistosoma mansoni* infections in Senegal due to resistance? an overview of the evidence. Trop Med Int Health 6(11):864–873. https://doi.org/10.1046/j.1365-3156.2001.00811.x
56. Paveley RA, Mansour NR, Hallyburton I, Bleicher LS, Benn AE, Mikic I, Guidi A, Gilbert IH, Hopkins AL, Bickle QD (2012) Whole organism high-content screening by label-free, image-based Bayesian classification for parasitic diseases. PLoS Negl Trop Dis 6:e1762. https://doi.org/10.1371/journal.pntd.0001762
57. González Cabrera D, Keiser J, Spangenberg T (2024) Analysis of the physicochemical properties of anti-schistosomal compounds to identify next-generation leads. ACS Med Chem Lett 15(5):626–630. https://doi.org/10.1021/acsmedchemlett.4c00026

From Small Molecule to Drug Product: The Chemical Journey of Praziquantel

4

David Maillard, Thomas Spangenberg, and Yara Sylla

Abstract

Building on a prior discussion of medicinal chemistry and the role of small molecules, this chapter focuses on praziquantel. Its chemical classification as a tetrahydroisoquinoline is detailed, alongside an analysis of key molecular descriptors and physicochemical properties. Additionally, the industrial-scale synthesis of praziquantel and its characterization are examined in the context of its development as a drug product. These insights provide a comprehensive overview of praziquantel's attributes and its relevance in medicinal chemistry.

Keywords

Praziquantel · Physicochemical properties · Enantiomer · Synthesis · Drug substance · Drug product

D. Maillard
Central Process Development – Downstream Processing Services, Merck Electronics KGaA, Darmstadt, Germany

T. Spangenberg (✉)
Global Health R&D of Merck Healthcare, Ares Trading S.A., (a subsidiary of Merck KGaA, Darmstadt, Germany), Eysins, Switzerland
e-mail: thomas.spangenberg@merckgroup.com

Y. Sylla
Global Project Leadership Department of Merck Healthcare KGaA, Darmstadt, Germany

T. Spangenberg (ed.), *Praziquantel*,
https://doi.org/10.1007/978-3-031-97397-0_4

A Brief History of the Birth of Medicinal Chemistry and What It Does?

Modern chemistry began to take shape in the eighteenth century, driven by the groundbreaking contributions of Carl Wilhelm Scheele and Joseph Priestley, who prepared and studied oxygen (O_2) and identified nitrogen (N_2) as key components of air. Building on these discoveries, Antoine Lavoisier advanced the field further by establishing the principles of mass conservation and distinguishing elements from compounds [1]. The nineteenth century marked a shift toward practical applications, as chemists focused on extracting and purifying active molecules from natural sources, particularly plants. This period also saw the chemical modification of natural dyes and the use of coal tar by-products, innovations that led to the creation of synthetic dyes [2, 3] (Fig. 4.1).

Organic chemistry, the branch of chemistry devoted to carbon-based compounds, gradually shaped the work of chemists and pharmacists during the nineteenth century. A pivotal moment occurred in 1828, when Friedrich Wöhler, while attempting to prepare ammonium cyanate, accidentally synthesized urea (Fig. 4.2). This marked the first time an organic compound was created from inorganic materials. Around the same period, chloroform emerged as the first organic molecule used for its therapeutic properties. It was independently discovered in 1831 by three chemists: Eugene Soubeiran, Justus von Liebig, and Samuel Guthrie [1]. A major advance in the medicinal application of organic chemistry occurred in 1856, when William Perkin, while attempting to synthesize quinine, inadvertently discovered mauveine, the first synthetic dye. This discovery not only revolutionized the dye industry but

Fig. 4.1 Collection of alkaloids and other isolated natural substances from the Chemical Laboratory of Emanuel Merck, Darmstadt, 1838, MA X1-1873

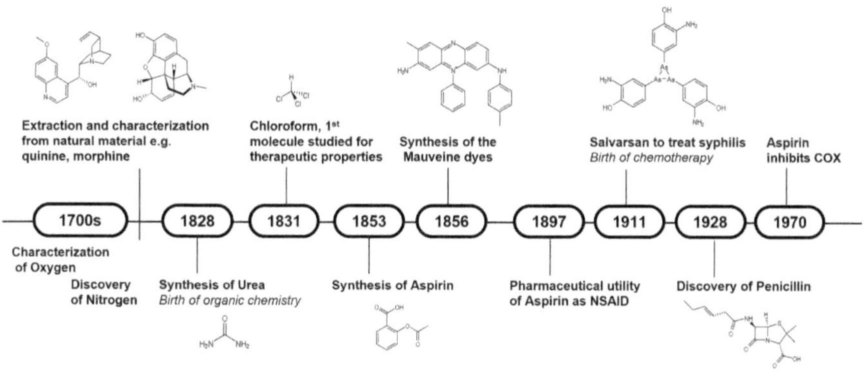

Fig. 4.2 Timeline of key discoveries pertaining to molecules of pharmaceutical interest spanning from the eighteenth to the twentieth century. COX: Cyclooxygenase; NSAID: nonsteroidal anti-inflammatory drug

also laid the foundation for the development of antiseptic and anti-infective drugs in subsequent decades [1, 4].

In 1906, building on experimental work with vaccines and antitoxins, and the molecular structures of dyes, Paul Ehrlich introduced the concept of the "magic bullet," that links chemical structure to pharmaceutical activity. This idea laid the foundation for modern medicinal chemistry. Just four years later, in 1910, Ehrlich's arsenic-based compound, Salvarsan, developed in collaboration with Hoechst, became the first systematically designed and effective treatment for syphilis [1, 5, 6]. The development of aspirin (acetylsalicylic acid) further illustrates the evolution of medicinal chemistry from natural compounds to synthesized drugs. Although, salicylic acid, the active ingredient, was first extracted from plants and characterized by Raffaele Piria in the 1830s it was not until 1853 that Charles Frédéric Gerhardt successfully synthesized acetylsalicylic acid, marking a significant milestone in organic chemistry. Several decades later, Bayer optimized its production, transforming aspirin into a widely used therapeutic agent. Its lasting role in pain relief and anti-inflammatory treatment highlights the importance of continued research into drug mechanisms and applications [1] (Fig. 4.2).

What Is Medicinal Chemistry?

Medicinal chemistry is deeply rooted in the history of organic chemistry and integrates principles from biological, medical, and pharmaceutical sciences. This multidisciplinary field focuses on the invention, discovery, and design of biologically active compounds, together with their synthesis and characterization. It also involves studying their metabolism, elucidating mechanisms of action, and establishing structure–activity relationships [7]. By bridging chemistry and biology, medicinal chemistry plays a critical role in the development of new therapies and deepening our understanding of drug behavior at the molecular level.

What Is a Small Molecule?

While 90 elements occur in nature in significant amounts, living organisms are primarily composed of six key elements: hydrogen (H), carbon (C), nitrogen (N), oxygen (O), phosphorus (P), and sulfur (S). Atoms can organize in various ways and quantities, resulting in an infinite diversity of shapes, weights, and properties. In a therapeutic context, biologically active agents range from small molecules to larger biologics. Biologics are complex entities that often provide exquisite specificity, as their size and structure limit unintended binding. In contrast, small therapeutics—low molecular weight organic compounds—are generally easier to synthesize and can be absorbed orally, penetrate cells, and be widely distributed throughout the body. However, their smaller size allows them to interact with multiple targets, which can sometimes lead to "off-target" toxicity (Fig. 4.3).

Small-molecule drugs have been pivotal in advancing human health and longevity, serving as the foundation of the pharmaceutical industry for nearly a century. Despite their central role, the discovery and development of these drugs is a time-consuming, costly, and complex process. Orally administered candidates, in particular, face significant challenges and a high attrition rate (Chap. 1) [1].

To guide the medicinal chemistry community in drug design, several easily calculated "rules of thumb" and efficiency metrics have been developed. These tools assess whether a compound with specific pharmacological or biological activity has physicochemical properties likely to make it an orally active drug in humans. The "rules" are based on physicochemical descriptors such as molecular weight (MW), lipophilicity (logarithm of the octanol-water partitioning coefficient, LogP, or the distribution coefficient, LogD), topological polar surface area (TPSA), and counts of hydrogen bond donors and acceptors (HBDs and HBAs), rotatable bonds (RBs),

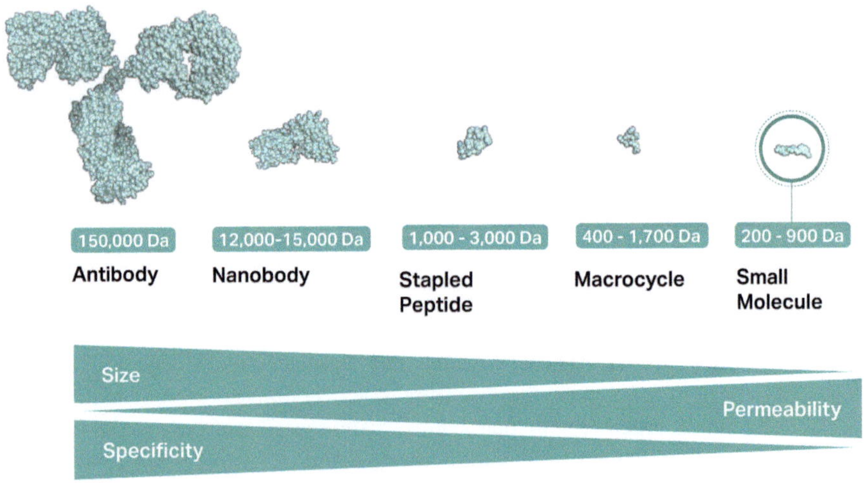

Fig. 4.3 From biologics to small molecules. Overview of sizes (in Dalton) and consequences on target specificity and permeability across membranes

and the number of aromatic rings. These are factors that chemists can intuitively manipulate in a molecule. In addition, efficiency metrics are used to evaluate how effectively a compound leverages its structural features to bind to the target. These metrics are calculated using formulas that incorporate binding affinity, molecular size, and lipophilicity [8–11].

A classic example is Lipinski's Rule of Five, which states that an orally active drug should ideally have no more than one violation of the following criteria. Exceeding these limits may hinder the drug's ability to cross biological membranes, reducing its bioavailability. Specifically, the criteria are: no more than five hydrogen bond donors (sum of nitrogen–hydrogen and oxygen–hydrogen bonds), no more than 10 hydrogen bond acceptors (nitrogen or oxygen atoms), a molecular weight (MW) under 500 Daltons (g/mol), and a LogP (the water/octanol partitioning coefficient) that does not exceed 5 [12]. These criteria were derived from an analysis of the molecular characteristics of orally administered drugs developed throughout the twentieth century, particularly in response to the decline in drug approvals during the late 1990s. Since then, additional simple rules of thumb have been developed, offering further guidance for evaluating compounds [9–11].

Multiparameter optimization is a critical aspect of the drug discovery and development process. It aims to identify small bioactive compounds that achieve an optimal balance between affinity, selectivity, and efficacy for the target, while also ensuring optimal pharmacokinetics, toxicity profiles, solubility, and permeability. Achieving this balance is inherently challenging, as each characteristic must be carefully tuned to ensure the compound's effectiveness and safety.

Where Does Praziquantel Come From?

Nature plays a crucial role in drug development, with approximately two-thirds of approved medications derived from natural sources, while the remaining third are entirely synthetic [13]. Notably, many natural products and synthetic drugs share common molecular fragments or moieties. One such example is the 1,2,3,4-tetrahydroisoquinoline (THIQ) scaffold - an important class of isoquinoline alkaloids found in various natural products. THIQ-based molecules exhibit a wide range of biological activities, including antiviral, antibacterial, antifungal, antiparasitic, and anticancer effects, and can serve as ligands for G protein–coupled receptors (GPCRs), nuclear receptors, and glucocorticoids [14, 15].

The widespread use of the THIQ core in therapeutic development stems from its structural versatility and ease of chemical modification. This adaptability has attracted significant interest among researchers, leading to the creation of numerous novel analogues with potent biological activities. The focused application and optimization of THIQ-based compounds gained significant momentum during the rise of modern medicinal chemistry in the twentieth century. During this time, researchers systematically investigated the structure–activity relationships of natural products and their derivatives, resulting in the design and synthesis of compounds with enhanced or tailored pharmacological profiles[14, 15]. Selected examples of THIQ-based drugs include the tranquilizers tetrabenazine [16], the antidepressant nomifensine [17], and the anticancer drug trabectedin [18]. Another notable example of a synthetic drug featuring the THIQ core is praziquantel (PZQ) (Fig. 4.4). Originally synthesized during Merck's drug discovery efforts to identify novel antipsychotic compounds, molecules from this program were later evaluated by Bayer for alternative therapeutic applications - a strategy known as drug repurposing (Chaps. 1 and 6). In 1975, Merck and Bayer collaborated to codevelop and

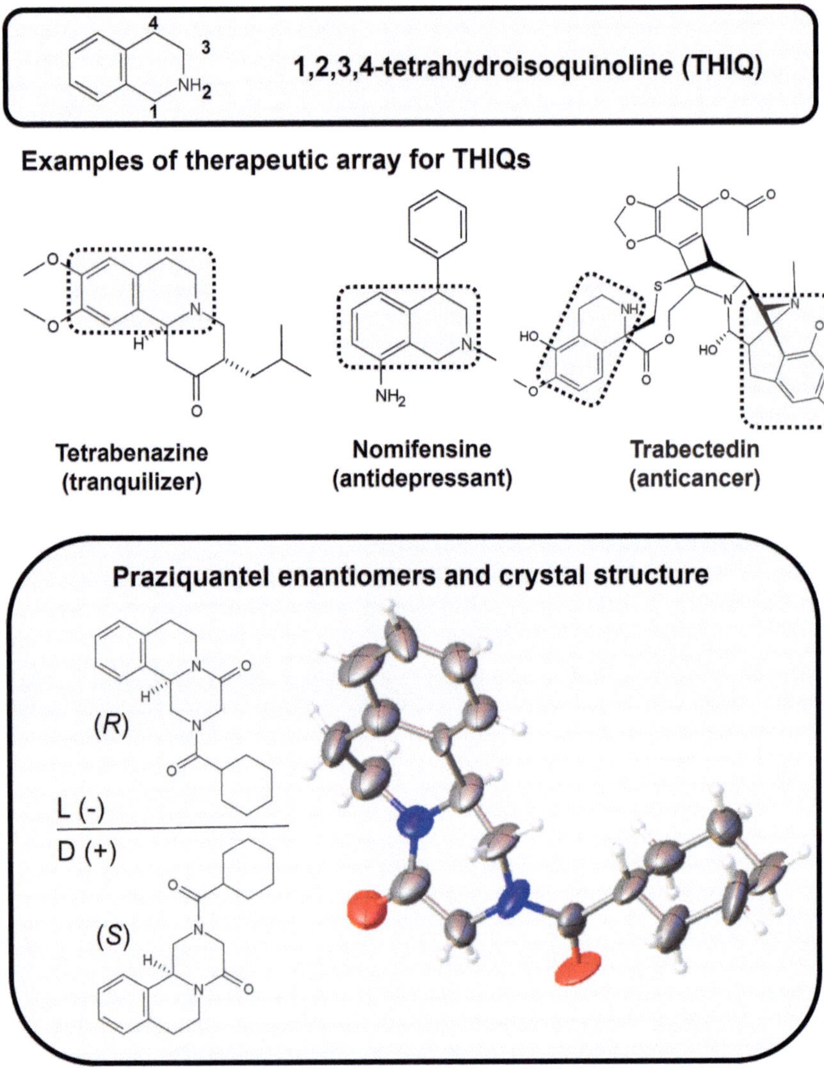

Fig. 4.4 Structure of 1,2,3,4-tetrahydroisoquinolines (THIQs) exemplified with tetrabenazine, nomifensine, trabectedin, and praziquantel enantiomers

patent PZQ as a novel anthelmintic. From over 400 analogues, PZQ was selected for its remarkable effectiveness against the blood flukes of *Schistosoma* strains, as well as other parasites such as liver flukes, intestinal tapeworms, and cestodes (Chap. 5) [19]. Its selection was also supported by its favorable tolerance in infected animals and its desirable molecular descriptor values and physicochemical characteristics. It is important to note that the optimal values for molecular descriptors and physicochemical properties are determined by the specific goals of the project, including the target, desired pharmacokinetic properties, route of administration, and therapeutic indication (Chap. 3).

What Are the Molecular Descriptors and Physicochemical Properties of Praziquantel?

A key feature of PZQ is its chirality, meaning it contains a carbon atom bonded to four different substituents. This property makes PZQ a racemate, as it exists as a mixture of equal amounts of two enantiomers—mirror-image forms of a chiral molecule. While enantiomers have identical physical properties, they differ in their interaction with plane-polarized light. There are several systems for describing enantiomers of a chiral molecule: (1) the R/S system, which assigns configurations based on the Cahn–Ingold–Prelog priority rules; (2) the (+) and (−) system, referring to the enantiomer's optical rotation, where (+) rotates plane-polarized light clockwise and (−) rotates it counterclockwise; and (3) the D (Dextrorotatory)/L (Levorotatory) system, which classifies enantiomers according to their structural relationship to the enantiomers of glyceraldehyde the reference compound (Fig. 4.4) [20].

The following section examines the key molecular descriptors and physicochemical properties of PZQ, highlighting how they collectively influence the molecule's behavior in biological systems. Molecular descriptors are numerical values derived from a molecule's structure, such as its size, shape, and electronic properties, while physicochemical properties refer to the compound's inherent physical and chemical traits, such as solubility, stability, and melting point. Together, these factors provide essential insights into the molecule's behavior, particularly in terms of absorption, distribution, metabolism, and excretion (ADME). They also play a decisive role in determining the drug's interactions with biological targets and its therapeutic efficacy (Chap. 7) (Table 4.1).

Solubility

Solubility reflects a substance's ability to dissolve in a solvent and form a homogeneous solution. Solubility screening assays typically measure either the concentration of the dissolved compound or the portion that remains undissolved. In kinetic solubility assays, the compound is initially dissolved in an organic solvent, typically dimethyl sulfoxide (DMSO), and then gradually added to an aqueous buffer.

Table 4.1 Summary of key molecular descriptors and physicochemical properties of praziquantel

Molecular descriptors and physicochemical properties	Value
Kinetic solubility phosphate-buffered saline (PBS)/ Thermodynamic solubility	190 μM / < 1mg/mL
LogP	2.5
Molecular weight (MW)	312 g/mol
Number of aromatic rings	1
Melting point (mp)	136–140 °C
Cell permeability (Caco-2, Papp A → B) Efflux ratio	39×10^{-6} cm/s 1
Topological polar surface area (TPSA)	41 Å^2
Hydrogen bond donors (HBDs) Hydrogen bond acceptors (HBAs)	0 4

However, this method may overestimate the compound's true solubility, as DMSO can artificially enhance the solubility of poorly water-soluble molecules. In contrast, thermodynamic solubility assays involve adding an aqueous buffer to a solid compound and agitating the mixture for 24–48 h until equilibrium is reached, at which point the amount of dissolved compound stabilizes, providing a more accurate measurement of the molecule's intrinsic solubility [21]. Solubility is crucial for orally administered drugs, as it determines the fraction of the dose available for absorption through gastrointestinal membranes into systemic circulation (Chap. 7). Low solubility (e.g., <20 μM) can also lead to inaccurate estimations in bioassays, potentially affecting the reliability of pharmacokinetic data and drug efficacy predictions. Several factors, including lipophilicity, molecular weight (MW), and molecular structure (such as the number of aromatic rings) influence a compound's solubility.

- *Lipophilicity* refers to a compound's affinity for dissolving in nonpolar solvents (such as octanol) compared to polar solvents (such as water). This preference arises from weaker intermolecular forces like Van der Waals and dipolar interactions, in contrast to stronger hydrogen bonds formed with water. The conventional method for evaluating lipophilicity involves measuring the partition coefficient (LogP), which represents the concentration ratio of the unionized species of a compound between water and octanol. When both ionized and unionized forms are considered at a specific pH, the term "distribution coefficient" (LogD) is used instead [21]. Lipophilicity, which governs a compound's solubility and membrane permeability, is a critical determinant of its absorption, distribution, metabolism, and excretion (ADME) properties. High lipophilicity, typically indicated by LogP and LogD values above 3, is often associated with reduced aqueous solubility, which can pose challenges for drug formulation and bioavailability [8, 22, 23].

  Another important property closely related to solubility is a compound's melting point. Generally, there is an inverse relationship between melting point and solubility: higher melting points, which indicate stronger intermolecular forces such as hydrogen bonding, typically correspond to lower solubility. Compounds with melting points above 150 °C often show reduced solubility in aqueous environments [8].

- *Molecular weight (MW)* is the sum of the atomic masses of all atoms in a molecule. While not rigorous, medicinal chemists typically express molecular weight in g/mol, rather than in atomic mass units (amu), which is less frequently used in this context. It has been suggested that molecules with a molecular weight below 300 g/mol are more likely to exhibit high solubility (e.g. around 250 μM). As molecular weight increases, solubility generally decreases; however, this is not always the case. Solubility is also influenced by other factors, including molecular structure and the presence of functional groups. Indeed, compounds with molecular weights exceeding 700 g/mol can still be soluble, particularly if they incorporate polar or ionizable groups [24].
- *The number of aromatic rings* refers to the total count of both aromatic and heteroaromatic rings in a molecule. In fused aromatic systems, each ring is counted individually, even if atoms are shared. As the number of aromatic rings increases, solubility generally decreases, although other factors may modulate this effect. The developability of orally administered drugs is often compromised when a molecule contains more than three aromatic rings [25]. Planar configuration of aromatic rings is key to their electronic stability and reactivity, as it facilitates π–π stacking interactions with other flat molecules. These non-covalent interactions arising when electron-rich π orbitals align in close proximity, can reduce solubility by promoting aggregation, increasing hydrophobicity, and affecting crystal packing as well as solvation [26].

The racemic form of PZQ displays a relatively high kinetic solubility of 190 μM but limited thermodynamic solubility of <1 mg/mL. This solubility profile can be attributed to a combination of factors, including its moderate lipophilicity (LogP = 2.5), low molecular weight (312 g/mol), and the presence of only one aromatic ring. Additionally, PZQ's nonplanar structure likely minimizes π–π stacking interactions, which are known to negatively impact solubility in more planar aromatic systems.

Permeability

Permeability describes a substance's ability to cross biological barriers or membranes, an essential property in numerous physiological and pharmacological processes (Chap. 7). It plays a critical role in in vitro activity testing, as compounds must traverse the cellular lipid bilayer to reach their targets. For orally administered drugs, adequate permeability is essential for absorption across the intestinal epithelium into systemic circulation. This movement can occur via passive transcellular diffusion, paracellular diffusion, or transporter-mediated mechanisms. Two commonly used in vitro screening methods are employed to assess permeability:

1. *Cell monolayer method*: This technique typically employs either Caco-2 (Caucasian Colon Adenocarcinoma, second clone) or MDCK (Madin-Darby Canine Kidney) cell lines to model intestinal absorption. Caco-2 cells are particularly valuable because they express both efflux and uptake transporters, facilitating the evaluation of permeation through various mechanisms. In contrast, MDCK cells primarily reflect passive transcellular diffusion, as they express significantly fewer transporters.
2. *Parallel Artificial Membrane Permeability Assay (PAMPA)*: This assay evaluates permeability by measuring compound diffusion through an artificial membrane, mimicking transcellular and paracellular routes. Its main advantages include low cost and suitability to efficiently screen a large number of compounds.

   The relationship between permeability and lipophilicity is well established, as higher lipophilicity often correlates with increased membrane permeability. However, this correlation is not absolute, as factors such as ionization state, polarity, and MW can significantly influence permeability independently of a compound's lipophilicity [8, 21, 22, 27].

   - The *ionization state* of a molecule—whether neutral, acidic, basic, or zwitterionic—greatly influences its permeability and is determined by its pKa, the base-10 logarithm of the acid dissociation constant. A helpful guideline is that when the pKa of a compound equals the pH of the solution, approximately 50% of the molecules are ionized, and 50% remain neutral. Neutral species typically display the highest permeability, followed by basic molecules, which are positively charged in acidic conditions and may experience diffusion restrictions due to interactions with anionic membranes. Acidic and zwitterionic molecules generally exhibit lower permeability because their charges lead to electrostatic repulsion with negatively charged lipid membranes [21, 22, 27].
   - *Polarity*: Permeation typically increases as polarity decreases, since less polar compounds can more readily cross lipid membranes. This effect is influenced by factors such as polar surface area (PSA) and the number of hydrogen bonds, including both hydrogen bond donors (HBDs) and hydrogen bond acceptors (HBAs) [27–29].
   - *Topological polar surface area* (TPSA, Å^2) is a widely used measure of PSA in drug discovery. It accounts for the cumulative contributions of polar atoms (e.g., oxygen and nitrogen) and their attached hydrogens to the overall molecular surface area. As a result, the primary contributors to TPSA are HBDs and HBAs). A TPSA value of 100 Å^2 or more is generally considered high [8, 27–29]. It has been suggested that the principles of central nervous system (CNS) drug development, which favor a lower TPSA for enhanced brain penetration, are also applicable to schistosomiasis drug research. High polarity can hinder a compound's ability to cross the blood–brain barrier (BBB), reinforcing the preference for lower TPSA in CNS drug development. As a guideline, compounds with TPSA below 70 Å^2 are more likely to exhibit high permeability and effective CNS penetration [30].

A compound's polarity is typically driven by its HBAs. Therefore, when addressing polarity-related limitations in passive permeability, modifications to HBAs should be prioritized over hydrogen bond donors (HBDs). Nonetheless, removing unnecessary HBDs can be an effective strategy for optimizing permeability as this generally causes smaller increases in lipophilicity than reducing HBAs [31].

- As *molecular weight (MW)* increases, a compound's permeability typically decreases because larger size and greater steric hindrance can hinder the molecule's ability to pass through the membrane's channels or pores.

PZQ has a moderate LogD of 2.5, a low TPSA of 41 Å^2, and a moderate MW of 312 g/mol. Its low TPSA is primarily due to the absence of hydrogen bond donors and the presence of only four hydrogen bond acceptors. These properties make PZQ highly permeable, as reflected by its Caco-2 permeability (Papp A $\rightarrow$ B = 39×10^{-6} cm/s, efflux ratio = 1) which support effective oral drug absorption and penetration into the parasite.

How Is Praziquantel Synthesized and Manufactured?

PZQ is a small synthetic molecule with anthelmintic activity, functioning as the drug substance (DS), also known as the active pharmaceutical ingredient (API). In pharmaceutical formulations, the API is combined with excipients—typically inactive ingredients that improve stability, facilitate drug delivery, or optimize the medication's physical and sensory properties. Together, these components form the drug product (DP), which is administered to patients. Drug substances are generally produced through multistep chemical synthesis, transforming commercially available starting materials into the final targeted chemical entity. The entire chemical process, encompassing all synthetic steps, must be carried out in accordance with current Good Manufacturing Practice (cGMP) regulations. These regulations outline the minimum requirements for analytical methods, facilities, and controls used in the manufacturing, processing, and packaging of drug substances and drug products. To obtain a marketing authorization—and thereby become licensed to distribute, sell, and commercialize a medical product—the producer must register the manufacturing process with health authorities (Chaps. 11 and 12). Once authorization is granted, the validated manufacturing process must be executed as described in the registration dossier. Deviations to the manufacturing process must be explicitly documented, justified, and assessed for their potential impact on the quality, safety, and efficacy of the product.

In the pharmaceutical industry, production costs extend far beyond raw materials. The manufacturing costs of a drug substance or product include both material and conversion costs. Material costs cover all starting materials, solvents, reagents, catalysts, excipients, packaging, and auxiliary materials (e.g., lubricants or cleaning agents) used during manufacturing and cleaning processes. Conversion costs refer to the expenses associated with transforming raw materials into the final product.

These include operational expenses such as electricity, rent, and depreciation, as well as costs for plant insurance, equipment repairs and maintenance, analytical testing, and labor. Reducing production costs requires a focus on both material and conversion costs. For material costs, sourcing raw materials at competitive prices is a key priority. Additionally, improving the yield of the synthetic process is critical, as it increases efficiency and reduces material waste. Manufacturing strategy also plays a pivotal role in cost optimization. For instance, increasing batch sizes to enhance space–time yield and implementing production campaigns with successive batches—thereby minimizing cleaning and downtime—can significantly reduce conversion costs.

Merck's synthesis of PZQ was developed in the late 1960s, starting from isoquinoline and involving six steps, labeled A through F. The key Reissert reaction, commonly used for synthesizing substituted quinolines, transforms quinoline to quinaldic acid. In step A, isoquinoline reacts with acetyl chloride and sodium cyanide to give intermediate **1**, under carefully controlled temperature and pH. It is important to note that the reaction with cyclohexanecarboxylic acid chloride does not proceed efficiently due to steric hindrance. Intermediate **2** is generated through high-pressure hydrogenation of the nitrile group using a metal catalyst (e.g., Raney Nickel), followed by rearrangement of the acetyl group (step B). In step C, intermediate **2** undergoes acetylation with chloroacetyl chloride, followed by ring closure in step D using potassium *tert*-butoxide (KO-*t*-Bu) to yield intermediate **4**. Steps C and D could be telescoped to avoid the isolation of intermediate **3**. Intermediate **4** is then saponified by phosphoric acid to form praziquanamine **5** (step E, depicted in red in Fig. 4.5). Finally, in step F, praziquanamine is acylated with cyclohexanecarboxylic acid chloride forming an amide bond and resulting in PZQ **6** as a racemate. The overall yield of this six-step synthesis, starting from isoquinoline, is approximately 37%.

In the mid-1980s, the South Korean pharmaceutical company Shin-Poong developed a more cost-effective manufacturing process for PZQ, streamlining the synthesis to just four steps (A*–D*, F). In this approach, steps C* and D* are telescoped, eliminating the need to isolate intermediates **9** and **10**. In step A*, phenylethylamine reacts with chloroacetyl chloride to form intermediate **7**. Step B* involves the aminoalkylation of intermediate **7** with 2,2-dimethoxyethan-1-amine **8** (a protected 2-aminoacetaldehyde) in the presence of sodium hydroxide (NaOH), yielding compound **9**. This intermediate is then further transformed without isolation, and the aminoacetaldehyde dimethyl acetal reagent used in excess is recovered through work-up and distillation. Concentrated sulfuric acid (step D*) induces a Pictet-Spengler reaction of **9**, forming praziquanamine **5** via ring closure of the non-isolated intermediate **10**. Finally, in step F, praziquanamine **5** is converted to PZQ **6** (as a racemate) via acylation with cyclohexanecarboxylic acid chloride. The overall yield of this four-step process is approximately 61%, starting from phenylethylamine. There are several factors that enhance the economic viability of the Shin-Poong process. First, it achieves a higher overall yield while utilizing relatively

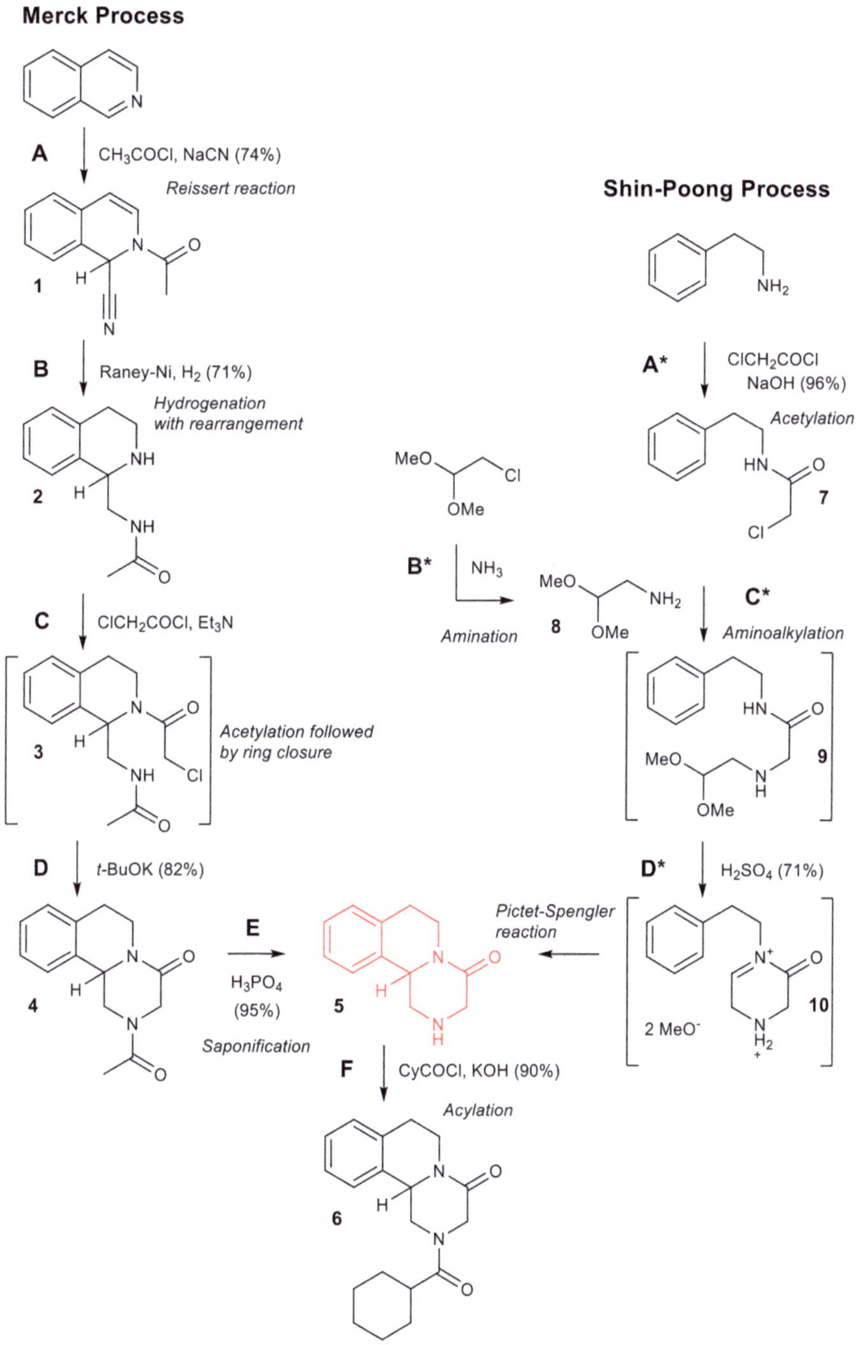

Fig. 4.5 The two primary manufacturing processes for praziquantel, known as the Merck process and the Shin-Poong process

inexpensive chemicals. Although the aminoacetaldehyde dimethyl acetal reagent is an exception, it can be recycled, further reducing costs by relying on readily available reagents. This results in a significant reduction in material costs. Second, the synthesis is more straightforward, with fewer isolated intermediates and shorter processing times, including fewer crystallizations, separations, and drying steps. This reduces analytical costs associated with investigating and releasing intermediates. Third, the type of chemistry involved also impacts the process significantly. The Merck process includes a high-pressure hydrogenation step (B), which requires specialized equipment and limits batch size, as large-volume autoclaves are uncommon and typically restricted to 2000–2500 L. In contrast, the Shin-Poong process uses simpler, easily scalable chemistry, enabling the use of larger vessels and higher batch sizes, contributing to a reduction in conversion costs.

PZQ is a racemate, containing equal amounts of both the (*R*)- and (*S*)-enantiomers. Since (*S*)-praziquantel ((*S*)-PZQ) exhibits lower activity compared to its (*R*)-enantiomer ((*R*)-PZQ), efforts have been focused on producing the pure (*R*)-enantiomer for pediatric applications. From a cost perspective, chiral separation via diastereomeric salt formation of intermediate amines in the PZQ synthesis is more practical and straightforward than pursuing enantioselective routes. The latter would require longer development time, for instance by establishing an enantioselective Pictet-Spengler reaction from compound **9** or performing enantioselective hydrogenation of **1** or **2** to obtain the (*R*)-enantiomer of intermediates **3**.

Building on this approach, praziquanamine **5** (depicted in red in Fig. 4.5) was selected for resolution as it is a key intermediate in both PZQ manufacturing processes (Fig. 4.6). Screening of chiral acids as resolving agents revealed an interesting finding: (*R*)-praziquanamine forms an insoluble salt with two molecules of naproxen, a nonsteroidal anti-inflammatory drug produced in large volumes and commercially available exclusively as the (*S*)-enantiomer. This co-crystal or diastereomeric salt, termed *(RSS)-11* in Fig. 4.6, precipitates from solution, while (*S*)-praziquanamine remains in the mother liquor-form of praziquanamine remains dissolved, while the salt precipitates from the mother liquor. This provides a significant advantage: the mother liquor contains virtually no naproxen requiring recycling, and since naproxen is insoluble in water, its recovery is straightforward. Instead of using an alkaline extraction method, the *(RSS)-11* salt is suspended in water, and a mineral acid is added. This process keeps the mineral salt of (*R*)-praziquanamine in the aqueous solution, while naproxen is efficiently removed by filtration. Thorough washing of the wet residue allows for the recovery of nearly pure naproxen. For (*R*)-PZQ preparation, isolating (*R*)-praziquanamine from its aqueous ammonium salt is unnecessary because the compound is highly water-soluble, making organic extraction labor-intensive. In a two-phase system of dichloromethane and water, the ammonium salt solution is slightly alkalized, followed by addition of cyclohexanecarboxylic acid chloride. The pH is maintained with alkali hydroxide, enabling an extractive Schotten–Baumann reaction.

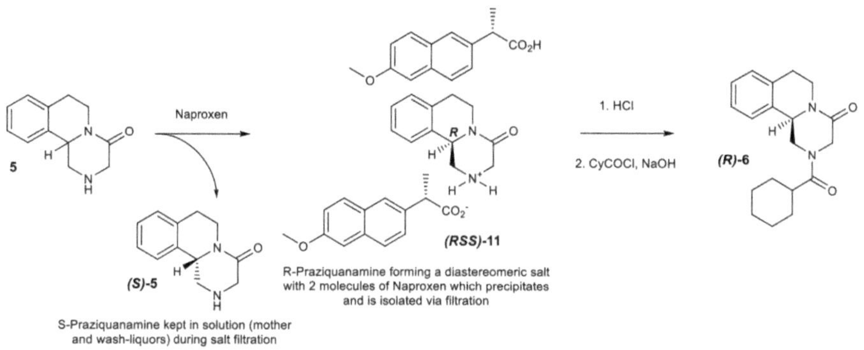

Fig. 4.6 (*R*)-PZQ manufacturing process. Naproxen is added to the racemic praziquanamine allowing the (*R*)-enantiomer to precipitate out of the solution. The salt is recovered by filtration. Treatment of the salt with an acid in water allows to keep (*R*)-praziquanamine in solution and remove the water-insoluble naproxen before proceeding to the final acylation step

APIs Properties

The physicochemical properties of APIs, especially solubility, stability, in vitro dissolution, and bioavailability, are of critical importance to enable the drug substance to be formulated into a finished dosage form, ensuring its suitability to patients. The choice of excipients, the manufacturing process, and the formulation design are all influenced by the physicochemical properties, which also impact the performance of the dosage form. Their evaluation begins during the development of a new drug in the preclinical phase and continues throughout the clinical phases. This helps monitor the impact of changes, such as those in API synthesis, on these properties. A physical property is a characteristic of a substance that can be observed or measured without changing its chemical composition, including color, solubility, polymorphic form, particle size, morphology, surface area, and melting point. In contrast, chemical properties describe the inherent ability of a material to react and form new substances. Various parameters were evaluated to assess the physical characteristics of APIs, with the results for (*R*)-PZQ presented in Fig. 4.7.

During drug substance development, analytical assessment includes impurity profiling and elucidation of degradation pathways. This is accomplished by performing forced degradation studies, where the drug substance is exposed to various stress conditions, including varying temperatures, humidity, pH (neutral, acidic, and basic), oxidative conditions (oxidants added to the solution), and light exposure. Impurity evaluation is carried out using liquid chromatography-mass spectrometry (LC-MS) to identify and quantify impurities. All potential organic impurities must also be assessed for mutagenic potential. This is accomplished through literature reviews, in-silico structure–activity relationship analyses, and the use of two complementary prediction models, in accordance with the International Council for Harmonisation (ICH) Guideline M7, which provides guidance for assessing and controlling DNA-reactive (mutagenic) impurities in pharmaceuticals. This process

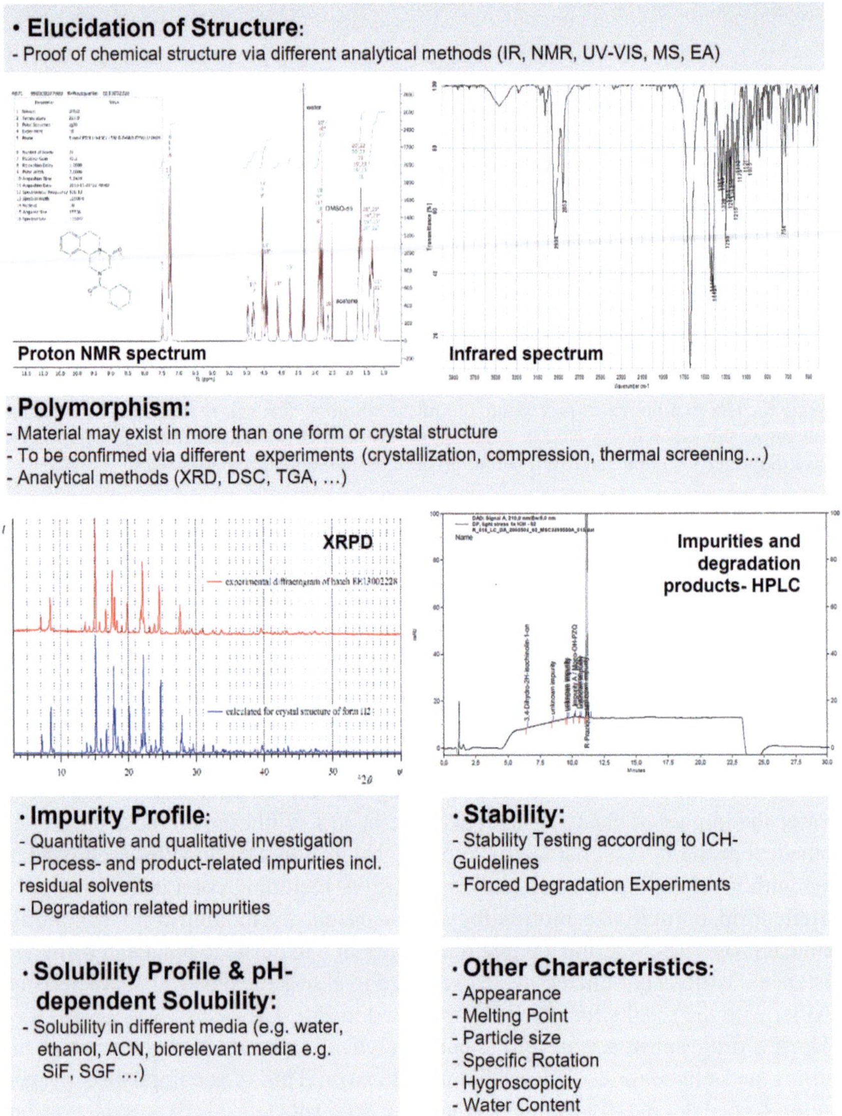

Fig. 4.7 Physical properties of (*R*)-PZQ and typical set of tests conducted for APIs

helps mitigate potential carcinogenic risks. Additionally, the evaluation includes assessing the risk of forming *N*-nitrosamines, which are recognized as potent carcinogens [32]. If any concern appears additional in vitro and in vivo models such as Ames and micronucleus tests are used to exclude potential risks (Chap. 8).

Drug Product

A drug product (DP), also referred to as a finished dosage form (FDF), is a formulation that contains one or more active and/or inactive ingredients. Drug products are available in various dosage forms, such as tablets, capsules, liquids, creams, and patches, and can be administered through different routes, including oral, intravenous infusion, or as ear or eye drops. Their manufacture, including all the steps and In-Process Control (IPC) testing, must be carried out in compliance with current cGMP regulations.

Both PZQ and (*R*)-PZQ DPs are available as formulations (Fig. 4.8), produced using a state-of-the-art wet granulation process. This method enables local production in endemic countries where schistosomiasis is prevalent. Standard excipients, which are well-established and easy to use, were selected to ensure accessibility for local manufacturing in these regions. According to the Biopharmaceutical Classification System (BCS), these compounds are classified as Class II, characterized by low solubility and high permeability (Chap. 7). For BCS II APIs, in vivo absorption is primarily influenced by solubility [33]. Therefore, disintegration time and dissolution rate are critical parameters to monitor during formulation development. The choice of disintegrants, fillers, and the compatibility of excipients with the API significantly affects disintegration and dissolution. In a wet granulation process, selecting an appropriate binder for granule formation is also critical.

PZQ was originally developed in the 1980s as a 600 mg film-coated tablet (Cesol 600®). To create a suitable pediatric formulation for children under 6 years of age, a 150 mg (*R*)-PZQ (Ceskid®) tablet was developed. By utilizing only the

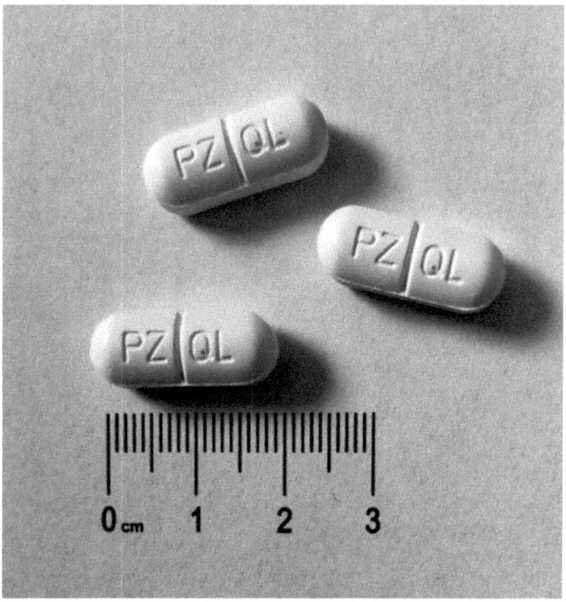

Fig. 4.8 Praziquantel (Cesol 600®) tablets

pharmaceutically active enantiomer, the tablet strength was reduced, resulting in a more adequate tablet size for children. To enhance adherence in young children, the tablet was designed as a dispersible formulation that dissolves in water within seconds, making it easy to swallow and drink. A sweetener is added to improve the taste. Both drug products are packaged in high-density polyethylene (HDPE) containers for mass treatment in endemic countries and are stable at temperatures up to 30 °C. The shelf life is determined through stability studies conducted in accordance with ICH guidelines, considering the climatic zones in which the drug products are distributed. PZQ (Cesol 600®) has a shelf life of 5 years, which is the maximum allowed under current guidelines, while (*R*)-PZQ (Ceskid®) has a shelf life of 4 years, with the possibility of extension based on additional stability data.

Summary

PZQ is a racemic synthetic small molecule with a tetrahydroisoquinoline (THIQ) core moiety, a structural motif also found in some natural products. It has a moderate molecular weight of 312 g/mol, a moderate lipophilicity with a LogD value of 2.5, and a low polar surface area of 41 $Å^2$. Two commercially viable large-scale methodologies for synthesizing the active pharmaceutical ingredient (API) have been developed, each involving six to four linear steps with yields ranging from approximately 37–61%. The (*R*)-enantiomers of PZQ can be efficiently obtained on a large scale via diastereomeric salt formation with naproxen. PZQ (Cesol 600®) is a stable API formulated as a 600 mg film-coated tablet, with a shelf life of 5 years and a storage temperature of up to 30 °C.

Competing Interests The authors are employees of Merck KGaA or affiliates.

References

1. Chast F (2008) A history of drug discovery. In: Wermuth CG (ed) The practice of medicinal chemistry, 3rd edn. Elsevier, Oxford, pp 3–62
2. Drews J (2000) Drug discovery: a historical perspective. Science 287(5460):1960–1964. https://doi.org/10.1126/science.287.5460.1960
3. Malerba F, Orsenigo L (2015) The evolution of the pharmaceutical industry. Bus Hist 57(5):664–687. https://doi.org/10.1080/00076791.2014.975119
4. Wainwright M (2008) Dyes in the development of drugs and pharmaceuticals. Dyes Pigments 76(3):582–589. https://doi.org/10.1016/j.dyepig.2007.01.015
5. Ehrlich P (1906) Über den jetzigen Stand der Chemotherapie. Berliner Klinische Wochenschrift 43:169–172. https://doi.org/10.1002/cber.19090420105
6. Wang ML (2009) The modern pharmaceutical industry: history, current position and challenges. In: Global health partnerships. Palgrave Macmillan, London. https://doi.org/10.1057/9780230582873_2
7. Wermuth CG (ed) (1992) Medicinal chemistry for the 21st century, Chemistry for the 21st century monographs. Blackwell, Oxford, pp 25–45

8. Gleeson MP (2008) Generation of a set of simple, interpretable ADMET rules of thumb. J Med Chem 51(4):817–834. https://doi.org/10.1021/jm701122q
9. Meanwell NA (2011) Improving drug candidates by design: a focus on physicochemical properties as a means of improving compound disposition and safety. Chem Res Toxicol 24(9):1420–1456. https://doi.org/10.1021/tx200211v
10. Meanwell NA (2016) An update on recent applications of efficiency metrics, strategies for replacing problematic elements, and compounds in nontraditional drug space. Chem Res Toxicol 29(4):564–616. https://doi.org/10.1021/acs.chemrestox.6b00043
11. Mignani S, Rodrigues J, Tomas H, Jalal R, Pal Singh P, Majoral J-P, Vishwakarma RA (2018) Present drug-likeness filters in medicinal chemistry during the hit and lead optimization process: how far can they be simplified? Drug Discov Today 23(3):605–615. https://doi.org/10.1016/j.drudis.2018.01.010
12. Lipinski CA, Lombardo F, Dominy BW, Feeney PJ (1997) Experimental and computational approaches to estimate solubility and permeability in drug discovery and development settings. Adv Drug Deliv Rev 46(1–3):3–26. https://doi.org/10.1016/s0169-409x(00)00129-0
13. Newman DJ, Cragg GM (2016) Natural products as sources of new drugs from 1981 to 2014. J Nat Prod 79(3):629–661. https://doi.org/10.1021/acs.jnatprod.5b01055
14. Muthukrishnam I, Sridharan V, Menéndez JC (2019) Progress in the chemistry of tetrahydroquinolines. Chem Rev 119(8):5057–5191. https://doi.org/10.1021/acs.chemrev.8b00567
15. Faheem Kumar BK, Sekhar KVGC, Chander S, Kunjiappan S, Murugesan S (2021) Medicinal chemistry perspectives of 1,2,3,4-tetrahydroisoquinoline analogs – biological activities and SAR studies. RSC Adv 11(20):12254–12287. https://doi.org/10.1039/d1ra01480c
16. Brossi A, Lindlar H, Walter M, Schnider O (1958) Syntheseversuche in der Emetin-Reihe. 1. Mitteilung. 2-Oxo-hydrobenzo[a]chinolizine. Helv Chim Acta 41(119–139):114. https://doi.org/10.1002/hlca.660410117
17. Hunt P, Kannengiesser M, Raynaud J (1974) Nomifensine: a new potent inhibitor of dopamine uptake into synaptosomes from rat brain corpus striatum. J Pharm Pharmacol 26(5):370–371. https://doi.org/10.1111/j.2042-7158.1974.tb09294.x
18. Sigel MM, Lichter W, Ghaffar A, Wellham LL, Weinheimer AJ (1980) Substances from marine organisms influencing tumor growth and immune responses. In: Escobar MR, Friedman H (eds) Macrophages and lymphocytes. Advances in experimental medicine and biology, vol 121B. Springer, Boston. https://doi.org/10.1007/978-1-4684-3593-1_48
19. Seubert J, Pohlke R, Loebich F (1977) Synthesis and properties of praziquantel, a novel broad spectrum anthelmintic with excellent activity against schistosomes and cestodes. Experientia 33(8):1036–1037. https://doi.org/10.1007/BF01945954
20. Moss GP (1996) Basic terminology of stereochemistry (IUPAC recommendations 1996). Pure Appl Chem 68(12):2193–2222. https://doi.org/10.1351/pac199668122193
21. Gleeson MP, Leeson PD, Waterbeemd HVD (2015) Physicochemical properties and compound quality. In: Davis A, Ward SE (eds) The handbook of medicinal chemistry, RSC eTextbook collection, pp 1–31. https://doi.org/10.1039/9781782621836-00001
22. Kerns EH, Di L (2004) Physicochemical profiling: overview of the screens. Drug Discov Today Technol 1(4):343–348. https://doi.org/10.1016/j.ddtec.2004.08.011
23. Waring MJ (2010) Lipophilicity in drug discovery. Expert Opin Drug Discov 5(3):235–248. https://doi.org/10.1517/17460441003605098
24. Miller RR, Madeira M, Wood HB, Geissler WM, Raab CE, Martin IJ (2020) Integrating the impact of lipophilicity on potency and pharmacokinetic parameters enables the use of diverse chemical space during small molecule drug optimization. J Med Chem 63(21):12156–12170. https://doi.org/10.1021/acs.jmedchem.9b01813
25. Walker MA (2015) Improving solubility via structural modification. In: Meanwell N (ed) Tactics in contemporary drug design, topics in medicinal chemistry. Springer, Berlin/Heidelberg, pp 69–106
26. Ritchie TJ, Macdonald SJF (2009) The impact of aromatic ring count on compound developability – are too many aromatic rings a liability in drug design? Drug Discov Today 14(21–22):1011–1020. https://doi.org/10.1016/j.drudis.2009.07.014

27. Lovering F, Bikker J, Humblet C (2009) Escape from flatland: increasing saturation as an approach to improving clinical success. J Med Chem 52(21):6752–6756. https://doi.org/10.1021/jm901241e
28. Yang W, Lipert M, Nofsinger R (2023) Current screening, design, and delivery approaches to address low permeability of chemically synthesized modalities in drug discovery and early clinical development. Drug Discov Today 28(9):1–13. https://doi.org/10.1016/j.drudis.2023.103685
29. Prasanna S, Doerksen RJ (2009) Topological polar surface area: a useful descriptor in 2D-QSAR. Curr Med Chem 16(1):21–41. https://doi.org/10.2174/092986709787002817
30. Rankovic Z (2014) CNS drug design: balancing physicochemical properties for optimal brain exposure. J Med Chem 58(6):2584–2608. https://doi.org/10.1021/jm501535r
31. Kenny WP (2022) Hydrogen-bond donors in drug design. J Med Chem 65(21):14261–14275. https://doi.org/10.1021/acs.jmedchem.2c01147
32. Nudelman R, Kocks G, Mouton B, Ponting DJ, Schlingemann J, Simon S, Smith GF, Teasdale A, Werner A-L (2023) The nitrosamine "Saga": lessons learned from five years of scrutiny. Org Process Res Dev 27(10):1719–1735. https://doi.org/10.1021/acs.oprd.3c0010
33. Amidon GL, Lennernäs H, Shah VP, Crison JR (1995) A theoretical basis for a biopharmaceutic drug classification: the correlation of in vitro drug product dissolution and in vivo bioavailability. Pharm Res 12(3):413–420. https://doi.org/10.1023/A:1016212804288

The Molecular Drug Target of Praziquantel

5

Sang-Kyu Park, Daniel J. Sprague, and Jonathan S. Marchant

Abstract

This chapter describes work that led to the identification of the target of praziquantel (PZQ) in parasitic flatworms. The molecular target of this drug is an ion channel of the transient receptor potential family, named $TRPM_{PZQ}$. $TRPM_{PZQ}$ is activated by PZQ to cause a rapid depolarization of excitable cells in schistosomes, and the properties of this recently discovered ion channel align well with all known facets of PZQ action. The discovery of this elusive target has rapidly enabled progress to illuminate how PZQ works, understand scenarios in which PZQ is poorly effective, and unlock a new phase of target-based discovery to realize new anthelmintic chemotypes and useful tools for probing parasitic flatworm biology.

Keywords

Parasitic flatworm · Drug development · Transient receptor potential ion channel · Molecular modeling · Target-based screening

Introduction

Any reader of this book will realize how important praziquantel (PZQ) has been in combating the scourge of schistosomiasis. One could argue that PZQ is the most important drug that people have never heard of. As a cheap, effective medication (with over 1.5 billion tablets of PZQ donated to date), it has played a crucial role in reducing the prevalence of schistosomiasis worldwide [1, 2]. It is therefore

S.-K. Park · D. J. Sprague · J. S. Marchant (✉)
Department of Cell Biology, Neurobiology and Anatomy, Medical College of Wisconsin, Milwaukee, WI, USA
e-mail: jmarchant@mcw.edu

T. Spangenberg (ed.), *Praziquantel*,
https://doi.org/10.1007/978-3-031-97397-0_5

surprising that over 40 years of clinical usage, the target of PZQ has remained undefined. The answer of "we don't know" to the question of "how does this drug work?" is a surprising retort in the modern era of medicine. While knowledge of a drug target is by no means essential for drug development, very few drugs lack defined targets given the inherent value that knowledge of a molecular target provides for candidate optimization through the preclinical pipeline. Developing a new drug can take decades and cost in excess of a billion US dollars (Chap. 1), so definition of a drug target is advantageous for mitigating failures in the lengthy drug development process [3, 4].

PZQ was, however, discovered a long time ago and was quickly recognized as highly effective (Chap. 4) [5, 6]. Exposure of parasitic flatworms to PZQ in vitro was patently deleterious—PZQ caused an immediate spastic contraction and blistering of the worm surface [7]. These adverse phenotypes were consistent with in vivo data, where PZQ effectively resolved schistosome infections (Chap. 6). PZQ showed activity against many types of parasitic flatworm and host side effects were minimal, positive outcomes that were juxtaposed with an obvious veterinary utility and a pressing clinical need [5, 6]. Target identification was therefore unnecessary to immediately realize the potential of this exciting, new drug. This also belied the reality that parasite target discovery is challenging, even today. PZQ was approved for human use in a pre-genomic area where no facile route to candidate targets, their sequences, or expression profiles existed. Starkly, there was no clear PZQ-evoked cellular response from which to distill the underlying molecular actor, no immortalized flatworm cell culture model for screening, no native PZQ-evoked current for expression cloning, and no radioligand binding signature for biochemical fractionation. And so, it came to pass that PZQ became approved for clinical use without an associated target, as typical for many approved drugs in this era (Chaps. 10 and 11). This situation persisted for decades until the unmasking of a target in 2019—a transient receptor potential ion channel of the melastatin subfamily, $TRPM_{PZQ}$ [8].

Despite the longevity of this mystery, identification of the parasite target of PZQ was an important problem to solve. This is because the lack of knowledge of the target was a roadblock to realizing several opportunities that knowledge of the target would catalyze. First, target-based drug development. As definition of the target realizes the molecular solution to how PZQ works, this would enable opportunity to discover other ligands that engage the same target. Mapping the ligand binding site empowers rational approaches toward engineering new molecules to optimize available drug–receptor interactions through structure-based design. Target identification also catalyzes target-based screening capabilities to discover other active ligands. This can be accomplished in vitro using high-throughput screening (HTS) of libraries containing hundreds of thousands of compounds, or in silico using an array of computational and artificial intelligence-based methods. Therefore, with a target in hand, the potential for discovering new derivatives or entirely different ligand classes (chemotypes) is enabled. Second, monitoring and improving PZQ effectiveness. Knowledge of the drug target provides capacity to appraise drug performance. How might natural variation in the sequence of this target impact the therapeutic effectiveness of PZQ? This is of obvious interest in the context of assessing PZQ resistance in natural schistosome populations, and surveillance for emergence of target variants with lower PZQ sensitivity in response to cycles of drug exposure associated with mass drug administration campaigns. Also, whether sequence differences between the same target in different parasitic flatworms underpin poorer treatment outcomes for infections? This would provide an opportunity to develop better treatments tailored for specific diseases, and potentially identify ligands with broad spectrum activity. Target identity also enables functional studies of the target, in order to understand how target activity is regulated and how target expression is controlled. All of this knowledge could potentially illuminate strategies for monitoring and improving PZQ efficacy. Finally, furthering our understanding of worm biology, target identification provides impetus to understand the endogenous role of the target, as well as to highlight target interactors and closely related proteins as additional candidates for drug development. Illumination of the underlying biology is important, as if worm vulnerabilities to chemotherapeutic attack are better understood, and this will help prioritize the selection and validation of novel targets.

This chapter illustrates how these advantages have begun to be realized following the discovery of $TRPM_{PZQ}$. We first address the discovery of $TRPM_{PZQ}$ and subsequent work to validate this target. Then we discuss how this breakthrough rapidly enabled identification of new $TRPM_{PZQ}$ ligands, novel chemotypes active against parasites refractory to PZQ, as well as additional insights into worm biology. The goal is to underscore the inherent value of knowledge of "the drug target" by showing how such knowledge catalyzes scientific progress.

Discovery of $TRPM_{PZQ}$

Praziquantel (PZQ) is a small molecule (MW = 312 g/mol), with a chiral carbon atom defining two enantiomers (*R*)-PZQ and (*S*)-PZQ (Chap. 4). It is classified as an "antiparasitic" on account of its bioactivity against various types of parasitic flatworm infections. When applied to isolated schistosome worms, it evokes (1) a rapid, spastic paralysis. This contraction is closely coupled with (2) a blistering of the worm surface (the tegument) and (3) a change in the permeability of various cations that results in a protracted depolarization paralleling worm contraction (Chap. 6). This triad of phenotypes is seen in various flukes and tapeworms exposed to PZQ, consistent with the resolution of the associated infections by PZQ. For each of these actions, (*R*)-PZQ is effective at much lower concentrations than (*S*)-PZQ. (*R*)-PZQ is therefore considered the eutomer (the more potent enantiomer) and (*S*)-PZQ the distomer (the less potent enantiomer). (*R*)-PZQ causes contraction of adult schistosomes in the tens of nanomolar range, with the effective concentration of (*R*)-PZQ causing a half-maximal response (EC_{50} ~ 50 nM) lower than the EC_{50} for (*S*)-PZQ (EC_{50} ~ 1 μM) [8]. The target therefore displays stereoselectivity toward the enantiomers, and the difference in potency between (*R*)-PZQ and (*S*)-PZQ is referred to as the eudismic ratio (~20-fold for the example of schistosome contraction above). Traditionally, the clinical formulation of PZQ has been a

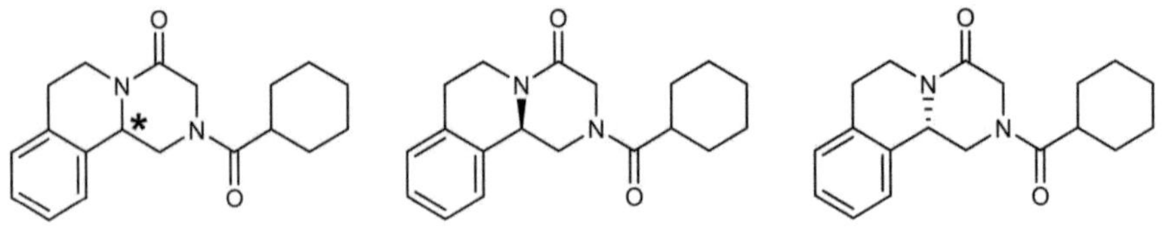

praziquantel (PZQ) M_r: 312.4 **(*R*)-PZQ** **(*S*)-PZQ**

Fig. 5.1 Stereochemistry of PZQ. Structure of PZQ, highlighting the chiral center (asterisked, *). The individual enantiomers—(*R*)-PZQ (middle) and (*S*)-PZQ (right)—are also shown

racemic mixture of both enantiomers, but a recent enantiopure formulation of (*R*)-PZQ has been developed for treating schistosomiasis [9] (Fig. 5.1).

While these PZQ-evoked effects provide no epiphany as to the molecular target, these phenotypes do provide clues. First, and most obvious, is the sustained contractile paralysis evoked by PZQ [7, 10], which implies a protracted disruption of ion homeostasis in muscle. The lack of this effect in Ca^{2+}-free media [7], underpinned a long-standing "Ca^{2+} hypothesis" for PZQ action pointing to a PZQ-evoked disruption of worm Ca^{2+} homeostasis [11]. The kinetics of the PZQ-evoked depolarization and contraction suggested a dysregulation of a signaling pathway associated with a rapid ion flux—such as an ion channel integral or indirectly coupled to the PZQ target. This appeared more likely than a Ca^{2+} pump or exchanger, which typically exhibit slower kinetics. The ionic signature of fluxes associated with PZQ action was broad—an influx of Na^+ and Ca^{2+} and efflux of K^+ [7]—suggesting the involvement of a nonselective cation current(s). Finally, the sensitivity to PZQ (tens of nanomolar) and the stereoselectivity of the response to the enantiomers (*R*)-PZQ being more potent than (*S*)-PZQ established clear characteristics to be reflected in the properties of the target.

Another benchmark was the broad activity of PZQ against different types of parasitic flatworm. PZQ is known to be effective against different types of flukes (blood, intestinal, lung, and liver) and tapeworms [5, 12]. The molecular target would need to be present in all these different parasites. But noteworthy is a genus of liver fluke (*Fasciola* species), which are unaffected by exposure to high concentrations of PZQ (50 μM) [13]. There needed to be an explanation as to the lack of efficacy of PZQ on these parasites and for treatment of acute and chronic fascioliasis [14, 15]. All these relevant clues were relevant for appraisal of a list of molecular suspects.

Progress with assembly and annotation of the *Schistosoma mansoni* genome (and other parasitic flatworm genomes) proved essential in the search for candidate targets [16]. The importance of investments in establishing these genomic resources for the field cannot be understated. The genomes of parasitic flatworms [16, 17] revealed a broad diversity of ion channels and G protein coupled receptors (GPCRs), known regulators of cellular Ca^{2+} homeostasis. However, there was little functional insight into the properties of these proteins, as relatively few ion channels and

GPCRs had been cloned and functionally expressed beyond the efforts of a few, pioneering laboratories. Cloning and functionally profiling all these potential candidates in an unbiased manner would have been a daunting task. The line up of potential suspects needed to be narrowed. This prioritization was achieved by taking a step back from parasite models and scrutinizing possible PZQ targets in humans where functional screening panels and methods had long been optimized. While arguably counterintuitive to search for targets of an antiparasitic therapy within the human host, this strategy was revealing as this work unveiled human targets for PZQ that, while displaying low affinity (EC_{50}s in the micromolar range, consistent with the lack of side effects of PZQ), these targets exhibited selectivity toward PZQ over closely related family members. For example, one host target was the *5-hydroxytryptamine 2B receptor* (5-HT_{2B}R, EC_{50} for PZQ ~8 μM), [18, 19] which exhibited selectivity toward (*R*)-PZQ compared with the closely related 5-HT_{2A} or 5-HT_{2C} receptors [18, 19]. The human transient receptor potential melastatin subfamily M (melastatin) member 8 (TRPM8) critically emerged as another host PZQ target [20], activated by micromolar concentrations of (*S*)-PZQ (EC_{50} for (*S*)-PZQ ~20 μM) compared with a lack of PZQ activity at other host TRPM channels [21].

These insights were revealing as the human data prioritized specific gene families, which if scrutinized in the parasites might unveil members that replicated an ability to bind PZQ. Both these particular families had already been viewed as intriguing anthelmintic targets, based on the long-evidenced role of serotonergic signaling in flatworm motility [22, 23], and TRP channels in sensory physiology [24, 25]. Therefore, following this logic, our laboratory began a screening campaign of parasite bioaminergic GPCRs and TRP channels to look for responsiveness to PZQ. This approach was based on heterologous expression, which entailed functional profiling of parasite gene products individually expressed in mammalian cell lines. This allowed testing each parasite protein for PZQ responsiveness using various second messenger reporter assays sensitive to elevations in cellular Ca^{2+} or cyclic adenosine monophosphate (cAMP). While based on a sound premise, this was nevertheless a blunt and chancy approach, as there was no known activator of many of these proteins to serve as a positive expression control. Months of negative data collection in our laboratory ensued, until dramatically a "hit" was found, with the expressed protein being a *S. mansoni* TRPM channel (*Smp_246790*). The data from this original experiment are reproduced in Fig. 5.2, showing a real-time fluorescent reporter assay in human embryonic kidney cells (HEK293) loaded with a fluorescent Ca^{2+} indicator where increased fluorescence signals reflect an increased cytoplasmic Ca^{2+} concentration. In non-transfected cells, increasing concentrations of PZQ, (*R*)-PZQ or (*S*)-PZQ (vertically, 1 nM to 50 μM), failed to elevate Ca^{2+} (Fig. 5.2). However, in cells expressing this particular TRPM channel, PZQ caused a concentration-dependent elevation in Ca^{2+} (Fig. 5.2), with a sensitivity consistent with the sensitivity of schistosomes to PZQ. The PZQ-evoked Ca^{2+} elevation was also rapid and sustained mimicking worm contraction (Fig. 5.2). (*R*)-PZQ was more potent than (*S*)-PZQ (Fig. 5.2), consistent with the known eudismic ratio for the action of both enantiomers on schistosomes. These features checked off previously discussed benchmarks for the expected target. Finally, this channel could also be

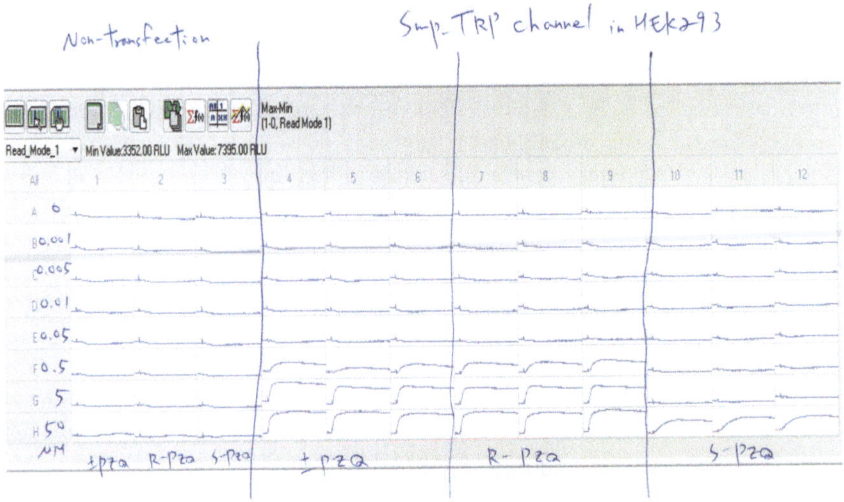

Fig. 5.2 PZQ activates a *S. mansoni* TRPM ion channel. Raw assay data from the original experiment performed by Sang-Kyu Park revealed PZQ activity at $TRPM_{PZQ}$. Data in a 96-well plate represent the real-time fluorescence of a Ca^{2+} reporter dye in HEK293 cells expressing $TRPM_{PZQ}$ compared with non-transfected HEK293 cells (left). Cells were challenged with increasing concentrations (top to bottom) of PZQ, or (*R*)-PZQ or (*S*)-PZQ. The stereoselectivity and submicromolar sensitivity of the response is clear

found in all types of parasitic flatworm sensitive to PZQ (including *Fasciola* sp.), with the properties of these orthologs aligning with PZQ sensitivity of the parasites [8, 26].

The excitement from these initial screening data was obvious. A parasitic flatworm target of PZQ had finally been identified, with the properties of this particular TRP channel closely mirroring the facets of PZQ action on parasitic flatworms. This ion channel was named $TRPM_{PZQ}$, tagged based upon its activator, with an initial characterization of this ion channel reported by Park et al. [8].

Validation of $TRPM_{PZQ}$

While these first impressions were alluring, there was a clear need to study $TRPM_{PZQ}$ further. Any potential target must be fully prosecuted to prove that it mediates the action of the drug in question, and that manipulations of the presumed target yield predicted outcomes. Therefore, there was much caution in our laboratory over grammar and use of "definite" and "indefinite" articles, taking careful to describe $TRPM_{PZQ}$ as "a" parasite target of PZQ rather than "the" target of PZQ until such additional lines of evidence could be accrued. This was appropriate as there are plenty of examples of presumed targets turning out to be "false" even when drugs are in very late stages of the development pipeline. Because of their size, small molecules are notoriously promiscuous, having many different target interactions that span a range of affinities. The ability of a single ligand to engage different targets is the phenomenon of drug polypharmacology. Therefore, further effort to validate $TRPM_{PZQ}$ as "the" relevant primary target was needed. Unambiguous validation could derive from a clean, genetic knockout of $TRPM_{PZQ}$, which would be predicted to result in a loss of responsivity to PZQ; however, this approach remains challenging in schistosomes and especially for an ion channel target expressed at low levels in excitable cells. Instead, other lines of experimental evidence were prioritized and each added weight to the contention that $TRPM_{PZQ}$ behaves as "the" clinically relevant target of PZQ. Readers are referred to a recent review collating how the progressive accrual of experimental evidence through diverse approaches has supported target validation [27]. In particular, work from Tim Anderson's laboratory provided independent support for $TRPM_{PZQ}$ as a critical determinant of schistosome sensitivity to PZQ through genetic mapping studies [28, 29]. Also work using electrophysiology to identify an endogenous nonselective cation current in schistosomes evoked by PZQ, which mirrored the characteristics of $TRPM_{PZQ}$ currents recorded in vitro [30]. Additionally, critical insights were gleaned from work mapping the binding site of PZQ in $TRPM_{PZQ}$ [31]. This was important for target validation as our initial study did not illuminate how PZQ activated $TRPM_{PZQ}$.

The first step in understanding the properties of the PZQ binding site was made by profiling the ability of a series of PZQ analogs to activate $TRPM_{PZQ}$ [31]. This established a unique ranking or "fingerprint"—a structure–activity relationship (SAR)—within this analog series reflective of the unique drug–receptor interactions made by different analogs within the same binding pocket. One would expect the SAR of analogs at the relevant target to match the SAR of the observed phenotype (e.g., worm contraction). For PZQ, hundreds of derivatives have been synthesized to date and the SAR for PZQ analogs on worm contraction is known to be "tight," with the majority of PZQ derivatives displaying poorer activity than PZQ itself (Chap. 6). A collaboration with Merck Healthcare KGaA was critical for our laboratory in providing access to a blinded panel of PZQ analogs, allowing profiling of a series of PZQ derivatives [31]. The goal was to compare the relative potency of $TRPM_{PZQ}$ activation (EC_{50} of 150 ± 50 nM for (*R*)-PZQ) versus schistosome contraction [31]. This general approach is described in Fig. 5.3. Analogs were profiled in different assays measuring either $TRPM_{PZQ}$ activation or worm contraction, and the potency of responses compared. Across the entire series of PZQ analogs, the fingerprint for channel activation and the fingerprint for worm contraction were almost identical (represented as "heat maps" in Fig. 5.3), consistent with $TRPM_{PZQ}$ activation underpinning contraction [31]. The SAR data also proved "tight," suggesting the structural requirements for PZQ engagement within the $TRPM_{PZQ}$ binding pocket were similarly stringent. Unblinding the structures of these analogs revealed which regions of the PZQ molecule were important for activity—the essential drug pharmacophore—and reciprocally, which regions showed more tolerability for modification.

However, insight from SAR analyses of PZQ analogs provided only half the story. The properties of the ligand pharmacophore (the "key") must correlate with the architecture of the PZQ binding pocket (the "lock") in $TRPM_{PZQ}$. And from a

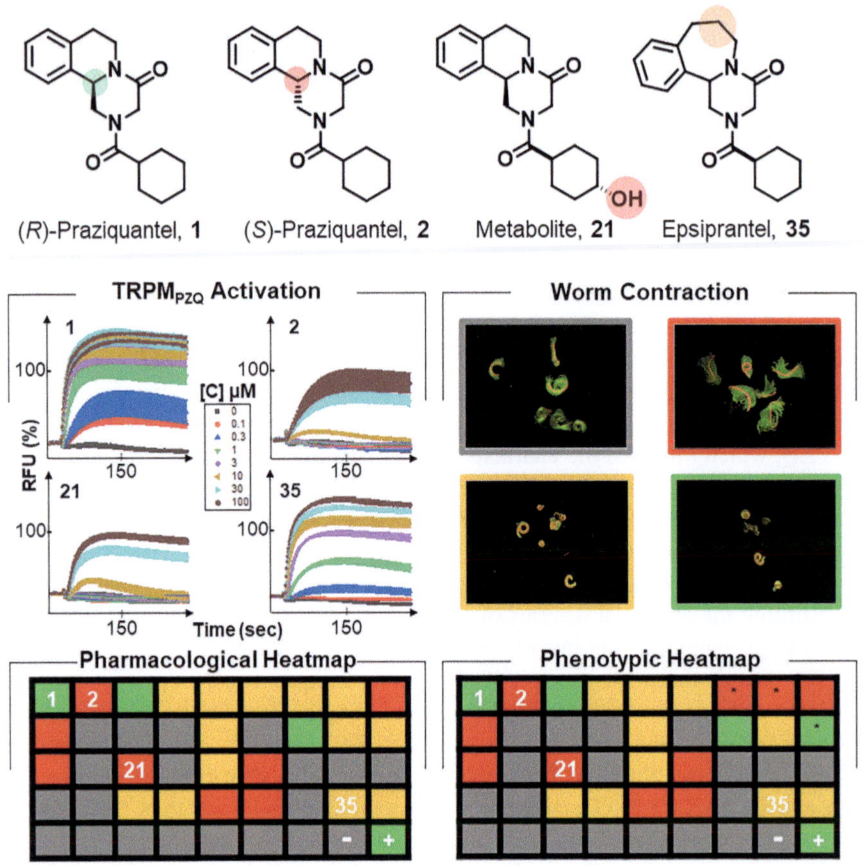

Fig. 5.3 Comparison of structure–activity relationships of PZQ analogs on the target versus the parasite. *Top*: examples of four different PZQ analogs from a screening set of 43 ligands (identified by numbers). *Middle*: strategy to compare results from an in vitro Ca^{2+} reporter assay (to measure $TRPM_{PZQ}$ activation) with the activity of the same analogs on adult schistosomes (to measure worm contraction). One would expect the SAR at the relevant target ("pharmacological SAR") to match that of the drug-evoked response ("phenotypic SAR"). *Bottom*: these heatmaps proved almost identical, consistent with $TRPM_{PZQ}$ mediating the PZQ-evoked paralysis [31]. (Figure modified with permission from [31])

target perspective, there was yet no insight into the molecular determinants of affinity (how well the key "fits" into the lock, an assessment of binding strength) or efficacy (how well the key "opens" the lock, an ability to evoke conformational changes that open the channel pore) as the identity of the PZQ binding site in $TRPM_{PZQ}$ was unknown. $TRPM_{PZQ}$ is a large protein (each monomer is >2200 amino acids, Fig. 5.4a), with extensive $–NH_2$ and –COOH terminal cytoplasmic regions extending from the transmembrane spanning region in the middle [8, 11]. The first four transmembrane helices (S1–S4) comprise a module known as the voltage-sensor-like domain (VSLD), based on the voltage sensing role of this region in

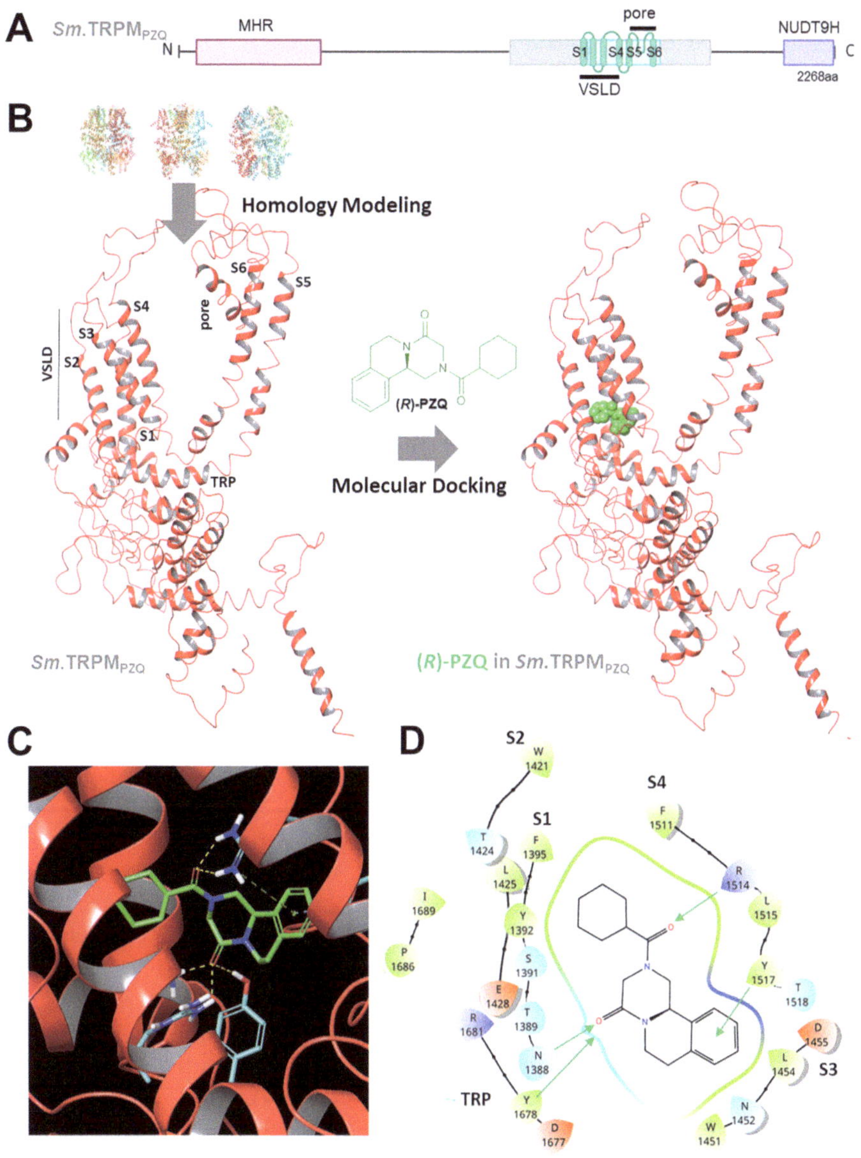

Fig. 5.4 Definition of the PZQ binding site in $\mathrm{TRPM_{PZQ}}$. (**a**) Schematic showing the modular organization of *Sm*.$\mathrm{TRPM_{PZQ}}$ depicting the cytoplasmic NH_2 terminal TRPM homology region (MHR, pink), the TRPM-like domain (gray) containing six transmembrane helices (S1–S6, green) that form the voltage-sensor like domain (VSLD) and pore (S5–S6) and the cytoplasmic –COOH terminal domain with the NUDT9-homology region (NUDT9H, purple). (**b**) *Left*: a predicted monomeric structure for *Sm*.$\mathrm{TRPM_{PZQ}}$ (residues 1100–1800) was generated by homology modeling using solved TRPM structures (TRPM2 (PDB: 6CO7), TRPM4 (PDB: 6BCJ), and TRPM8 (PDB: 6NR3)). *Right*: molecular docking was used to identify a likely binding pose for (*R*)-PZQ within the homology model. (**c**) Three-dimensional view of PZQ docked within the VSLD of *Sm*.$\mathrm{TRPM_{PZQ}}$. Different interactions between PZQ and residues on S1, S4, and the TRP helix are highlighted (dashed lines). (**d**) A cartoon representation of the same interactions reproduced with permission from [31]

voltage-operated ion channels. The VSLD is separated from the pore-domain, which is formed from a domain containing transmembrane helices S5 and S6 (Fig. 5.4b). The TRP helix, a general feature of the TRP superfamily runs parallel to the membrane on the cytosolic side. The cytoplasmic NH_2 terminal domain contains four regions resembling TRPM homology regions (MHRs), establishing these channels as TRPM-like, and the –COOH terminal domain contains a region resembling a mitochondrial hydrolase, NUDT9. The role of the NUDT9-homology domain in $TRPM_{PZQ}$ function is currently unknown. Where did the PZQ binding pocket reside in $TRPM_{PZQ}$? Again, key insight came using knowledge extrapolated from vertebrate channels. While overall sequence homology of $TRPM_{PZQ}$ to vertebrate TRPM channels was low, sequence alignment of the different TRP channels revealed that regions of higher homology existed within the VSLD of $TRPM_{PZQ}$. Residues known to be important in a well-characterized ligand binding pocket found in the VSLD of the human TRPM8 channel were conserved at the same positions in the transmembrane helices of $TRPM_{PZQ}$ [31]. Point mutagenesis of these residues in $TRPM_{PZQ}$ ablated PZQ activity, whereas mutations at residues outside this pocket did not impair the ability of PZQ to activate $TRPM_{PZQ}$ [31]. To define possible ligand binding poses in the VSLD pocket of $TRPM_{PZQ}$, a homology model of the $TRPM_{PZQ}$ transmembrane region was constructed (Fig. 5.4b) using known human TRPM channel structures as templates [31]. A homology model refers to an atomic resolution model that is the output from computationally fitting the amino acid sequence of the $TRPM_{PZQ}$ transmembrane region into known experimental three-dimensional structures of related proteins. In the absence of an actual structure for $TRPM_{PZQ}$ (which would require considerable experimental optimization), the homology model allows a route to molecular docking analyses where the most likely (*R*)-PZQ binding confirmations can be predicted by ranking each ligand binding pose by fit and affinity. Different binding poses were evaluated experimentally by point mutagenesis, where the specific residues predicted to be involved in binding PZQ were individually mutated to impair the specific drug–receptor interactions [31]. The top ranked PZQ binding pose supported by mutagenesis data is shown in Fig. 5.4b. This PZQ binding pose is shown in detail in both a three-dimensional (Fig. 5.4c) and a schematic projection (Fig. 5.4d). Docking simulations with (*R*)-PZQ revealed that PZQ bridged two hydrophobic cavities separated by a region of polar contacts where the internal and external carbonyl groups of PZQ form hydrogen bonding interactions with residues in the S1 (N1388), S4 (R1514), and TRP domain [31]. The cyclohexyl ring of PZQ projected into a hydrophobic cleft defined by three hydrophobic residues: F1395 (S1), W1421 (S2), and L1425 (S2). The second hydrophobic cavity accommodates the aromatic ring system of PZQ and is framed by six residues: W1451, N1452, L1454, D1455, R1514, and Y1517. A salt bridge between D1455 and R1514 shaped this cavity, in which the unsaturated part of the PZQ scaffold interacted with Y1517 via π–π stacking and with R1514 via cation–π interactions. From a ligand perspective, both carbonyl groups were known to be important for PZQ activity. From a target perspective, mutation of their predicted interacting residues also caused a loss of activity. The fact that reciprocal changes in the "lock" and "key" yield the same outcome supports the binding pose prediction. Another example relates to the cyclohexyl ring of PZQ. From our analysis of PZQ analogs, this ring emerged as a key part of the ligand pharmacophore [31]. Modifications to the cyclohexyl ring were poorly tolerated, leading to a loss of efficacy [31]. Evaluation of the predicted binding pose in the TRPM_{PZQ} homology model revealed why. The cyclohexyl ring inserted within a tight hydrophobic pocket at the apex of the ligand binding pocket (Fig. 5.4c). Any additions to the ring structure would likely clash with binding pocket residues or disrupt the orientation of PZQ such that formation of critical drug–receptor interactions needed for channel activation would likely be impaired. Again, the reciprocity of "fit" could be shown from the target perspective through targeted mutagenesis to introduce bulkier residues into the vicinity of the cyclohexyl ring. In the homology model, residue L1425 was predicted to project above the cyclohexyl ring. Increasing the size of this overhanging residue (L1425F or L1425W) to progressively decreased PZQ potency [31], most likely because the increased residue size sterically hindered the ability of the cyclohexyl ring of PZQ to occupy the hydrophobic cavity. Again, the concurrence between ligand- and target-based manipulations proved reassuring. Resolution of the PZQ binding site in TRPM_{PZQ} allowed scrutiny of the characteristics of this pocket. Twenty-three amino acid residues in the transmembrane S1, S2, S3, S4 helices and the cytoplasmic TRP helix were predicted to fall within 5 Å of the PZQ binding pose (Fig. 5.4d) [31].

Again, the availability of genomic data from other parasitic flatworms proved immensely helpful, as it enabled comparative bioinformatics to compare this schistosome binding pocket with sequences of TRPM_{PZQ} orthologs in other parasites. This analysis revealed high conservation of these 23 residues [26], suggesting that a similar PZQ binding pocket existed in the other TRPM_{PZQ} orthologs. Figure 5.5 shows an alignment of these residues, illustrating a complete conversation of all 23 residues in other flukes known to be sensitive to PZQ. These included TRPM_{PZQ} orthologs from other schistosome species (*S. haematobium* and *S. japonicum*), as well as other trematodes (e.g., *Clonorchis sinensis*, *Opisthorchis viverrini*, and

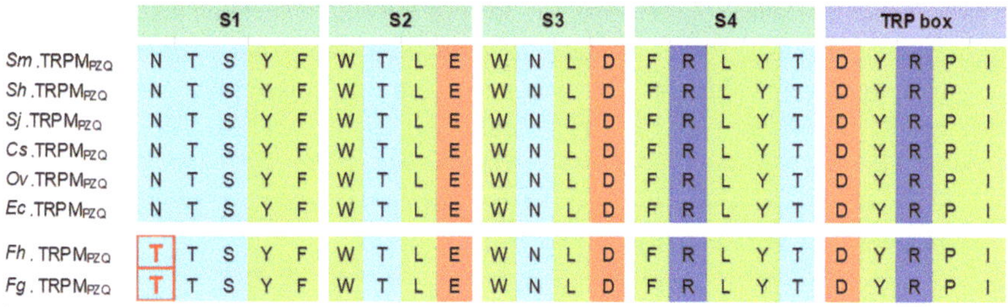

Fig. 5.5 TRPM_{PZQ} binding pocket residues in close proximity to PZQ. Alignment of 23 binding pocket residues that lie with 5 Å of the predicted PZQ binding pose in schistosome TRPM_{PZQ} (*S. mansoni*, *Sm*.TRPM_{PZQ}; *S. haematobium*, *Sh*.TRPM_{PZQ}; and *S. japonicum*, *Sj*.TRPM_{PZQ}) as well as other trematode TRPM_{PZQ} channels (*Clonorchis sinensis*, *Cs*.TRPM_{PZQ}; *Opisthorchis viverrini*, *Ov*.TRPM_{PZQ}; and *Echinostoma caproni*, *Ec*.TRPM_{PZQ}). The one amino acid difference (a residue in S1) with *Fasciola* TRPM_{PZQ} (*Fasciola hepatica*, *Fh*.TRPM_{PZQ} and *Fasciola gigantica*, *Fg*.TRPM_{PZQ}) is highlighted. (Data are modified from [26, 31])

Echinostoma caproni). Functional profiling of $TRPM_{PZQ}$ from these other schistosome species and the other parasitic flatworms confirmed that they were sensitive to PZQ [26, 32].

But what about *Fasciola*? A $TRPM_{PZQ}$ ortholog was annotated within the genomes of *Fasciola hepatica* (*Fh.* $TRPM_{PZQ}$) and *Fasciola gigantica* (*Fg.* $TRPM_{PZQ}$), so the presumed drug target was present in these liver fluke. If $TRPM_{PZQ}$ is expressed, why are these parasitic flukes not sensitive to PZQ? Sequence alignment showed that the residues lining the binding pocket of these *Fasciola* channels appeared well conserved, with 22 of the 23 residues being identical [31]. If you had to bet based on these data, you would wager that *Fasciola* $TRPM_{PZQ}$ should show sensitivity to PZQ. However, functional expression of *Fh.*$TRPM_{PZQ}$ or *Fg.*$TRPM_{PZQ}$ yielded no sensitivity to PZQ [26, 31], unlike results with all the other $TRPM_{PZQ}$ orthologs. Scrutiny of the single residue difference revealed it was a conservative difference (Fig. 5.5)—an asparagine (N1388), the S1 helix of *Schistosoma mansoni* $TRPM_{PZQ}$, was represented by a threonine at the same position (T1270) in the S1 helix of *Fasciola* $TRPM_{PZQ}$ [31]. However, this small difference proved critical. Replacement of the asparagine with a threonine (N1388T) was a loss-of-function mutation for *Sm.*$TRPM_{PZQ}$ annulling PZQ activity [31]. In contrast, the reciprocal mutation in *Fasciola* $TRPM_{PZQ}$ (T1270N in *Fh.*$TRPM_{PZQ}$ and T1271N in *Fg.*$TRPM_{PZQ}$) was a gain-of-function mutation, with the mutant *Fasciola* $TRPM_{PZQ}$ channels displaying robust responsivity to PZQ [26, 31]. This was a remarkable result, advancing a simple but elegant molecular explanation for the insensitivity of these particular liver flukes to PZQ. This amino acid difference occurred at a residue that is predicted in the binding pose to form a critical hydrogen bond with the internal carbonyl of PZQ [31]. Loss of this drug–receptor interaction when a threonine residue is present in *Fasciola* spp. ablated PZQ action. This evidence was a smoking gun for validating the candidacy of $TRPM_{PZQ}$ as the relevant clinical target of PZQ. The definition of the $TRPM_{PZQ}$ binding pocket had provided a clear answer as to why *Fasciola* are insensitive to PZQ. The lack of phenotypic effects of PZQ when applied to these flukes would argue against other molecular targets being relevant. The fact that a single, conservative amino acid difference dictates the insensitivity of these liver flukes to PZQ is quite sobering and presents obvious anxiety whether de novo variation within this binding pocket could impact PZQ sensitivity and thereby the effectiveness of PZQ as a clinical therapy in other parasites.

Target-Based Screening of $TRPM_{PZQ}$

With a molecular target in hand, target-based drug screening approaches become available to discover novel agonists (activators) and antagonists (inhibitors) of $TRPM_{PZQ}$. This could provide new tools for manipulating worm biology, and potentially novel anthelmintic leads. The chances of finding new $TRPM_{PZQ}$ regulators would obviously be maximized by screening a large number of compounds, sourced from different chemical libraries. The number of chemicals profiled can be in the hundreds of thousands for biological assays ("physical" screens), or an even greater number for in silico assays ("virtual" screens). Chemical libraries can be focused on particular themes (e.g., cellular proliferation), flavors of target (e.g., kinases), types of ligands (e.g., natural products), or simply collated as "samplers" to represent "diversity" of chemical structures to better the odds of sampling different ligands that regulate the target. Given the current lack of an experimental structure of $TRPM_{PZQ}$ (a homology model is simply a prediction of structure), we decided to perform a physical screen to search for new $TRPM_{PZQ}$ regulators. The logical first step was to conduct a pilot screen of a manageable number of ligands to optimize methods, before upscaling this approach into a larger, higher throughput screening effort using fully automated methods. A high-throughput screen (HTS) can profile over 100,000 individual assays per day. Such analyses necessitate careful, prior optimization of the screening conditions to ensure HTS results are robust and reproducible. A key challenge is successful miniaturization of the assay into small volumes, as a typical screen is done in multi-well plates with a large number of wells (e.g., 1586-well plates) to allow economical screening of many compounds. These plates afford an assay volume of typically a few microliters per well, necessitating a good signal versus background ("signal-to-noise ratio") within this small sample volume. One widely used measure of assay robustness is the Z′ factor, which takes into account the mean values of both the positive and negative assay controls as well as their standard deviations [33]. A criterion of $Z' > 0.5$ is considered an excellent starting point. Thus, effort must be invested to optimize the performance of the miniaturized assay before launching a HTS. Such assay development is critically important, integral to the success of the subsequent screen. HTS screens can be tailored for identifying activators and/or inhibitors, both involve carefully selected cut-off criteria to prioritize likely "hits" (or in the case of agonists, "actives") over background signals. An initial screen—often referred to as a primary screen—is typically done at a fixed, ligand concentration before full concentration–response curves are run on the initial hits in a secondary screen that assesses potency through measurements of EC_{50} values for agonists (Half Maximal Effective Concentration) or IC_{50} values for antagonists (Half Maximal Inhibitory Concentration). Hit rates in a HTS even using diverse libraries can be low—single digit, or lower, percentages across the entire library. This hit rate is then further decreased as putative hits that exceed the cut-off thresholds in the screen must also be subject to subsequent validation, one component of which is counter-screening to ensure that hits do not result from activity in control cells, interference with assay reporters, or adverse effects on cell viability. These activities separate out real positives that can be further progressed, from false positives. There is lengthy pipeline between a HTS hit and a chemical lead: a treacherous journey that decreases the diversity of chemical matter but increases confidence in specific leads as the overall diversity decreases. Readers are referred to excellent reviews on the topic of target-based screening [34, 35].

For *S. mansoni* $TRPM_{PZQ}$, we optimized the Ca^{2+} reporter-based assay for miniaturization (into a 6 μL volume with a 1536-well plate, Fig. 5.6a). Assay parameters were evaluated with different reporter dyes at different cell densities and temperatures [36]. Thankfully, the large, sustained Ca^{2+} signal resolved with high affinity Ca^{2+} indicators consequent to PZQ-evoked $TRPM_{PZQ}$ activation facilitated this

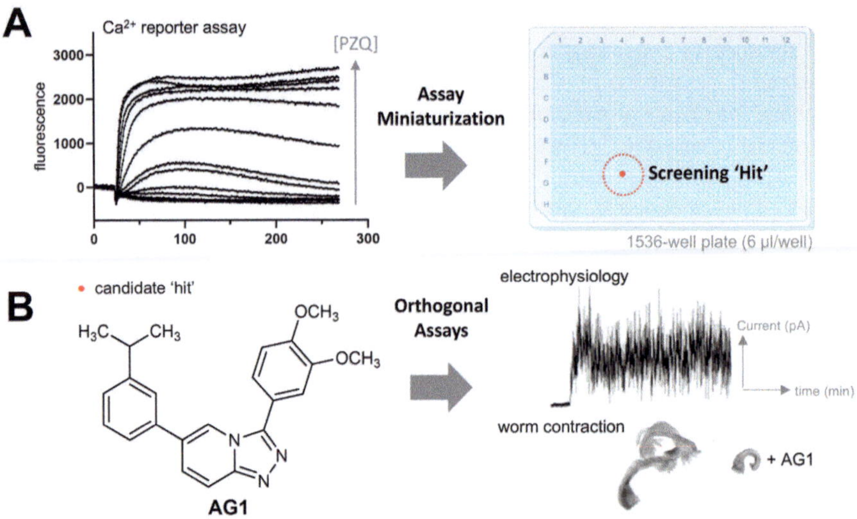

Fig. 5.6 Miniaturization of the reporter assay for a pilot screening campaign. (**a**) The Ca^{2+} reporter assay (left) for profiling *Sm*.TRPM$_{PZQ}$ activity was optimized for a miniaturized screening format (right). (**b**) *Left*: a candidate "active" from this screen that represented a different chemotype to PZQ was named **AG1** (agonist-1). *Right*: the activity of **AG1** was validated following compound resynthesis in different assays (electrophysiology and contraction of adult schistosomes) than used in the original screen

process, and a Z′ score of >0.7 was achieved [36]. A pilot screen was run against a total of ~16,000 compounds, and in this primary agonist screen (each compound screened at 5 μM) a total of three hits were resolved [36]. Two of these hits proved to be PZQ, present as a test compound in two of the chosen libraries. The resulting novel "hit" rate was therefore 1/15,934, or ~0.006%, continuing the theme of a "tight" SAR profile at TRPM$_{PZQ}$. The sole "hit" (agonist-1, **AG1**) comprised a different chemotype from PZQ (Fig. 5.6b). **AG1** was then profiled in a secondary screen, yielding an EC_{50} of ~2 μM. Resynthesis and reevaluation in a different laboratory confirmed activity (EC_{50} of ~9 μM) [36]. The activity of **AG1** was also validated using electrophysiology (Fig. 5.6b), exploiting an entirely different method as an orthogonal approach to confirm activity [36]. Finally, when added to adult schistosome worms, **AG1** also caused worm contraction, phenocopying PZQ [36]. This supported correct target validation, as the predictable phenotype (contraction) resulted from TRPM$_{PZQ}$ activation effected by entirely different chemotype. Remarkably, the antischistosomal activity of **AG1** had previously been recognized [37, 38]: it was one of several "hits" reported from a large (300,000 compounds) phenotypic screen against various schistosome life cycle stages [37]. The fact that a target-based screen and an antischistosomal phenotypic screen should independently converge on the same molecule, revealed as a TRPM$_{PZQ}$ agonist, underscored the viability of TRPM$_{PZQ}$ as a novel druggable target to realize new anthelmintic leads.

High-Throughput Screening of $TRPM_{PZQ}$ Yielded a Broad Spectrum Flukicide

Next, we upscaled the $TRPM_{PZQ}$ screen to >600,000 compounds. This required automated drug dispensing and plate handling robotics in collaboration with a dedicated HTS screening facility. With the higher number of compounds, and a different selection of drug libraries, more putative agonist "hits" were resolved, representing several novel chemotypes. With this larger portfolio of chemical matter, we were intrigued whether any of these new analogs would activate *Fasciola* $TRPM_{PZQ}$ as these different chemotypes might display unique drug–receptor interactions and be able to engage the *Fasciola* VSLD ligand binding pocket. Therefore, we counter-screened hits against *Fasciola hepatica* $TRPM_{PZQ}$ (*Fh*.$TRPM_{PZQ}$) and identified a benzamidoquinazolinone that activated both channels, *Sm*.$TRPM_{PZQ}$ (EC_{50} = ~1.2 μM) and *Fh*.$TRPM_{PZQ}$ (EC_{50} = 3.0 μM) [13]. A "hit" from a screening library is not a final product, but a substrate for further optimization by medicinal chemistry to identify derivatives with improved potency. Using this initial hit as a start, an analog was identified—named BZQ (Fig. 5.7a)—with improved potency at *Sm*.$TRPM_{PZQ}$, (~13-fold, EC_{50} = ~90 nM) and *Fh*.$TRPM_{PZQ}$ (~threefold, EC_{50} = ~1 μM) (Fig. 5.7b) [13]. Data with other fluke $TRPM_{PZQ}$ orthologs revealed BZQ acted as a broad spectrum $TRPM_{PZQ}$ activator [13]. However, the critical experiment was to add BZQ to liver flukes and observe whether BZQ caused similar effects on *Fasciola* as seen with PZQ on schistosomes. After all, BZQ had been identified solely from target-based screening, so if $TRPM_{PZQ}$ was correctly identified as the mediator of PZQ action, one would expect a different activator at the same target to yield similar effects. This experiment was performed by Simone Häberlein's laboratory and the results revealed that BZQ phenocopied the effects of PZQ (Fig. 5.7c) [13]. In the presence of BZQ, the liver fluke rapidly contracted into a sustained, spastic paralysis with concomitant evidence of surface damage [13], a further boost for correct target validation.

While BZQ engaged the VSLD of *Fh*.$TRPM_{PZQ}$, reproducing common interactions needed for $TRPM_{PZQ}$ activation (discrete interactions with S1, S4, and the TRP domain), the broad spectrum activity of BZQ resulted from an interaction with a different S1 residue than PZQ. BZQ engaged a common residue found in the S1 helix—a threonine residue localized immediately adjacent to the variant residue, N$\underline{T}$ in *Sm*.$TRPM_{PZQ}$ and T$\underline{T}$ in *Fh*.$TRPM_{PZQ}$. This particular threonine residue is conserved in the S1 helix of all fluke $TRPM_{PZQ}$ sequences (Fig. 5.5), supporting broad spectrum activity [13]. So, identification of $TRPM_{PZQ}$ rapidly catalyzed the discovery of novel ligands through HTS. This pool of $TRPM_{PZQ}$ activators realized novel chemotypes that had utility for treating diseases where current therapeutic options are limited.

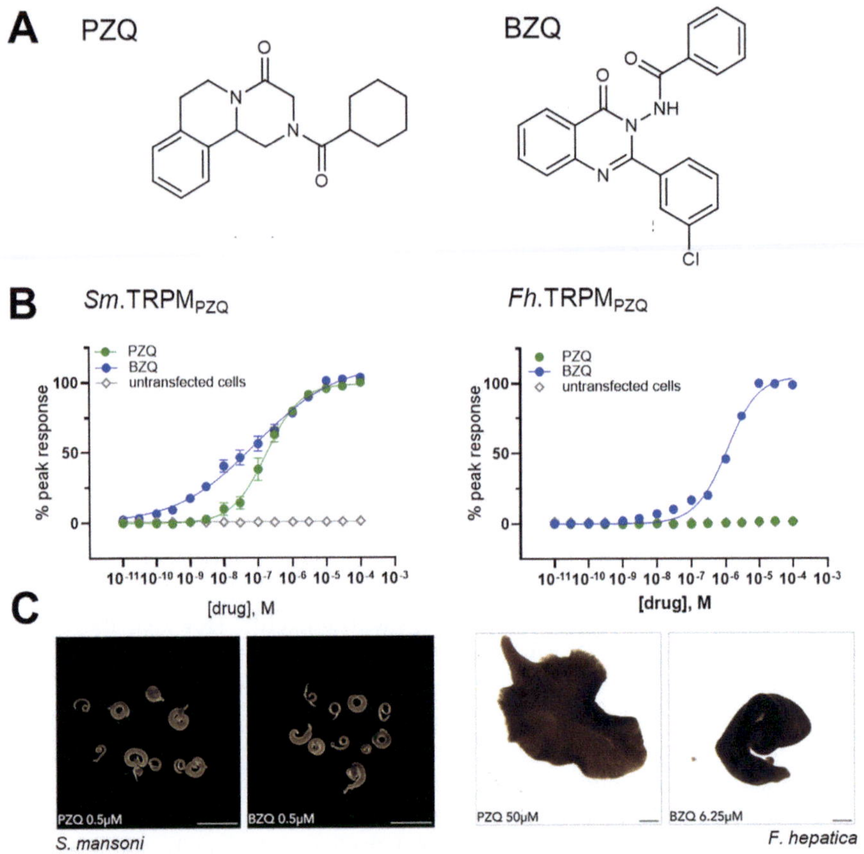

Fig. 5.7 Discovery of a broad spectrum flukicide in a high-throughput screening campaign. (**a**) Structures of PZQ (left) and BZQ (right). (**b**) Concentration response relationship for PZQ (green) and BZQ (blue) at *Sm*.TRPM$_{PZQ}$ (left) and *Fh*.TRPM$_{PZQ}$ (right) compared with control (non-transfected) cells (open symbols). (**c**) Action of PZQ and BZQ on adult *S. mansoni* worms (left) and *F. hepatica* (right). (Data are reproduced with permission from [13])

Illuminating New Aspects of Parasite Biology

In terms of clinical relevance, identification of TRPM$_{PZQ}$ permits surveillance of the extent of variation in TRPM$_{PZQ}$ sequence and expression within natural schistosome populations. This is now a focus of several groups engaged in fieldwork worldwide. Sequence variants associated with lower PZQ sensitivity could potentially prove problematic in terms of an increased likelihood that clinical dosing of PZQ fails to eliminate worms harboring TRPM$_{PZQ}$ variants with lower sensitivity. Surveillance of such "variants of concern" would be a priority to assess their frequency and persistence in areas known as "hot spots" where cycles of PZQ treatment have failed to decrease the incidence of infection. While drug resistance toward PZQ has not yet proved problematic in the field, surveillance of $TRPM_{PZQ}$ variants can only be helpful as one component of a strategy to monitor the ongoing clinical effectiveness of PZQ. Commonly, the effectiveness of many antiparasitic therapies has been curtailed by drug resistance.

Basic research into the endogenous role of $TRPM_{PZQ}$ is also important. Drug target identification provides an opportunity to understand the role of $TRPM_{PZQ}$ in schistosome biology. Is $TRPM_{PZQ}$ essential for schistosome viability? Is $TRPM_{PZQ}$ required throughout the parasitic life cycle? To better understand the role of $TRPM_{PZQ}$, it will be important to understand how the channel is endogenously activated. Is there an endogenous ligand that also occupies the VSLD binding pocket? Or are other sensory cures more important—for example, $TRPM_{PZQ}$ has been shown to be activated by membrane stretch [39], suggesting a possible link with worm motility. $TRPM_{PZQ}$ is expressed in different types of neurons and PZQ disrupts endogenous neuronal oscillations [30]. Here, there is likely opportunity based on studying $TRPM_{PZQ}$ antagonists discovered in the target-based screen. Pharmacological $TRPM_{PZQ}$ blockade should reveal the endogenous functions of this channel during various stages of the parasite life cycle. Identification of $TRPM_{PZQ}$ has also focused attention on the broader TRPM family in parasitic flatworms. While $TRPM_{PZQ}$ appears uniquely sensitive to PZQ, there are seven predicted members of the TRPM family in schistosomes [40, 41]. Knowledge of the properties of these other channels is scant. TRPM have a long evolutionary pedigree, so it is possible that these other TRPM family members also retain druggable VSLD binding pockets that can be exploited as an Achilles heel. Chemotherapeutic vulnerability may therefore extend throughout the TRPM family, providing untapped anthelmintic potential. This possibility has recently been realized with our discovery that another member of the TRPM family is targeted by a different anthelmintic chemotype, meclonazepam. Meclonazepam, a benzodiazepine, causes spastic worm paralysis and surface damage and has been recently shown to engage the VSLD binding pocket of a different TRPM channel, tagged TRP_{MCLZ} [42]. TRP_{MCLZ} is not activated by PZQ [42]. Meclonazepam is effective at killing juvenile worms, which in contrast recover from PZQ treatment. Therefore, identification of $TRPM_{MCLZ}$ advances an entirely new anthelmintic drug target for the field, and the discovery of $TRPM_{MCLZ}$ was only realized in the wake of $TRPM_{PZQ}$ identification that prioritized this particular ion channel family for study [40]. Like the game of "Battleship," a "hit" rapidly narrowed the search for the next strike.

Summary

This chapter has described the work that identified $TRPM_{PZQ}$ as the parasite target of PZQ, the mechanism by which PZQ interacts with this ion channel, as well as the structural requirements for $TRPM_{PZQ}$ activation. The discovery of $TRPM_{PZQ}$ has rapidly catalyzed our understanding of how PZQ works and why also it can be ineffective. It has also defined a critical molecular marker for surveillance of drug resistance. Already target-based screening of $TRPM_{PZQ}$ has yielded new antischistosomal chemotypes as well as tools, such as $TRPM_{PZQ}$ antagonists, useful for probing schistosome biology. The recent award of a Nobel Prize for studies of TRP channel signaling in sensory physiology provided timely impetus for exploring the function and druggability of TRP channels in disease-causing flatworms, where these channels likely fulfill sensory roles throughout the parasitic life cycle. Indeed, a related schistosome TRPM channel has recently been defined as another druggable target, advancing further screening opportunities. But perhaps the most important conclusion from this work is simply the speed at which all this progress has occurred. Knowledge of the target of PZQ has rapidly unlocked all these new insights, spurring new opportunities for anthelmintic development.

Acknowledgments The authors thank past and present members of the Marchant Laboratory who contributed to deciphering the mechanism of action of PZQ. Particular thanks are due to Jutta Reinhard-Rupp, Lukas Friedrich, and Thomas Spangenberg (Merck Healthcare KGaA) for their generosity and help catalyzing our work, which has been a joyful collaboration. Several schematics in this chapter were created in part using BioRender.

Competing Interests The work in the Marchant Laboratory has been supported by NIH grants (AI145781 and AI155405 to J.S.M.) and the Falk Foundation.

References

1. Kokaliaris C, Garba A, Matuska M, Bronzan RN, Colley DG, Dorkenoo AM, Ekpo UF, Fleming FM, French MD, Kabore A, Mbonigaba JB, Midzi N, Mwinzi PNM, N'Goran EK, Polo MR, Sacko M, Tchuem Tchuente LA, Tukahebwa EM, Uvon PA, Yang G, Wiesner L, Zhang Y, Utzinger J, Vounatsou P (2022) Effect of preventive chemotherapy with praziquantel on schistosomiasis among school-aged children in sub-Saharan Africa: a spatiotemporal modelling study. Lancet Infect Dis 22(1):136–149. https://doi.org/10.1016/S1473-3099(21)00090-6
2. Bradley M, Taylor R, Jacobson J, Guex M, Hopkins A, Jensen J, Leonard L, Waltz J, Kuykens L, Sow PS, Madeja UD, Hida T, Ole-Moiyoi K, King J, Argaw D, Mohamed J, Polo MR, Yajima A, Ottesen E (2021) Medicine donation programmes supporting the global drive to end the burden of neglected tropical diseases. Trans R Soc Trop Med Hyg 115(2):136–144. https://doi.org/10.1093/trstmh/traa167
3. Paul SM, Mytelka DS, Dunwiddie CT, Persinger CC, Munos BH, Lindborg SR, Schacht AL (2010) How to improve R&D productivity: the pharmaceutical industry's grand challenge. Nat Rev Drug Discov 9(3):203–214. https://doi.org/10.1038/nrd3078
4. Emmerich CH, Gamboa LM, Hofmann MCJ, Bonin-Andresen M, Arbach O, Schendel P, Gerlach B, Hempel K, Bespalov A, Dirnagl U, Parnham MJ (2021) Improving target assessment in biomedical research: the GOT-IT recommendations. Nat Rev Drug Discov 20(1):64–81. https://doi.org/10.1038/s41573-020-0087-3
5. Andrews P, Thomas H, Pohlke R, Seubert J (1983) Praziquantel. Med Res Rev 3(2):147–200. https://doi.org/10.1002/med.2610030204
6. Gonnert R, Andrews P (1977) Praziquantel, a new board-spectrum antischistosomal agent. Z Parasitenkd 52:29–150
7. Pax R, Bennett JL, Fetterer R (1978) A benzodiazepine derivative and praziquantel: effects on musculature of *Schistosoma mansoni* and *Schistosoma japonicum*. Naunyn Schmiedeberg's Arch Pharmacol 304:309–315

8. Park SK, Gunaratne GS, Chulkov EG, Moehring F, McCusker P, Dosa PI, Chan JD, Stucky CL, Marchant JS (2019) The anthelmintic drug praziquantel activates a schistosome transient receptor potential channel. J Biol Chem 294:18873–18880
9. N'Goran EK, Odiere MR, Assande Aka R, Ouattara M, Aka NAD, Ogutu B, Rawago F, Bagchus WM, Bodding M, Kourany-Lefoll E, Tappert A, Yin X, Bezuidenhout D, Badenhorst H, Huber E, Dalken B, Haj-Ali SO (2023) Efficacy, safety, and palatability of arpraziquantel (L-praziquantel) orodispersible tablets in children aged 3 months to 6 years infected with *Schistosoma* in Cote d'Ivoire and Kenya: an open-label, partly randomised, phase 3 trial. Lancet Infect Dis 23:867–876
10. Fetterer RH, Pax RA, Bennett JL (1980) Praziquantel, potassium and 2,4-dinitrophenol: analysis of their action on the musculature of *Schistosoma mansoni*. Eur J Pharmacol 64:31–38
11. Park SK, Marchant JS (2020) The journey to discovering a flatworm target of Praziquantel: a long TRP. Trends Parasitol 36:182–194
12. Chai JY (2013) Praziquantel treatment in trematode and cestode infections: an update. Infect Chemother 45:32–43
13. Sprague DJ, Park SK, Gramberg S, Bauer L, Rohr CM, Chulkov EG, Smith E, Scampavia L, Spicer TP, Haeberlein S, Marchant JS (2023) Target-based discovery of a broad spectrum flukicide. bioRxiv 2023.2009.2022.559026
14. Farid Z, Kamal M, Mansour N (1989) Praziquantel and *Fasciola hepatica* infection. Trans R Soc Trop Med Hyg 83:813
15. Farid Z, Trabolsi B, Boctor F, Hafez A (1986) Unsuccessful use of praziquantel to treat acute fascioliasis in children. J Infect Dis 154:920–921
16. Howe KL, Bolt BJ, Shafie M, Kersey P, Berriman M (2017) WormBase ParaSite - a comprehensive resource for helminth genomics. Mol Biochem Parasitol 215:2–10
17. Berriman M, Haas BJ, LoVerde PT, Wilson RA, Dillon GP, Cerqueira GC, Mashiyama ST, Al-Lazikani B, Andrade LF, Ashton PD, Aslett MA, Bartholomeu DC, Blandin G, Caffrey CR, Coghlan A, Coulson R, Day TA, Delcher A, DeMarco R, Djikeng A, Eyre T, Gamble JA, Ghedin E, Gu Y, Hertz-Fowler C, Hirai H, Hirai Y, Houston R, Ivens A, Johnston DA, Lacerda D, Macedo CD, McVeigh P, Ning Z, Oliveira G, Overington JP, Parkhill J, Pertea M, Pierce RJ, Protasio AV, Quail MA, Rajandream MA, Rogers J, Sajid M, Salzberg SL, Stanke M, Tivey AR, White O, Williams DL, Wortman J, Wu W, Zamanian M, Zerlotini A, Fraser-Liggett CM, Barrell BG, El-Sayed NM (2009) The genome of the blood fluke *Schistosoma mansoni*. Nature 460:352–358
18. Yahya NA, Lanham JK, Sprague DJ, Palygin O, McCorvy JD, Marchant JS (2023) Molecular and cellular basis of praziquantel action in the cardiovascular system. Am J Physiol Cell Physiol 324:C573–C587
19. Chan JD, Cupit PM, Gunaratne GS, McCorvy JD, Yang Y, Stoltz K, Webb TR, Dosa PI, Roth BL, Abagyan R, Cunningham C, Marchant JS (2017) The anthelmintic praziquantel is a human serotoninergic G-protein-coupled receptor ligand. Nat Commun 8:1910
20. Babes RM, Selescu T, Domocos D, Babes A (2017) The anthelminthic drug praziquantel is a selective agonist of the sensory transient receptor potential melastatin type 8 channel. Toxicol Appl Pharmacol 336:55–65
21. Gunaratne GS, Yahya NA, Dosa PI, Marchant JS (2018) Activation of host transient receptor potential (TRP) channels by praziquantel stereoisomers. PLoS Negl Trop Dis 12:e0006420
22. Mansour TE (1984) Serotonin receptors in parasitic worms. Adv Parasitol 23:1–36
23. Patocka N, Sharma N, Rashid M, Ribeiro P (2014) Serotonin signaling in *Schistosoma mansoni*: a serotonin-activated G protein-coupled receptor controls parasite movement. PLoS Pathog 10:e1003878
24. Wolstenholme AJ, Williamson SM, Reaves BJ (2011) TRP channels in parasites. Adv Exp Med Biol 704:359–371
25. Bais S, Churgin MA, Fang-Yen C, Greenberg RM (2015) Evidence for novel pharmacological sensitivities of transient receptor potential (TRP) channels in *Schistosoma mansoni*. PLoS Negl Trop Dis 9:e0004295

26. Rohr CM, Sprague DJ, Park SK, Malcolm NJ, Marchant JS (2023) Natural variation in the binding pocket of a parasitic flatworm TRPM channel resolves the basis for praziquantel sensitivity. Proc Natl Acad Sci USA 120:e2217732120
27. Marchant JS (2024) Progress interrogating $TRPM_{PZQ}$ as the target of praziquantel. PLoS Negl Trop Dis 18:e0011929
28. Le Clec'h W, Chevalier FD, Mattos ACA, Strickland A, Diaz R, McDew-White M, Rohr CM, Kinung'hi S, Allan F, Webster BL, Webster JP, Emery AM, Rollinson D, Djirmay AG, Al Mashikhi KM, Al Yafae S, Idris MA, Mone H, Mouahid G, LoVerde P, Marchant JS, Anderson TJC (2021) Genetic analysis of praziquantel response in schistosome parasites implicates a transient receptor potential channel. Sci Transl Med 13:eabj9114
29. Chevalier FD, Le Clec'h W, Berriman M, Anderson TJ (2024) A single locus determines praziquantel response in *Schistosoma mansoni*. Antimicrob Agents Chemother e0143223
30. Chulkov EG, Rohr CM, Marchant JS (2023) Praziquantel activates a native cation current in Schistosoma mansoni. Front Parasitol 2. https://doi.org/10.3389/fpara.2023.1285177
31. Park SK, Friedrich L, Yahya NA, Rohr CM, Chulkov EG, Maillard D, Rippmann F, Spangenberg T, Marchant JS (2021) Mechanism of praziquantel action at a parasitic flatworm ion channel. Sci Transl Med 13:eabj5832
32. Chulkov EG, Palygin O, Yahya NA, Park SK, Marchant JS (2023) Electrophysiological characterization of a schistosome transient receptor potential channel activated by praziquantel. Int J Parasitol 53:415–425
33. Zhang JH, Chung TD, Oldenburg KR (1999) A simple statistical parameter for use in evaluation and validation of high throughput screening assays. J Biomol Screen 4:67–73
34. Hughes JP, Rees S, Kalindjian SB, Philpott KL (2011) Principles of early drug discovery. Br J Pharmacol 162:1239–1249
35. Inglese J, Johnson RL, Simeonov A, Xia M, Zheng W, Austin CP, Auld DS (2007) High-throughput screening assays for the identification of chemical probes. Nat Chem Biol 3:466–479
36. Chulkov EG, Smith E, Rohr CM, Yahya NA, Park SK, Scampavia L, Spicer TP, Marchant JS (2021) Identification of novel modulators of a schistosome transient receptor potential channel targeted by praziquantel. PLoS Negl Trop Dis 15:e0009898
37. Mansour NR, Paveley R, Gardner JM, Bell AS, Parkinson T, Bickle Q (2016) High throughput screening identifies novel Lead compounds with activity against larval, juvenile and adult *Schistosoma mansoni*. PLoS Negl Trop Dis 10:e0004659
38. Maccesi M, Aguiar PHN, Pasche V, Padilla M, Suzuki BM, Montefusco S, Abagyan R, Keiser J, Mourao MM, Caffrey CR (2019) Multi-center screening of the pathogen box collection for schistosomiasis drug discovery. Parasit Vectors 12:493
39. Chulkov EG, Isaeva E, Stucky CL, Marchant JS (2023) Use the force, fluke: ligand-independent gating of *Schistosoma mansoni* ion channel TRPM(PZQ). Int J Parasitol 53:427–434
40. Sprague DJ, Rohr CM, Marchant JS (2024) TRP drop, TRP drop: a steady patter of anti-schistosomal target illumination. Front Parasitol 3
41. Bais S, Greenberg RM (2020) Schistosome TRP channels: an appraisal. Int J Parasitol Drugs Drug Resist 13:1–7
42. Park SK, Sprague DJ, Rohr CM, Chulkov EG, Petrow I, Kumar S, Marchant JS (2024) The anthelmintic meclonazepam activates a schistosome transient receptor potential channel. J Biol Chem 300:105528

In Vitro and In Vivo Activity of Praziquantel Against Schistosome Worms

6

Thomas Spangenberg and Jennifer Keiser

Abstract

When molecular targets are lacking, a phenotypic-based drug discovery approach may prove useful. This is how PZQ's anthelmintic activity was discovered by inducing characteristic effects to the worms in vitro that could be translated into in vivo efficacy. In this chapter we portray how compounds are tested against *Schistosoma mansoni* in vitro and in vivo. As an example for successful phenotyping screening, we describe the structure–activity relationship of praziquantel.

Keywords

PZQ · Phenotypic assay · Efficacy · In Vitro · In Vivo · Structure-activity relationship · Drug resistance

Introduction

Between 1999 and 2008, most (62%) of the first-in-class drugs approved by the United States Food and Drug Administration (FDA) were discovered empirically without a drug target hypothesis [1]. This is also true for praziquantel (PZQ). While one approach consists in discovering new drugs based on established causal

T. Spangenberg (✉)
Global Health R&D of Merck Healthcare, Ares Trading S.A., (a subsidiary of Merck KGaA, Darmstadt, Germany), Eysins, Switzerland
e-mail: thomas.spangenberg@merckgroup.com

J. Keiser
Department of Medical Parasitology and Infection Biology, Swiss Tropical and Public Health Institute, Allschwil, Switzerland

University of Basel, Basel, Switzerland

T. Spangenberg (ed.), *Praziquantel*,
https://doi.org/10.1007/978-3-031-97397-0_6

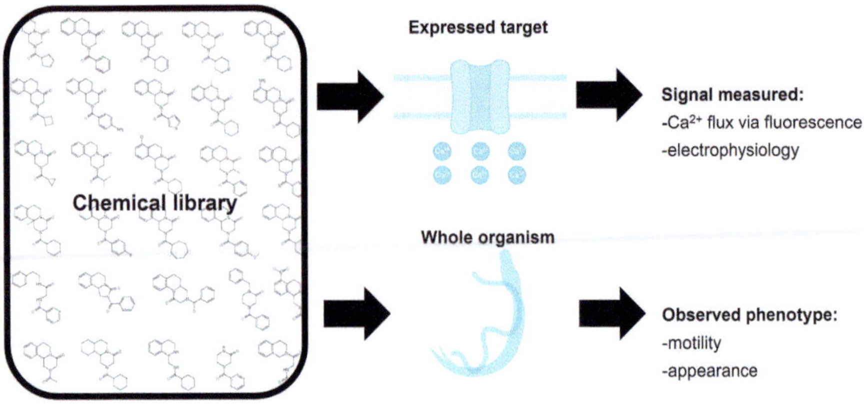

Fig. 6.1 Target-based versus phenotype-based drug screening approaches

relationship between a molecular target and a disease state, termed "target-based drug discovery" (Chap. 5), an alternative relies on chemical interrogation of a disease-relevant biological system in a molecular target-agnostic fashion, termed "phenotype-based drug discovery." The latter strategy is pragmatic and biology-driven that provides tool molecules to link therapeutic biology to previously unknown signaling pathways, molecular mechanisms, and drug targets. This approach termed "polypharmacology" also provides the opportunity to potentially engage more than one target simultaneously (Fig. 6.1).

Another potential benefit of phenotypic screening, given its focus on overall biological responses, is that it may improve the translatability of the activity of a given drug candidate from in vitro to in vivo. In other words, from a preclinical setting to a clinical setting as the approach involves testing compounds in whole-cell or whole-organism systems, allowing for a more comprehensive understanding of a compound's effects on complex biological systems. This approach captures the integrated response of the target and its environment, offering insights into both intended and unintended effects of the drug candidate. Importantly, drug candidates stemming from a phenotypic-based drug discovery program can transition to the clinic in the absence of knowledge of the molecular target. So, it is not a prerequisite to know the target to advance a drug candidate. This was also the case for PZQ.

Here, we will discuss the various steps that were undertaking to allow the testing of PZQ in in vitro and in vivo phenotypic assays.

What Will Be Used in the Phenotype-Based Assay?

As the aim is to find antischistosomal molecules, i.e., compounds that impact the viability of schistosome worms (Chap. 3), they would need to produce an observable effect against the worm. Depending on what molecules are targeting, which is unknown at that time, the induced phenotype may appear at a specific time point, leading to various observations, such as changes in worm motility, alterations in their appearance, or effects on their viability. Unlike single-celled malaria parasites, schistosome worms do not multiply in the human host (Chap. 2). Instead, their number and sex (gender) are determined by the cercariae that enter through the skin, after which they undergo several morphological changes as they migrate through different organs. Male and female worms then mate, with the females subsequently laying eggs.

Here, six morphologically characterized stages of worm development are discussed [2] (Fig. 6.2). After skin penetration, schistosomules mature in the lungs and migrate toward the portal system. **Stage 1** worms only display a light stain, indicating the beginning of the intestine (caecum). This stage has a short lifespan. Further maturation to **Stage 2** results in a darker stain, with the intestine bifurcating, but not bypassing the ventral sucker (acetabulum). In **Stage 3**, the dark bifurcating stain bypasses the ventral sucker and eventually links together. **Stage 4** is reached when the dark bifurcated stain extends to the end of the parasite but does not grow longer than the bifurcated intestine. In **Stage 5**, the linked intestine grows longer than the bifurcated part but remains under three times its length. Finally, **Stage 6** represents young and mature adults, with the linked caecum reaching three times the length of the bifurcated intestine. Notably, the size of the worms increases from less than 1 mm in **Stage 1** to approximately 1 cm for adult worms (primarily females), and their development is not synchronous, meaning that multiple stages can be observed at a given time point.

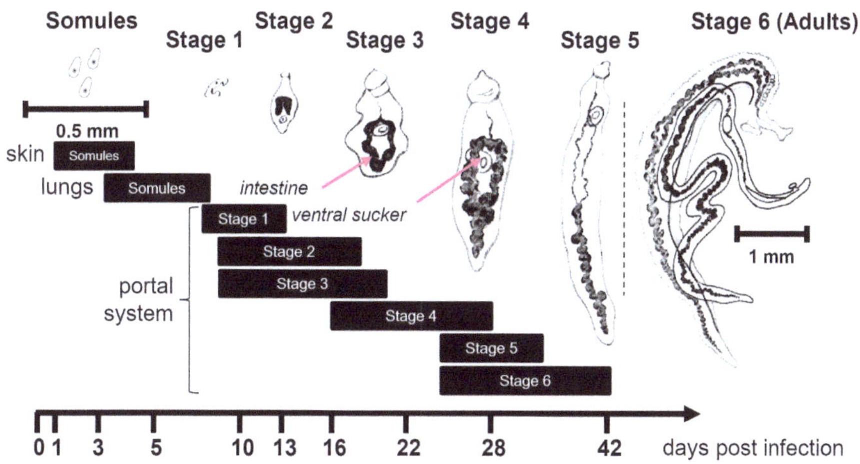

Fig. 6.2 Standardized reference stages used to describe the development of *S. mansoni*. (Adapted from Barbosa et al. [2])

How Are Parasites Accessed for Drug Testing?

To test a set of compounds from a chemical library, a reliable supply of worms is needed. For *S. mansoni*, eggs collected from the stool of infected patients can be obtained from various endemic areas (Chap. 2). These eggs hatch into miracidia, which are then used to infect the intermediate snail host, *Biomphalaria glabrata*, maintained in a water tank (Fig. 6.3). Several institutions maintain a laboratory life cycle of *S. mansoni*, often using hamsters as the mammalian host [3].

After 35 days of incubation in the snail, cercaria shedding can be triggered by light and collected. At this stage, they can be mechanically transformed into schistosomules by removing their tails. Once schistosomules are obtained, they are placed into in vitro culture using a medium that allows their maintenance (e.g., a slightly modified version of Basch's protocol) [4] supplemented with antibiotics (streptomycin/penicillin) to avoid contamination. Standard culture conditions are humidified air at 37 °C and 6% CO_2. A key limitation of in vitro culture is the difficulty in reaching the adult stage of the worm. As a result, schistosomules and juvenile stages can be maintained under these conditions, but they do not progress to adulthood [5]. An alternative approach is to perform an ex vivo experiment, where worms are collected directly from an infected animal (e.g., mouse or hamster) after reaching the desired maturation stage. These worms are then placed into an in vitro culture medium to assess drug-induced phenotypes. It is worth highlighting that only a low number of worms are typically obtained from an infected animal (e.g., around 30 worm pairs in mice), making this a low throughput approach. In this

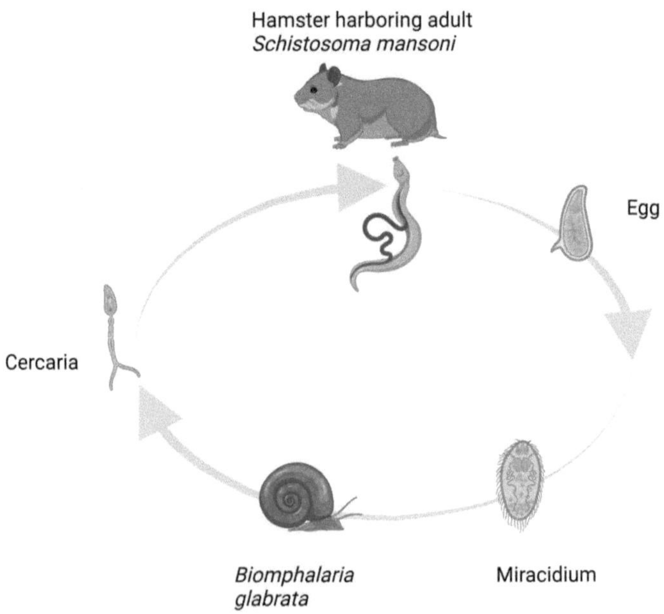

Fig. 6.3 Laboratory life cycle of *S. mansoni* in hamsters

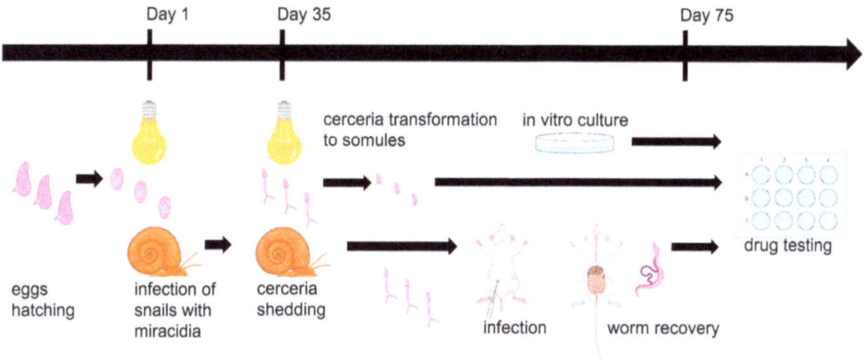

Fig. 6.4 Drug testing employing *S. mansoni* worms through generation of schistosomules or from perfusion of an infected mouse. (Made in part with BioRender®)

method, 80–120 cercaria are injected subcutaneously into a mouse (e.g., the Naval Medical Research Institute, NMRI, strain). After 7 weeks, the mice are dissected, and adult schistosomes are collected from the mesenteric veins. The worms are maintained in a 24-well culture plate with a culture medium (Roswell Park Memorial Institute Medium, RPMI, 1640), supplemented with proteins (10% fetal bovine serum, FBS), and kept from infections by means of antibiotics. Parasites are cultured at 37 °C with 5% CO_2 (Fig. 6.4).

How Are Compounds Tested In Vitro?

Now that the worm supply has been addressed, testing of compounds can be envisaged. Unlike organisms that replicate within the human host, schistosomes do not multiply but instead grow to adulthood, the relevant stage, in about 40 days before pairing and laying eggs. Consequently, we will examine this specific developmental stage of the worm, describing the effects of a compound incubated at a given concentration, rather than measuring growth inhibition as it is done with replicating parasites in other infectious diseases such as malaria.

So, what are we measuring? When testing compounds, worms exposed to molecules are observed and compared to unexposed control worms. The resulting phenotypes may vary, most commonly affecting motility (e.g., hypo/hyper), damaging the tegument (outer skin), or alterating the general appearance (e.g., shrinkage, darkening). To quantify the intensity of these effects, a score or grade is assigned. These scores are generally normalized to ensure the replication, reproducibility, and translatability of the results. Typically, compounds are first tested at a single concentration (e.g., 10 μM) and incubated with the worms for a specified period of time (e.g., 24 h, 48 h, or 72 h). This allows to perform a first triaging of active compounds. For instance, exposure of *S. mansoni* adult worms to PZQ leads to two striking phenomena in vitro: an almost instantaneous spastic paralysis of the parasite musculature and a rapid structural damage to the tegument (Fig. 6.5a) [6, 7].

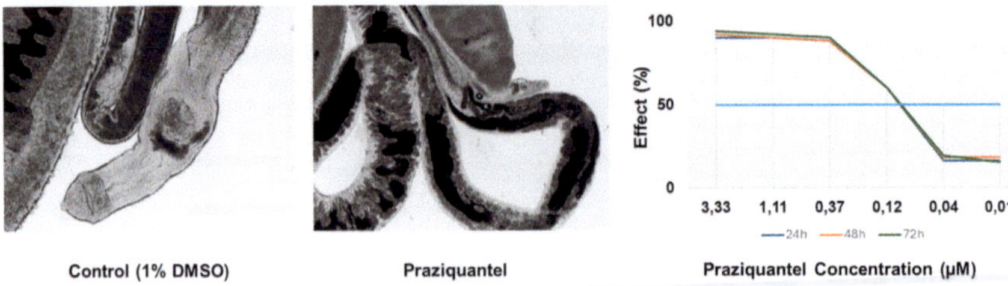

Fig. 6.5 DMSO control and PZQ phenotype on adult male and female worms seen by light microscopy. Dose–response curves (% effect as a function of PZQ concentration in micromolar) phenotype on adult *S. mansoni* male and female worms versus control

Table 6.1 Ex vivo EC_{50} pertaining to (*R*)-PZQ, (*S*)-PZQ, and the main metabolite *trans*-4-hydroxypraziquantel on adult *S. mansoni* and *S. haematobium* worms [8, 9]

Compound	Adult worm species	EC_{50} at 4 h (μM)	EC_{50} at 72 h (μM)	EC_{90} at 72 h (μM)
(*R*)-PZQ	*S. mansoni*	0.128	0.064	0.128
(*S*)-PZQ	*S. mansoni*	18	19	57
Trans-4-hydroxypraziquantel	*S. mansoni*	43	13	187
(*R*)-PZQ	*S. haematobium*	0.022	0.032	0.096
(*S*)-PZQ	*S. haematobium*	11	11	19
Trans-4-hydroxypraziquantel	*S. haematobium*	4.7	4.7	11

PZQ possesses an asymmetric center and consists of 1:1 mixture of two enantiomers (Chap. 4). In vitro assays confirmed that (*R*)-PZQ reproduces the characteristic phenotype, whereas (*S*)-PZQ at similar concentrations induces only transient hyperactivity in the worms. The main metabolite, *trans*-4-hydroxypraziquantel, is generated through hepatic oxidation and displays a less pronounced effect, indicating reduced potency (Chap. 7).

To obtain a more quantitative measurement of activity, dose–response curves were generated by assessing the motility of male and female worms as a function of drug concentration (Fig. 6.5b). The dose–response curves were established using different incubation times (24 h, 48 h, and 72 h) to capture the kinetics of the phenotype.

In Table 6.1, data on the PZQ enantiomers and its main metabolite are presented for the two most relevant adult *Schistosoma* species, i.e., *S. mansoni* and *S. haematobium*, which were obtained ex vivo [8, 9]. The EC_{50}, defined as the concentration required to achieve 50% of the maximal effect (in this case, motility), is 128 nM for (*R*)-PZQ, 18 μM for (*S*)-PZQ (resulting in an eudismic ratio of 142, which compares the potency of the two enantiomers), and 18 μM for the main metabolite, *trans*-4-hydroxypraziquantel, in *S. mansoni* worms at 4 h. Drug activity did not improve significantly with prolonged exposure times (72 h). Under similar experimental conditions for *S. haematobium*, the EC_{50} values for (*R*)-PZQ, (*S*)-PZQ, and *trans*-4-hydroxypraziquantel were 22 nM, 11 μM, and 4.7 μM, respectively, at 4 h, with no significant change after further incubation to 72 h. These data suggest that PZQ has a fast-acting mode of action with submicromolar activity. In a related experiment using juvenile worms (e.g., **Stage 4**, Fig. 6.2), a similar phenotype was observed; however, after washing out the drug and further maintaining the worms in vitro, they were able to recover from the drug's action indicating that the effect was reversible.

As only free (unbound) fraction of a drug can produce a pharmacological effect, it is important to understand to which extent PZQ is bound to proteins both in vitro and in vivo, since protein presence in both systems can reduce the concentration of free drug. Determining these values is essential for establishing effective free concentrations and for correlating results across different assays or species. For (*R*)-PZQ, the fraction unbound (F_u) was determined to be approximately 50% in the in vitro assay and around 20% in mouse plasma (Chap. 7).

In phenotypic testing, one might legitimately ask whether the molecule permeates adequately into the worm, how fast this process occurs, and to what extent? To address these questions, a radioactive ^{14}C-labeled PZQ was synthesized. Upon exposure, the drug was found to distribut evenly throughout the parasite's tissues, indicating the absence of receptor-mediated uptake and no formation of metabolites in the parasite. PZQ is taken up rapidly and reversibly, with apparent internal concentration equal to the external concentration within 2 min in schistosomes [10].

How Are Compounds Tested In Vivo?

In Vivo Efficacy Studies

The use of animal models remains a cornerstone in the study of schistosomiasis, with murine hosts, especially mice, being the most commonly used due to their small size, ease of handling, and cost-effectiveness. The mouse model is employed to investigate various aspects of human schistosomiasis, including parasite biology, pathogenesis, immunology, chemotherapy screening, and vaccine development. Other permissive species such as hamsters, gerbils, rabbits, or non-human primates have also been employed in in vivo studies. However, rats are considered semipermissive hosts for schistosomiasis, meaning they are not ideal for studying the full life cycle of the parasite, and thus are unlikely to be suitable for this purpose.

Animals are infected either through skin penetration (via the tail or shaved abdomen) or by subcutaneous (s.c.) or intraperitoneal (i.p.) injection of cercariae (Fig. 6.6). In mice, roughly 25% of the s.c. injected cercariae mature into adult worms, and the intensity of infection is correlated with the number of cercariae successfully becoming adults. Typically, the infection inoculum consists of around 100 cercariae [11]. As shown in Fig. 6.2, the parasites reach defined developmental stages at specific time points postinfection, generally between six and eight weeks [12]. Drugs can, therefore, be tested at various stages of worm development, typically around day 21 (3 weeks) for juvenile worms and around day 49 (7 weeks) for adult worms. Upon sacrifice of the animal, worms are recovered and counted either

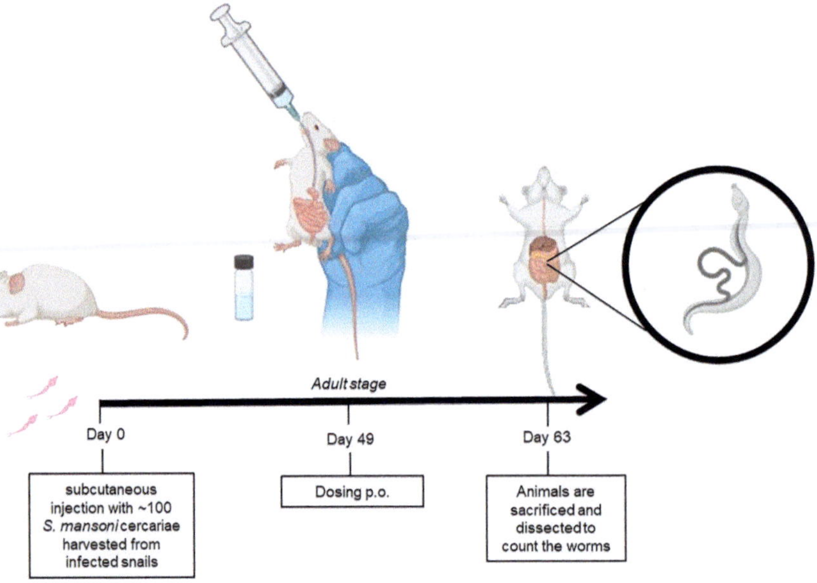

Fig. 6.6 *S. mansoni* infected mouse study scheme for oral drug testing at adult stage. (Made in part with BioRender®)

by (1) reverse perfusion of the portal venous system or (2) examination of the mesenteric veins, often combined with pressing the liver between glass plates to detect remaining schistosomes. These procedures are performed under a stereoscopic microscope. The number of worms recovered from the control and treated groups is compared, and the resulting worm burden reduction (WBR) serves as the primary measure of drug efficacy in this preclinical model (Table 6.2).

Three main variations influence the characteristics of the mouse model across different laboratories, complicating the direct comparability of experimental results: First, the use of mice of different strains and genders; Second, utilization of parasites of different isolate origins, infection route and dose; Third, variations in worm recovery methodology and drug efficacy assessment. Despite these differences, worm burden reduction (WBR) is widely accepted as the primary readout for evaluating drug effects.

Table 6.2 provides an overview of the in vivo activity of PZQ, (*R*)-PZQ, and (*S*)-PZQ in the aforementioned *S. mansoni* mouse model. Various factors, including dosing protocols, drug administration routes, worm developmental stages, gender, and conditions influencing systemic drug exposure, were evaluated. Efficacy was expressed as WBR. A single oral dose of PZQ at 400 or 500 mg/kg resulted in nearly full efficacy, with WBR ranging from 95% to 100%. Under similar conditions, (*R*)-PZQ achieved 98% WBR at 200 mg/kg, while the less potent (*S*)-PZQ enantiomer produced only 18% WBR at 400 mg/kg relative to the untreated control group [8]. Although pharmacokinetic–pharmacodynamic (PK–PD) studies are

Table 6.2 In vivo efficacy of PZQ and its enantiomers in *S. mansoni* infected mouse

Compound	Dose (mg/kg)	Route	Regimen	Worm development stage (day)	Number of mice	Mean WBR% (range or SD)	Comment	Reference
Vehicle	–	–	–	Adult	4–8	0	Control	–
PZQ	200	p.o.	q.d.	Adult	6	68 (65–71)	$AUC_{0\text{-}t}$ = 7.1 (1.2–11.1) μM/h ($P < 0.05$)	[13]
PZQ	400	p.o.	q.d..	Adult	6	97 (84–100)	$AUC_{0\text{-}t}$ = 11.4 (6.5–25.7) μM/h	[13]
PZQ	400	p.o.	q.d.	Adult	4	73 (46 ± 100)	With DEXA ($P < 0.05$)	[13]
PZQ	400	p.o.	q.d.	Adult	6	53 (0 ± 71)	With ABT, $AUC_{0\text{-}t}$ = 22.3 (10.0–39.2) μM/h	[13]
(*R*)-PZQ	400	p.o.	q.d.	Adult	3	100.0 (0)		[8]
(*R*)-PZQ	100	p.o.	q.d.	Adult	6	52.0 (30.8)		[8]
(*S*)-PZQ	400	p.o.	q.d.	Adult	4	18.0 (±21.4)		[8]
PZQ	250 (total)	p.o.	t.i.d.	Adult	4	100	2% Cremophor El in water	[14]
PZQ	500	i.m.	q.d.	Adult	6	87	Formulated in polyethylene glycol	[14]
PZQ	500	s.c.	q.d.	Adult	4	100	Formulated in olive oil	[14]
PZQ	500	p.o.	q.d.	Skin–lung stage (day 4)	35	41	2% Cremophor El in water	[14]

(continued)

Table 6.2 (continued)

Compound	Dose (mg/kg)	Route	Regimen	Worm development stage (day)	Number of mice	Mean WBR% (range or SD)	Comment	Reference
PZQ	500	p.o.	q.d.	Lung stage (day 7)	35	62	2% Cremophor El in water	[14]
PZQ	500	p.o.	q.d.	Lung stage (day 14)	35	21	2% Cremophor El in water	[14]
PZQ	500	p.o.	q.d.	Lung stage (day 21)	6	0	2% Cremophor El in water	[14]
PZQ	500	p.o.	q.d.	Lung stage (day 28)	7	0	2% Cremophor El in water	[14]
PZQ	500	p.o.	q.d.	Lung stage (day 35)	7	28	2% Cremophor El in water	[14]
PZQ	500	p.o.	q.d.	Adult (day 48)	15	92	2% Cremophor El in water	[14]
PZQ	500	p.o.	q.d.	Male adult (day 48)	15	100	2% Cremophor El in water	[14]
PZQ	500	p.o.	5x	Female adult (day 48)	15	41	2% Cremophor El in water	[14]

QD; WBR [%] = 100% − (100%/WBcontrol × WBtreatment); Statistical comparisons were done using the Kruskal–Wallis test at a significance level of $p < 0.05$. p.o.: per os (oral), s.c.: subcutaneous; i.m.: intramuscular; q.d.: quaque die (1 day); t.i.d.: ter in die (3 times a day). *S. mansoni* Liberian strain were used in all studies cited. DEXA control group AUC = **12_0** (9.8 ± 17.2) WBR = **84** (38 ± 92)

commonly used in drug discovery to identify factors driving drug efficacy, a clear PK–PD correlation for PZQ in *S. mansoni* rodent models has not yet been established. An important physiological consideration is the parasite's migratory route (Chaps. 2 and 7); juvenile worms travel through the inferior vena cava to the heart, then to the left lung and back to the heart, before entering the abdominal aorta and ultimately the portal vasculature, where the parasites mature [2]. Administering PZQ at a lower dose (e.g., 250 mg/kg) more frequently, such as dividing the total daily dose into three equal administrations (termed ter in die, t.i.d. meaning three times a day), also provides full efficacy [14]. This effect may result not only from more prolonged exposure of the worms to the drug but also from PZQ's physicochemical properties. As a Class II drug in the Biopharmaceutical Classification System (Chaps. 4 and 7) PZQ is highly permeable but poorly soluble, leading to slow absorption within the gut lumen. This slower absorption allows the drug to exert its effect in the mesenteric veins before undergoing extensive first-pass metabolism in the liver via the cytochrome P450 system (Chap. 7). Interestingly, administering 500 mg/kg of PZQ orally or subcutaneously produced full efficacy, whereas intramuscular administration, caused a slight reduction in efficacy, with a WBR of 87%.

Next, PZQ was administered to infected mice that had been pretreated with a cytochrome P450 inhibitor, aminobenzotriazole (ABT), to reduce metabolism (Chap. 7), thereby increasing systemic concentrations of PZQ. As a result, a single oral dose of 50 mg/kg achieved similar plasma concentrations to those of 400 mg/kg dose, but efficacy remained limited. Conversely, pretreatment with dexamethasone (DEXA) to induce liver enzymes did not diminish efficacy at a 400 mg/kg oral dose. These findings suggest that systemic exposure alone is not predictive of efficacy; rather, efficacy depends on the administered dose. This aligns with the fact that *S. mansoni* worms are located in the mesenteric vein, allowing PZQ to exert its effect before the drug is extensively metabolized by the liver. Therefore, the concentration of PZQ in the portal circulation may be a critical determinant of efficacy [13]. It is important to note that the exposure of *trans*-4-hydroxypraziquantel was not measured in this study, leaving open the possibility that elevated circulating levels of the less potent metabolite may have partially contributed to the overall drug efficacy [15]. To investigate stage-specific efficacy, infected mice were treated with an oral dose of 500 mg/kg, which had previously demonstrated full efficacy against adult worms at week 7. In this study, small to modest efficacy was observed during the first 2 weeks postinfection, with WBR ranging from 21% to 61%. However, there was a lack of efficacy between weeks 3 and 5. In experiments with single-sex infections, male worms were fully susceptible to the 500 mg/kg dose, while female worms showed significantly reduced susceptibility, even at 5 × 500 mg/kg, achieving only 41% WBR.

How Relevant Is the Mouse Model?

A key similarity between *S. mansoni* and *S. japonicum*, both of which cause human intestinal schistosomiasis, is that they inhabit the same anatomical location—the mesenteric veins—in both humans and mice. Conversely, *S. haematobium*, which typically resides in the venous plexus of the human bladder, cannot replicate this site specificity in mice, where it also localizes to the mesenteric veins. Despite its utility, notable differences exist between human schistosomiasis and the mouse model. For example, in humans, worm burdens rarely exceed five worm pairs per kilogram, whereas in animal models, this number often surpasses 50 worm pairs per kilogram, reflecting a much higher infection intensity in mice. Other differences are: (1) mice are typically infected via a single exposure, while humans acquire infections over time; (2) laboratory strains with homogeneous genetic backgrounds are used in mice, whereas humans are infected with natural parasite isolates, which have heterogeneous genetics; and (3) mice are usually mono-infected, while humans are often coinfected with other parasites.

While there are limitations, the mouse model remains pivotal for drug discovery and has been successfully used over the last decades.

What Is the Structure–Activity Relationship of Praziquantel?

The outstanding anthelminthic activity of PZQ was discovered during a screening campaign of over 400 analogues stemming from an antipsychotic research program. Medicinal chemists synthesized these analogues by derivatizing the core scaffold, 1,2,3,6,7,11β-hexahydro-4*H*-pyrazino(2,1-a)isoquinoline-4-one, primarily through simple amide bond formation to introduce new substituents. Generating biological data on a series of PZQ analogues or near neighbors enables the establishment of a structure–activity relationship (SAR), which describes how changes in chemical structure of a molecule affect biological activity. This approach allows to identify which elements can be replaced to modulate the properties of PZQ such as solubility (Chap. 4), metabolic stability (Chap. 7), toxicity (Chap. 8), and which elements are essential to maintain activity, i.e., the minimum pharmacophore. As shown in Fig. 6.7, selected PZQ analogues (all tested as racemic mixtures) exhibit modifications at various positions, including extensive alterations to the scaffold. These include the formation of isomeric structures, expansion or reduction of the heterocyclic system, changes in ring size, replacement of one or two rings with open-chain analogues, as well as substitutions and replacements involving the two amide bonds and the cyclohexane ring. The compounds were tested both in vitro and in vivo (in mice) against *S. mansoni* and evaluated based on the aforementioned in vitro phenotype as well as the worm burden in the mouse model: +++ (green) indicates full effect ≤1 μg/mL and 100% WBR ≤ 5 × 50 mg/kg; ++ (yellow) indicates full effect ≤100 μg/mL and 100% WBR ≤ 5 × 500 mg/kg; + (orange) indicates partial effect at 100 μg/mL and 90% WBR ≤ 5 × 500 mg/kg; 0 (red) indicates no effect at 100 μg/mL and no effect at 5 × 500 mg/kg.

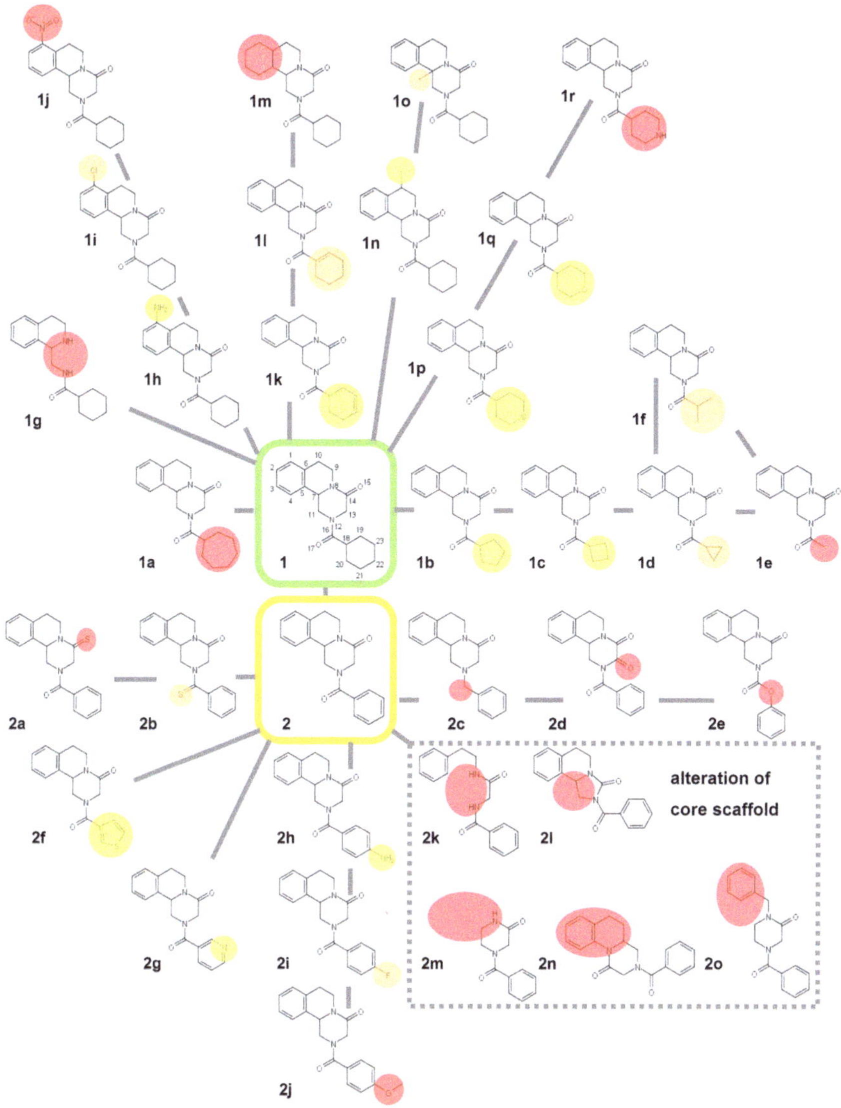

Score	In vitro	In vivo
+++	Full effect ≤1 µg/mL	100% WBR ≤ 5 x 50 mg/kg
++	Full effect ≤100 µg/mL	100% WBR ≤ 5 x 500 mg/kg
+	partial effect at 100 µg/mL	90% WBR ≤ 5 x 500 mg/kg
0	no effect at 100 µg/mL	No effect at 5 x 500 mg/kg

Fig. 6.7 Structure–activity relationship of praziquantel analogues

At first glance, the SAR appears narrow, with PZQ (Fig. 6.7, **1**) showing the strongest activity (+++) in vitro and in vivo. Substitution of the cyclohexyl group (positions 18–23) reveals that increasing the ring size to a seven-membered azepanyl (**1a**) results in a loss of activity. In contrast, reducing the ring size to a cyclopentyl (**1b**) or a cyclobutyl (**1c**) retains reasonable activity (++). Further reduction to a cyclopropyl (**1d**) or an isopropyl (**1f**) results in weak activity (+), while replacement with a simple methyl group (**1e**) leads to complete loss of activity. When introducing unsaturation by removing two hydrogen atoms to form a double bond in the cyclohexyl ring (**1l**, **1k**), activity varies (++; +) depending on the position of the double bond within the ring. This suggests that the typical low-energy chair-like conformation of the cyclohexyl ring is important for activity. Adding a heteroatom into the cyclohexyl ring (position 22) also diminishes activity, with potency decreasing from sulfur (**1p**, ++), to oxygen (**1q**, ++), and further to nitrogen (**1r**, +). Full saturation of the tetrahydroisoquinoline (Chap. 4) moiety (**1m**) renders the molecule devoid of activity. This suggests that the benzene ring in the tetrahydroisoquinoline moiety is important for activity (Chap. 5), while some substitutions at position 1 on the ring are tolerated to varying degrees. For example, the amino substituent ($-NH_2$, **1h**), an electron-donating group (EDG), confers greater activity than electron-withdrawing groups (EWGs) such as chloro (–Cl, **1i**) and nitro ($-NO_2$, **1j**). Furthermore, activity decreases progressively with increasing electron-withdrawing strength, with the nitro group (a stronger EWG) showing lower activity than the chloro substituent (a weaker EWG) (**1h** > **1i** > **1j**). Adding a methyl group to the middle ring of the tetrahydroisoquinoline moiety is tolerated to varying degrees (**1n** > **1o**), potentially altering the conformation of the fused tricyclic ring system. This effect is also observed when the piperazine ring is opened (**1g**, 0). Additionally, the oxygen atom of the acyl group at position 15 may play a crucial role in activity. In derivatives of compound **2** (+), where the cyclohexyl ring is replaced by a benzene ring, substituting the oxygen atoms at positions 15 and 17 with sulfur atoms (**2a**, **2b**) dramatically decreases activity (0, +), suggesting that hydrogen-bond interactions play a crucial role in activity. This is further supported by the removal of the oxygen atom at position 17 (**2c**), resulting in a complete loss of activity (0). Therefore, substantial anthelminthic activity requires the presence of an acyl or thioacyl group at position 17, and an acyl group at position 15. Similarly, inserting additional oxygen atoms at positions 13 and 18 (**2d**, **2e**) results in a loss of activity (0, 0). Comparable potency to compound **2** (++) is maintained by incorporating an amine, a strong EDG, at position 22 (**2h**, ++). However, activity is reduced when an EWG, such as a fluoride, is inserted (**2i**, +). Introducing a weak EDG, such as a methoxy group results in a complete loss of activity (**2j**, 0). Replacing the phenyl ring with a thiophene (**2f**) or a pyridine derivative (**2g**) retains activity (++, ++). Finally, scaffold derivatives (**2k**, **2l**, **2m**, **2n**, **2o**), where the core moiety is altered, are devoid of substantial activity (0), indicating that the anthelmintic property is limited to the pyrazinoisoquinoline system. Based on these results, PZQ was selected as the drug candidate and underwent further evaluation.

Can Worms Become Refractory to Praziquantel?

Like any other chemotherapeutic agent deployed against infectious diseases, drug resistance could arise. This may occur when parasites are under drug pressure, leading to a heritable loss of sensitivity to the drug, which is conferred to a parasite population that was previously susceptible to the same drug. Mechanisms of anthelmintic drug resistance may include upregulation of cellular efflux pathways, increased drug metabolism, alterations in drug receptor sites, and decreases in drug receptor abundance. The development of resistance is influenced by factors such as the host, the parasite, the type and usage of anthelmintic, animal management practices, and environmental conditions.

As schistosomes do not replicate within the human host, the number of parasites is limited. However, selection for PZQ-resistant parasites may still occur. Experimentally, resistant schistosomes were generated by treating infected snails with PZQ. The emerging parasites showed a 14-fold difference in the median effective concentration (EC_{50}), with the progenitor parasite population having an EC_{50} of 2.8 μM, compared to the progeny derived from PZQ selection, which had an EC_{50} of 41 μM [16]. Notably, the dose–response curve for the resistant population plateaus at 65% mortality, with the remaining 35% of parasites recovering even at high doses of PZQ (233 μM). These findings suggest that the parasite population is heterogenous, containing both PZQ-sensitive and PZQ-resistant worms.

Building on the observation that a mixed population of PZQ-sensitive and PZQ-resistant schistosomes exists, a genome-wide association study was conducted to map the loci underlying PZQ response. This analysis identified a transient receptor potential channel (Smp_246790) on chromosome 3 that is activated by submicromolar concentrations of PZQ (Chap. 5) [16]. The PZQ response showed recessive inheritance and marker-assisted selection of parasites at *Sm*.$TRPM_{PZQ}$ a single nucleotide polymorphism (SNP) that produced populations of PZQ-enriched resistant (PZQ-ER) and PZQ-enriched sensitive (PZQ-ES) parasites, exhibiting >377-fold difference in PZQ response. The PZQ-ER parasites survived treatment in mice at higher frequencies compared with PZQ-ES, and resistant parasites exhibited two-fold lower expression of *Sm*.$TRPM_{PZQ}$ relative to sensitive parasites. Identification of such molecular markers for characterizing PZQ-resistant alleles could be useful for parasite control programs and provide a framework for monitoring the evolution of PZQ resistance in schistosomiasis control efforts. Alternatively, *S. mansoni* insensitivity to PZQ was assessed by collecting samples from three patients in Senegal, in 1993. These patients failed to be cured, as they continued to pass eggs in their feces after receiving one to three treatments with the WHO-recommended single oral dose of 40 mg/kg PZQ. The isolates collected from these patients were then maintained by passaging them in *Biomphalaria glabrata* snails and outbred CD1 albino mice, following the protocols described previously. Mice were then infected with PZQ-susceptible parasites (positive control), lab-selected resistant parasites (negative control), and the Senegalese field isolates [17]. Starting from day 58 postinfection, mice were treated orally with PZQ at a dose of 200 mg/kg for 3 consecutive days. Worm burden reduction was 96.2% for the susceptible parasites,

59.6% for the lab-selected resistant parasites, and a mean of 68.7% for the field isolates, suggesting a reduced susceptibility to PZQ treatment in the latter.

Summary

Phenotypic testing of molecules is a successful approach for discovering new anthelmintic drugs, but due to the long life cycle of schistosomes, this method is labor-intensive and has a low throughput. PZQ was discovered from a set of 400 analogues and is active against adult schistosomes at submicromolar concentrations in vitro. Its onset of action is fast, resulting in two striking phenomena: almost instantaneous muscle paralysis and rapid structural damage to the worm tegument. PZQ has been shown to be fully effective in mouse models at a single oral dose of 400 mg/kg. Worms that have been subjected to drug selection with PZQ exhibit a marked reduction in susceptibility, highlighting the potential for loss of effectiveness over time.

Competing Interests The authors have no conflicts of interest to declare that are relevant to the content of this chapter.

References

1. Swinney DC, Anthony J (2011) How were new medicines discovered? Nat Rev Drug Discov 10(7):507–519. https://doi.org/10.1038/nrd3480
2. Barbosa MA, Pellegrino J, Coelho PMZ, Sampaio IBM (1978) Quantitative aspects of the migration and evolutive asynchronism of *Schistosoma mansoni* in mice. Rev Inst Med Trop S Paulo 20(3):121–132
3. Keiser J (2010) *In vitro* and i*n vivo* trematode models for chemotherapeutic studies. Parasitology 137(3):589–603. https://doi.org/10.1017/S0031182009991739
4. Basch PF (1981) Cultivation of *Schistosoma mansoni in vitro*. I. Establishment of cultures from cercariae and development until pairing. J Parasitol 67(2):179–185
5. Buchter V, Schneeberger PHH, Keiser J (2021) Validation of a human-serum-based *in vitro* growth method for drug screening on juvenile development stages of *Schistosoma mansoni*. PLoS Negl Trop Dis 15:e0009313. https://doi.org/10.1371/journal.pntd.0009313
6. Andrews P (1985) Praziquantel: mechanisms of anti-schistosomal activity. Pharmacol Ther 29(1):129–156. https://doi.org/10.1016/0163-7258(85)90020-8
7. Harder A, Andrews P, Thomas H (1987) Praziquantel: mode of action. Biochem Soc Trans 15(1):68–70. https://doi.org/10.1042/bst0150068
8. Meister I, Ingram-Sieber K, Cowan N, Todd M, Robertson MN, Meli C, Patra M, Gasser G, Keiser J (2014) Activity of praziquantel enantiomers and main metabolites against *Schistosoma mansoni*. Antimicrob Agents Chemother 58(9):5466–5472. https://doi.org/10.1128/AAC.02741-14
9. Kovac J, Vargas M, Keiser J (2017) In vitro and in vivo activity of R- and S- praziquantel enantiomers and the main human metabolite trans-4-hydroxy-praziquantel against Schistosoma haematobium. Parasit Vectors. https://doi.org/10.1186/s13071-017-2293-3
10. Andrews P, Thomas H, Weber H (1980) The *in vitro* uptake of ^{14}C-praziquantel by cestodes, trematodes, and a nematode. J Parasitol 66(6):920–925

11. Probst A, Biendl S, Keiser J (2022) Improving translational power in antischistosomal drug discovery. Adv Parasitol 117:47–73. https://doi.org/10.1016/bs.apar.2022.05.002
12. Xiao SH, Catto BA, Webster LT Jr (1985) Effects of praziquantel on different developmental stages of *Schistosoma mansoni in vitro* and *in vivo*. J Infect Dis 151(6):1130–1137. https://doi.org/10.1093/infdis/151.6.1130
13. Abla N, Keiser J, Vargas M, Reimers N, Haas H, Spangenberg T (2017) Evaluation of the pharmacokinetic-pharmacodynamic relationship of praziquantel in the *Schistosoma mansoni* mouse model. PLoS Negl Trop Dis 11:e0005942. https://doi.org/10.1371/journal.pntd.0005942
14. Gönnert R, Andrews P (1977) Praziquantel, a new board-spectrum antischistosomal agent. Z Parasitenkd 52(2):129–150. https://doi.org/10.1007/BF00389899
15. Friedrich L, Park SK, Ballard P, Ho Baeurle TH, Maillard D, Bödding M, Keiser J, Marchant JS, Spangenberg T (2023) Metabolism of (R)-praziquantel versus the activation of a parasite transient receptor potential melastatin ion channel. Chem Med Chem 18:e202300140. https://doi.org/10.1002/cmdc.202300140
16. Le Clec'h W, Chevalier FD, Mattos ACA, Strickland A, Diaz R, McDew-White M, Rohr CM, Kinung'hi S, Allan F, Webster BL, Webster JP, Emery AM, Rollinson D, Djirmay AG, Al Mashikhi KM, Al Yafae S, Idris MA, Moné H, Mouahid G, LoVerde P, Marchant JS, Anderson TJC (2021) Genetic analysis of praziquantel response in schistosome parasites implicates a transient receptor potential channel. Sci Transl Med 13:eabj9114. https://doi.org/10.1126/scitranslmed.abj9114
17. Liang Y-S, Coles GC, Doenhoff MJ, Southgat VR (2001) In vitro responses of praziquantel-resistant and -susceptible *Schistosoma mansoni* to praziquantel. Int J Parasitol 31(11):1227–1235. https://doi.org/10.1016/S0020-7519(01)00246-6

In Life Behavior of Praziquantel and Relationship Between Drug Exposure and Effect

7

Mathew Njoroge and Peter Ballard

Abstract

In this chapter, the pharmacokinetics of praziquantel will be discussed, covering key data and parameters related to how the drug is absorbed, distributed, metabolized, and excreted. Additionally, the relationship between the drug's exposure and its pharmacodynamic effects will be examined.

Keywords

Praziquantel · Metabolism · ADME · BCSII · PK/PD

Introduction

When drugs are used to treat a disease, correct a physiological imbalance or serve other therapeutic purposes, they must reach their site of action at sufficient concentrations and remain there long enough to achieve the intended effect. These factors, often called the pharmacokinetics (PK) of a drug, are as important as the activity at the target—the pharmacodynamics (PD)—for the achievement of therapeutic effects. For example, understanding the impact of praziquantel (PZQ) in patients with schistosomiasis requires considering both its PD—that is the effect against *Schistosoma mansoni* (*S. mansoni*) worms—and its PK—that is the concentrations achieved in the portal vein and other sites of infection, as well as the duration for which these exposures are maintained.

M. Njoroge (✉)
Holistic Drug Discovery and Development Centre (H3D), University of Cape Town, Rondebosch, South Africa
e-mail: MN.Njoroge@uct.ac.za

P. Ballard
PB DMPK Consulting Ltd., Chinley, UK

T. Spangenberg (ed.), *Praziquantel*,
https://doi.org/10.1007/978-3-031-97397-0_7

The PK of PZQ and other orally administered drugs can be analyzed by addressing key factors that determine the concentrations achieved in vivo and how long they are maintained. This involves understanding how the drugs are absorbed, distributed, metabolized, and excreted, usually abbreviated as ADME (Fig. 7.1).

Absorption marks the first phase of PK and describes how effectively an orally administered drug crosses biological membranes in the gut. This process depends on the drug's solubility, permeability, and other physicochemical properties (Chap. 4). Once absorbed into the bloodstream, the drug is distributed across various compartments and tissues, influenced by its permeability and physicochemical characteristics. Metabolism converts the parent compound into metabolites, which can alter the drug's activity, concentration, and duration in the body. This process depends on the drug's intrinsic interaction with metabolic enzymes, as well as its permeability to cellular sites of metabolism. Finally, the drug and/or its metabolites are eliminated through excretion, typically via renal or biliary pathways, involving organs such as the kidneys and liver. The interplay of these ADME processes defines the drug's PK, which together with its PD, determines the overall therapeutic effect of the drug in vivo. The importance of understanding the ADME of drugs is underlined by observations from the era before it was commonly implemented into drug

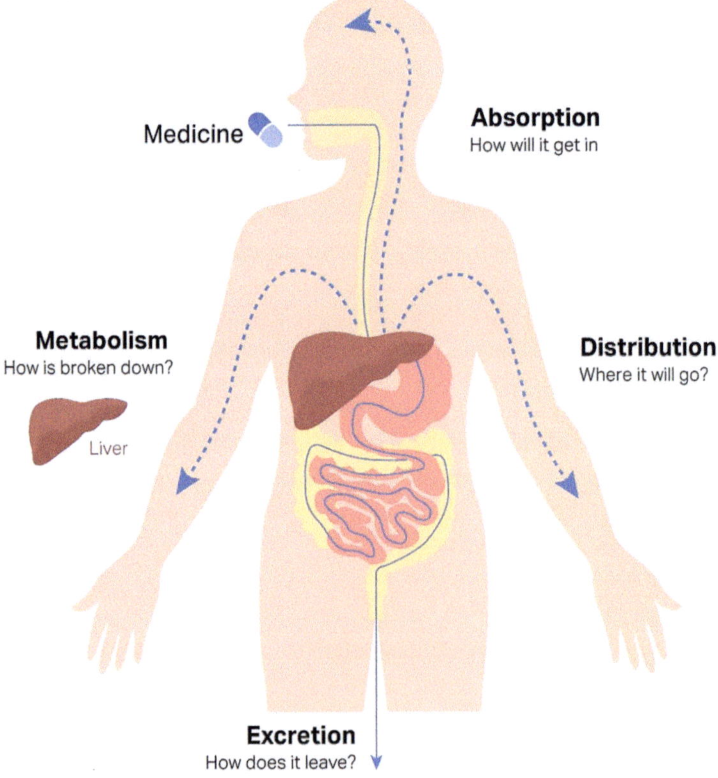

Fig. 7.1 ADME and in vivo drug exposure

discovery. During that period, up to 40% of failures were attributed to poor ADME/PK, with an even greater number potentially linked to how ADME/PK influences drug efficacy and toxicity [1, 2]. Therefore, the evaluation of ADME/PK properties has increasingly been integrated into the drug discovery and development process, playing a crucial role in identifying compounds with the potential to become successful drugs. In the early stages of drug discovery, this is achieved through iterative design-make-test-analyze (DMTA) cycles. During these cycles, in silico ADME properties are predicted through suitable models to inform compound design. After compounds are synthesized, in vitro and/or in vivo testing of their properties is conducted. Finally, the data is analyzed to determine whether optimal compounds have been achieved, or whether additional medicinal chemistry design is needed to address any liabilities identified [3, 4]. In vitro high- and medium-throughput studies dominate the progression of compounds through the drug discovery process. The assays must be selected based on their likely impact and relevance to the eventual human PK and toxicity profiles.

In vivo PK evaluation is conducted in different preclinical species, taking into account factors such as the relevance of the animal efficacy model to humans, the amount of compound required for evaluation, ethical and regulatory considerations, and other relevant factors. These studies generate critical data that support compound optimization in the DMTA cycle. Preclinical PK data also enhance confidence in human PK predictions by demonstrating the consistency and predictability of preclinical models [5]. When combined with other key data, such as efficacy and safety assessments, this information is vital for determining the appropriate human dose.

The Absorption Distribution Metabolism and Excretion of Praziquantel

Praziquantel (PZQ) is the primary drug for treating schistosomiasis and is typically administered orally at a dose of 40 mg/kg once daily. As noted in the introduction, the in vivo concentrations and time course of PZQ are critical for determining its efficacy against *Schistosoma* worms and other target organisms. These concentrations depend on the drug's ADME properties, which are described in the following sections. While concentrations at the site of action—such as the mesenteric veins for PZQ—are the most critical for determining exposure-effect relationships, they are often impractical to measure directly. Instead, blood or plasma concentrations are typically used as a surrogate, with modeling and simulation approaches employed to estimate site-specific drug levels. These methods are discussed in detail below.

Figure 7.2 shows a typical PK profile in human plasma following a single 40 mg/kg oral dose of PZQ administered to six healthy adult male volunteers. The figure displays data on a linear scale (left) and as a semilog plot (right). In PK, most processes follow first-order kinetics, meaning they are proportional to drug concentration. As a result, semilog plots of concentration versus time are often useful for interpreting the data and identifying any deviations from expected trends.

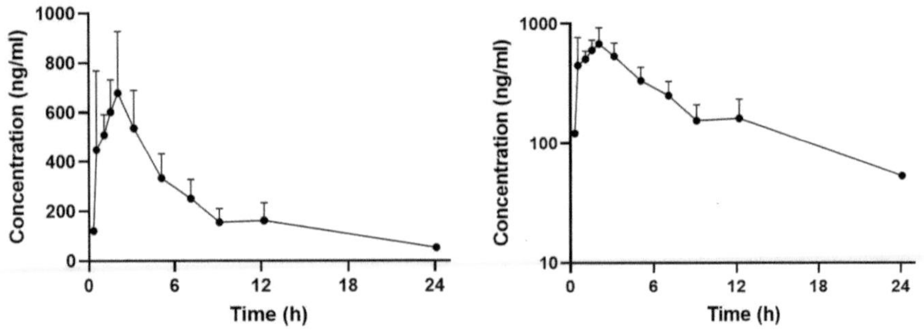

Fig. 7.2 Mean plasma concentration of praziquantel following a single oral 40 mg/kg dose to six healthy volunteers [6]

After administration of a single 40 mg/kg dose of PZQ, plasma concentrations rise as the drug is absorbed, reaching a maximum concentration (C_{max}) of approximately 700 ng/mL. C_{max} is a key parameter for evaluating both safety margins and efficacy, as it represents the peak drug exposure, which may approach or exceed the maximum tolerated concentration. The time to reach C_{max} (t_{max}), approximately 2 h, provides insight into the drug's absorption rate. The initial portion of the concentration–time curve, up to t_{max}, is primarily influenced by absorption. Beyond this point, the curve is shaped by the interplay between absorption, distribution, and clearance. The total area under the concentration–time curve (AUC), measured in units of h·ng/mL, reflects the drug's overall systemic exposure. For PZQ, the AUC is 5409 h·ng/mL. AUC is a crucial parameter for assessing both the therapeutic efficacy and toxicological risk of the compound. However, PZQ exhibits a high degree of interindividual variability, which can be attributed to differences in factors such as metabolism, absorption rates, and clearance among patients, leading to a wide range of plasma exposures.

In the following sections, we will examine how absorption, distribution, metabolism, and excretion of PZQ define its PK profile.

Absorption

To understand the absorption of PZQ, as well as other orally administered drugs, it is important to understand what occurs after ingestion. As shown in Fig. 7.3, the drug passes through the esophagus into the stomach, where it dissolves in the gut fluids, either in the stomach or the small intestine, depending on the formulation and the drug's physicochemical properties. The dissolved drug must then be sufficiently permeable to pass through the apical side of the enterocyte (epithelial cells lining the intestines) and exit through the basolateral side into the mesenteric veins (blood vessels that transport absorbed nutrients and drugs from the intestines to the liver), ultimately reaching systemic circulation.

From this description, we can delineate two important prerequisites for absorption: solubility and permeability.

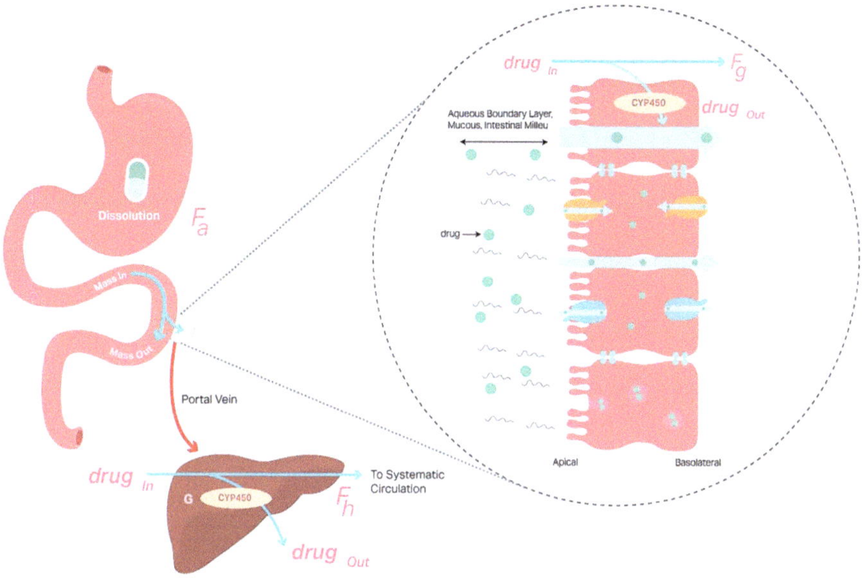

Fig. 7.3 Oral drug absorption

Solubility

In the context of oral absorption, the most important measure of solubility is how the solid form of the drug dissolves in intestinal fluids. This process can be measured in vitro using the solid drug and appropriate buffers. Solubility experiments, in which the drug is allowed to reach equilibrium in the buffer, are referred to as thermodynamic solubility and are considered the best method for evaluating solubility during drug discovery [7]. These experiments typically use simple buffers, such as phosphate or chloride salts, in the pH range of 1–8 to mimic the pH of different regions of the gastrointestinal tract (GIT). More complex buffers, such as simulated intestinal fluids, can also be used [8]. Examples include Fasted State Simulated Intestinal Fluid (FaSSIF, pH 6.5), Fed State Simulated Intestinal Fluid (FeSSIF, pH 5.5), and Fasted State Simulated Gastric Fluid (FaSSGF, pH 1.8), which have been used to excellent effect. During the early stages of drug discovery, performing thermodynamic solubility on each compound may be impractical and costly, as the compound requirements can be high, and the experiment is difficult to automate. In its place, the higher throughput kinetic solubility assay is used (Chap. 4) [9]. In this assay, stock solutions of the compound in dimethyl sulfoxide (DMSO)—a commonly used universal solvent in early drug discovery—are added directly to buffer solutions, and solubility is quantified after shaking for a specified period. This assay design provides insight into whether the compound has sufficient solubility to generate consistent data in in vitro experiments [10].

The solubility of PZQ in FaSSIF (pH 6.5) buffer is 0.3 mg/mL, which is similar to solubility values reported in other media, including in distilled water [11–13]. According to the United States Food and Drug Administration (FDA), for a compound to be considered highly soluble, the highest dose strength must dissolve completely in 250 mL of fluid across a pH range of 1–6.8 [14]. For example, a 600 mg tablet of PZQ must dissolve completely in 250 mL of GIT fluid, which requires a solubility of at least 2.4 mg/mL. When administered at doses up to 20 mg/kg three times a day, a 60 kg patient would require 1200 mg per dose, demanding a solubility of at least 4.8 mg/mL. PZQ is, therefore, considered a poorly soluble compound with only a small portion of the dose dissolving in the GIT, which potentially limits its absorption. Given that solubility is an inherent property of the compound, efforts to enhance PZQ's solubility have focused on developing alternative solid forms (Chap. 4) [15], cyclodextrin inclusion complexes [16, 17], and co-crystal approaches [18], among other strategies. However, none of these methods have achieved widespread clinical use.

Permeability

Once in solution, drugs must cross biological membranes to exert their activity, with enterocytes in the small intestine serving as the primary barrier to absorption, as shown in Fig. 7.3. As mentioned earlier, permeability and solubility are key prerequisites for absorption, and permeability can occur through several distinct mechanisms, as depicted in Fig. 7.4.

One of the primary mechanisms of permeability is passive diffusion through the cell membranes, driven by a concentration gradient. This process is predominant for compounds with moderate lipophilicity (LogD >1), and low polarity (TPSA <140 Å^2, Chap. 4) [19, 20]. Active transport involves specialized transporters located on either side of the membrane, which facilitate drug movement by utilizing energy. This mechanism requires the drug to bind with the transporter, enabling its transfer across the membrane. While active transport occurs alongside passive

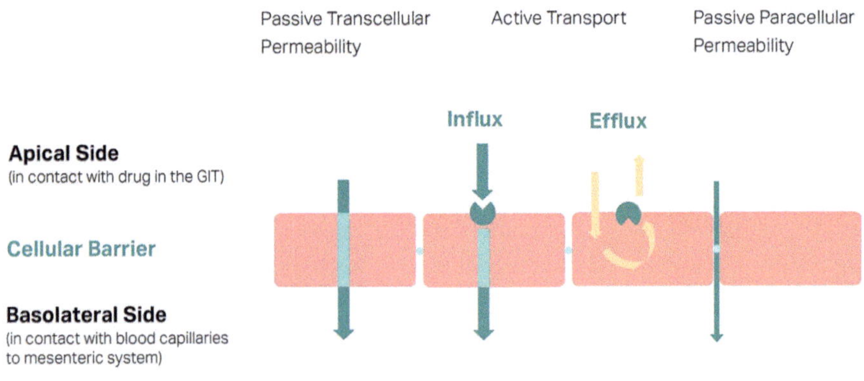

Fig. 7.4 Permeability mechanisms

diffusion, it has a greater impact on oral absorption and pharmacokinetics, particularly for compounds with low passive permeability [21, 22]. Passive permeability is frequently prioritized in drug discovery due to its simplicity and the fact that it relies solely on the concentration gradient across the membrane, making it easier to model once in vivo concentrations are understood [19]. In addition, unlike transporter-mediated mechanisms, which are subject to saturation and can result in nonlinear PK, passive permeability is not saturable, ensuring a more predictable absorption profile [22].

In drug discovery, permeability is assessed using either cellular systems, such as Caucasian colon adenocarcinoma cell line number 2 (Caco-2) or Madin-Darby Canine Kidney (MDCK) cells, or artificial membranes, such as those employed in the Parallel Artificial Membrane Permeability Assay (PAMPA) (Chap. 4). Cellular systems enable the evaluation of both active and passive transport mechanisms, as well as other processes like metabolism, while artificial membrane assays are limited to measuring passive permeability. Among these methods, the Caco-2 cell line is the most widely used for permeability studies [23]. The cells are cultured as a monolayer on permeable filters to simulate the epithelial lining of the gut wall. Permeability is assessed in both the apical-to-basolateral (A->B) and basolateral-to-apical (B->A) directions and is reported as the apparent permeability coefficient (P_{app}) in units of 10^{-6} cm/s or as LogP_{app}. Following permeability measurements, the ratio of transport in the B->A vs A->B directions is calculated to define the efflux ratio. A ratio of 1 indicates no active efflux, while values greater than 2 generally suggest the presence of active efflux. PZQ exhibits an A–>B permeability of 39×10^{-6} cm/s with an efflux ratio of 1, classifying it as a highly permeable compound without evidence of active efflux. This high permeability suggests that passive diffusion is the dominant transport mechanism, consistent with the physicochemical properties described in Chap. 4.

Permeability and solubility are key factors often considered together when evaluating oral drug absorption. Highly soluble compounds generate high drug concentrations in the intestinal lumen, creating a steep concentration gradient across the gut wall. When combined with high permeability, this can result in complete absorption. This interplay between solubility and permeability forms the basis of the Biopharmaceutical Classification System (BCS), which classifies drugs into categories based on these two properties (Fig. 7.5). The BCS framework is widely used to predict absorption potential and guide drug formulation strategies [24].

Praziquantel Absorption

PZQ is classified as a BCS Class 2 drug due to its low solubility and high permeability [25]. Its high permeability compensates for its low solubility, facilitating sufficient absorption. In fed human volunteers, absorption is generally rapid and complete, with more than 80% of the administered dose reaching the portal system—a network of veins that collects blood from the mesenteric veins and directs it to the liver. This is particularly important for PZQ's therapeutic effect, as the mesenteric veins are often a primary site of infection in intestinal schistosomiasis.

	High Solubility	Low Solubility
High Permeability	Class 1 e.g., Paracetamol	Class 2 e.g., Praziquantel
Low Permeability	Class 3 e.g., Ethambutol	Class 4 e.g., Acyclovir

Fig. 7.5 Biopharmaceutical Classification System (BCS)

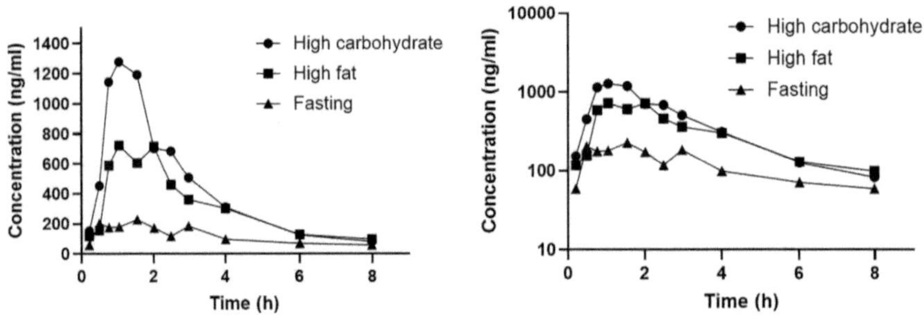

Fig. 7.6 Effect of food on praziquantel's systemic exposure (left panel: linear plot; right panel; semilog plot) [27]

As discussed previously, food can significantly affect absorption. As shown in Fig. 7.6, a high-carbohydrate meal increased C_{max} five-fold and AUC three-fold compared to fasting. The semilog plot helps visualize that the different conditions (high carbohydrate, high fat, or fasting) evaluated in the study are similar, while the linear plot shows greater variability. This pattern suggests an effect on absorption rather than distribution or clearance. While absorption exceeds 80% in the fed state, poor solubility likely reduces absorption in the fasted state, potentially leading to lower systemic exposure [26, 27]. Food effects are more pronounced for poorly soluble compounds [28], as changes in physiological conditions following a meal—such as alterations in gastric pH, increased blood flow, and the presence of lipid micelles in the gut—can enhance dissolution and absorption. These factors can shift the equilibrium between dissolved and undissolved compound, promoting greater dissolution or facilitating increased absorption, which ultimately leads to higher systemic concentrations.

As we return to discussing PZQ absorption, it is important to note that while absorption into the portal vein is high, the overall systemic bioavailability of unchanged PZQ remains low. Bioavailability is defined as the fraction of the unchanged drug that reaches systemic circulation. After oral administration, the drug must be absorbed across the gastrointestinal wall into the mesenteric veins, which drain into the portal vein and then the liver (Fig. 7.3). Enzymes in the gut or liver can metabolize the drug, reducing the amount available for systemic circulation—this process is known as first-pass metabolism. Therefore, bioavailability (F) is determined by the fraction absorbed (Fa), the fraction that escapes gut metabolism (Fg), and the fraction that escapes hepatic metabolism (Fh) (Eq. 7.1).

$$\mathrm{F} = \mathrm{Fa.Fg.Fh} \tag{7.1}$$

Although the fraction of PZQ absorbed is high, its first-pass metabolism in the liver (resulting in a low fraction escaping hepatic metabolism, or low Fh) reduces the amount of unchanged drug reaching systemic circulation. For many oral drugs, this would result in a loss of efficacy, as the unchanged drug is required to exert its pharmacological effect. However, for PZQ, the primary site of infection for *S. mansoni* is in the mesenteric and portal veins, so its rapid absorption ensures that high concentrations are still achieved at the site of infection. This has been demonstrated in a *S. mansoni* mouse model, where inhibition of PZQ metabolism in the liver resulted in increased systemic concentrations (as expected from Eq. 7.1, where the inhibitor increases Fh by blocking metabolism), but did not lead to an increase in efficacy (Chap. 6) [29]. We will examine other factors contributing to this observation in the subsequent metabolism section.

Distribution

When drugs enter systemic circulation, they are transported to various tissues, moving from the bloodstream to tissues and ultimately into cells, where an equilibrium is maintained between cellular and blood concentrations. This process is referred to as distribution. The concentration of a drug in the blood/plasma is influenced by the extent to which it distributes into tissues outside the vascular compartment (the portion of the circulatory system contained within blood vessels, including arteries, veins, and capillaries). Greater tissue distribution results in lower blood concentrations, while limited distribution keeps blood concentrations higher. This relationship is quantified by the volume of distribution (V_d), which can be calculated using the administered dose and the initial blood/plasma concentration (C_0) following an intravenous bolus dose (Eq. 7.2). V_d represents the apparent physiological volume into which a compound appears to distribute.

$$Vd = \frac{\text{Dose}}{C_0} \tag{7.2}$$

V_d is measured in units of volume (mL or L) and can also be expressed per unit body weight (L/kg). The calculation assumes that the entire administered dose reaches systemic circulation, as is the case with intravenous administration. However, oral absorption is often limited by various physiological barriers, making

100% bioavailability uncommon. When volume of distribution is calculated after oral administration, it is expressed as V_d/F to account for the systemic availability of the compound. If bioavailability is 100%, V_d/F simplifies to V_d. The volume of distribution is termed *apparent* because it does not correspond to any physiological volume, such as blood volume or total body water. For example, chloroquine, a drug used to treat malaria, has a V_d exceeding 100 L/kg, far greater than any physiological volume in the body. This large value arises because chloroquine extensively partitions to tissues, illustrating how the apparent volume of distribution reflects not the actual space the drug occupies but the extent of its distribution across compartments.

To explore this concept further, we can examine the volume of distribution at steady state (V_{SS}). Steady state is achieved when the amount of drug entering systemic circulation equals the amount leaving systemic circulation through clearance. At this point, V_{SS} represents the distribution of a drug between plasma and tissues. It is determined by the volume of plasma (V_p) and the volume of tissue (V_t), which depend on physiological factors such as tissue composition and blood flow. Additionally, V_{SS} is influenced by the difference in the fraction of the drug unbound in plasma ($f_{u,p}$) and in tissues ($f_{u,t}$). These unbound fractions are governed by the drug's physicochemical properties, such as lipophilicity and binding affinity.

$$V_{SS} = V_p + \sum \left(\frac{f_{u,p}}{f_{u,t}} \right) V_t \tag{7.3}$$

Equation 7.3 introduces an important concept: the fraction unbound. Drugs in the body exist in equilibrium between being bound to proteins or tissues and being unbound. Only the portion of the drug that is unbound can interact with targets at the site of action, cross cell membranes, and be subject to clearance. Binding is typically nonspecific and reversible, enabling the unbound portion to maintain equilibrium across different physiological compartments. In the vascular compartment, drugs primarily bind to plasma proteins such as albumin and alpha-1-acid glycoprotein. In tissues, binding is driven by interactions with cellular membrane phospholipids such as phosphatidylcholine and phosphatidylserine, which provide a large surface area for nonspecific interactions [30, 31].

Building on the concept of unbound fractions discussed earlier, Eq. 7.3 also highlights the role of physicochemical properties in determining drug distribution. Basic compounds typically exhibit low $f_{u,t}$ and high V_{SS} due to their strong affinity for acidic phospholipids in tissues. This affinity is driven by their lipophilicity and ionization state. Furthermore, in acidic cellular compartments, basic compounds can become ionized, leading to pH-dependent ion trapping that restricts their ability to diffuse back across membranes, thereby increasing their extravascular distribution. In contrast, acidic compounds generally have a low affinity for tissue phospholipids (high $f_{u,t}$) but bind strongly to plasma proteins (low plasma unbound fraction, $f_{u,p}$). This results in a lower V_{SS} as their distribution is more restricted to the vascular space.

Praziquantel Distribution

In preclinical studies performed in rats with radiolabeled PZQ, the compound and its metabolites were distributed across the body, with the highest levels detected in the liver and kidney—the main organs involved in its elimination [32]. This serves as an example of how the distribution of a drug can align with its therapeutic targets. Since schistosomiasis primarily affects the liver (*S. mansoni*) and the bladder (*S. haematobium*), this distribution pattern also contributes to its antiparasitic activity. The fraction unbound in plasma is about 0.2 (i.e., 80% bound) in humans and across most species reported, reflecting a balance between plasma protein binding and tissue distribution that supports its efficacy at target sites.

Metabolism

Drugs are xenobiotics meaning they are chemical substances not naturally produced or expected to be present within the organism. The body eliminates these compounds to prevent their accumulation, which can disrupt normal biological processes or cause toxicity. One of the primary mechanisms of drug elimination is metabolism, an enzymatic process that typically increases a compound's polarity, making it more water-soluble and facilitating its elimination. Metabolism is considered a clearance process because it permanently alters the compound's structure, often reducing (though not always eliminating) its pharmacological activity. The liver plays a central role in drug metabolism due to its high expression of metabolic enzymes, making hepatic clearance a major route of drug elimination. Hepatic clearance can be quantified by considering the volume of plasma completely cleared of drug per unit time, which is typically expressed in units such as mL/min or L/h. When normalized to body weight, it is expressed as mL/min/kg or L/h/kg. We can calculate clearance in relation to the dose and AUC as:

$$\mathrm{CL} = \frac{F.\mathrm{Dose}}{\mathrm{AUC}} \tag{7.4}$$

where F represents the bioavailability of the drug (Eq. 7.4). Rearranging this equation in terms of AUC yields the following expression (Eq. 7.5), where the observed systemic exposure (AUC) is directly proportional to bioavailability (F), dose and inversely proportional to the clearance of the drug.

$$\mathrm{AUC} = \frac{F.\mathrm{Dose}}{\mathrm{CL}} \tag{7.5}$$

According to the definition of clearance as the relationship between the drug elimination and the concentration in the body, an increase in clearance leads to a reduction in AUC, thereby decreasing the total exposure to the drug. This is important in drug discovery, as drugs with high clearance may require higher doses to maintain adequate systemic exposure compared to those with lower clearance.

Additionally, understanding the impact of hepatic clearance is essential when considering how pharmacogenetics can influence pharmacokinetics, as genetic variations can affect drug metabolism, drug elimination rates, and clearance. The expression and activity of drug-metabolizing enzymes are both influenced by genetics, with variability in an individual's genetic profile, leading to differences in enzyme levels or in how effectively drugs bind to the enzyme's active site for metabolism. Phenotypically, these genetic variations manifest as differences in the rate of drug metabolism. A *variant*—a specific form of a gene or enzyme that differs from the typical or "wild-type" version—may result in altered enzyme function. For example, if a drug-metabolizing enzyme variant has reduced activity, the drug's metabolism slows (lower clearance), leading to a higher AUC (greater drug exposure). Conversely, variants with increased activity accelerate metabolism (increasing clearance), resulting in a lower AUC. This variability is a key factor in interpatient differences during therapy, and the magnitude of these differences can significantly impact the efficacy and safety of a drug.

Changes in clearance not only materially affect the AUC but can also help to determine the duration of exposure as measured by the drugs elimination half-life ($t_{1/2}$) and usually expressed in units of hours. The half-life of a drug is dependent on both the clearance and volume of distribution in a simple expression (Eq. 7.6):

$$t_{1/2} = \frac{0.693.Vd}{CL} \tag{7.6}$$

In the example above, a drug-metabolizing enzyme variant with reduced activity leads to lower clearance. Since the volume of distribution is unlikely to be affected, the drug's half-life will increase in this situation. In healthy volunteers, PZQ has a half-life of around 3 h.

After exploring the role of hepatic clearance in pharmacokinetics (PK), the next focus is on the formation of metabolites. Drug metabolism is classified into distinct phases based on the type of reaction the compound undergoes. Phase I metabolism involves redox reactions that either introduce new functional groups (e.g., hydroxylation) or expose those already present on the compound (e.g., demethylation). Phase II involves conjugative reactions, where endogenous compounds such as glucuronic acid, sulfate, or glutathione are attached to functional groups formed in Phase I or already present on the molecule. Compounds can undergo Phase I metabolism, Phase II metabolism, or both. In some cases, depending on their physicochemical properties, they may bypass metabolism entirely and be excreted primarily as unchanged parent compounds. Metabolites can be pharmacologically active at the same site of action as the parent compound, active at a different site (producing either beneficial or adverse effects), inactive, or reactive. Reactive metabolites possess characteristics that enable irreversible binding to proteins or other macromolecules, which in some cases may lead to toxicity.

Metabolic clearance can be assessed in vitro using liver-derived cellular and subcellular fractions. Hepatocytes, the primary cells responsible for liver metabolism, can be isolated and cryopreserved for use in drug clearance studies. These cells retain both Phase I and Phase II enzymes, making them the gold standard for in vitro drug metabolism experiments [33]. Subcellular liver fractions are commonly used in vitro to study metabolic clearance. These fractions are prepared by homogenizing liver tissue, followed by centrifugation to obtain the S9 fraction, which contains both microsomal and cytosolic components. Further ultracentrifugation separates the S9 fraction into microsomes and cytosol. Liver microsomes, which are enriched in cytochrome P450 (CYP450) enzymes, are primarily used to study CYP450-mediated Phase I metabolism, particularly oxidation reactions. CYP450 enzymes are a large family of heme-containing enzymes that play a central role in Phase I metabolism by catalyzing the oxidation of many drugs. The major CYP450 families involved in drug metabolism include CYP1, CYP2, and CYP3, with CYP3A4 being particularly significant due to its role in metabolizing a large proportion of clinically used drugs. The S9 (Supernatant after 9,000 × g centrifugation) fraction, which contains a broader range of enzymes, is often used to evaluate both Phase I and Phase II metabolism, including conjugation reactions such as glucuronidation and sulfation. However, a key disadvantage of using subcellular fractions is that they do not contain the full complement of drug-metabolizing enzymes and other cellular components found in intact liver cells, making it unlikely that all in vivo metabolic pathways are adequately represented in vitro.

Praziquantel Metabolism

As alluded to in the discussion on bioavailability, PZQ is rapidly metabolized in the liver of humans and other species. In humans, within 4 h following oral administration, only 5% of the dose remains in plasma as the parent compound [34] (Fig. 7.7). PZQ is administered as a racemic mixture and undergoes enantioselective metabolism, with the (*R*)-enantiomer, which is the more potent, being metabolized more quickly (Chaps. 4, 5, and 6). As a result, the (*R*)-enantiomer is present at lower levels in vivo compared to the (*S*)-enantiomer.

The primary metabolite formed from both the (*R*)- and (*S*)-enantiomers is *trans*-4'-hydroxypraziquantel, which circulates at much higher concentrations than either parent enantiomer, as shown in Fig. 7.8 Other metabolites, including *cis*-4'-hydroxypraziquantel and 11ß-hydroxypraziquantel, have also been identified in humans, although they are present at significantly lower levels, with AUCs approximately 30-fold and 10-fold lower, respectively, compared to *trans*-4'-hydroxypraziquantel [36]. Furthermore, several other mono- and di-hydroxylated metabolites have been identified, though their clinical significance remains inadequately characterized [34].

In vitro studies using human hepatocytes demonstrate metabolic pathways that closely resemble those observed in vivo, leading to the formation of *trans*-4'-hydroxypraziquantel (Fig. 7.8). In contrast, human liver microsomes primarily produce *cis*-4'-hydroxypraziquantel as the main metabolite. This *cis*-metabolite is then converted to *trans*-4'-hydroxypraziquantel in hepatocytes, indicating that the differing metabolite profiles between hepatocytes and microsomes are due to the

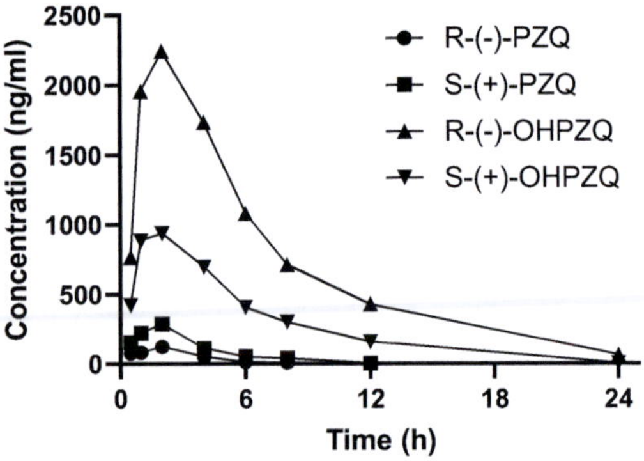

Fig. 7.7 Mean serum concentration of (*R*)- and (*S*)-praziquantel (PZQ) and their respective *trans*-4'-hydroxy metabolites (OH-PZQ) following oral administration of 600 mg PZQ to healthy volunteers [35]

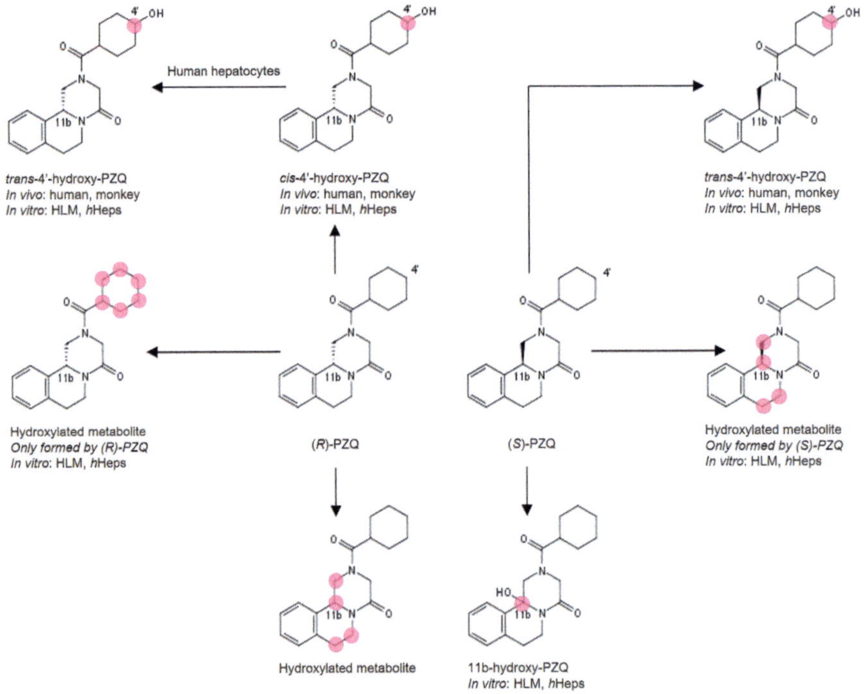

Fig. 7.8 In vitro and in vivo metabolism of praziquantel. (Adapted from Vendrell-Navaro et al. [38] with additional data). Pink dots: Possible hydroxylation/oxidation sites

presence of a cytosolic enzyme in hepatocytes, absent in the microsomal fraction [37]. PZQ metabolism is primarily mediated by CYP450 enzymes, with CYP1A2, CYP2C9, and CYP2C19 being the main enzymes involved. CYP3A4 also contributes to the metabolism of (*S*)-PZQ. As discussed previously, genetic variations can influence drug metabolism, and CYP2C19 is polymorphic, with several genetic variants observed at significant levels in different populations. Variants with reduced metabolic activity, such as CYP2C19*2 and CYP2C19*3, are associated with higher levels of PZQ and lower levels of *trans*-4-hydroxypraziquantel [38–40].

Excretion

Whether compounds undergo metabolism or remain unchanged, their ultimate removal from the body occurs primarily through excretion. This process can happen via several pathways: through the kidneys (urine), liver (bile), intestines (in feces), and, to a lesser extent, through other organs such as the skin (sweat, e.g., for alcohol) and the lungs (exhalation, e.g., for anesthetics). The total clearance of a drug can be represented as the sum of all these routes of clearance:

$$\mathrm{CL_s} = \mathrm{CL_r} + \mathrm{CL_h} + \ldots \tag{7.7}$$

Systemic clearance ($\mathrm{CL_s}$) of a drug is the sum of all routes of elimination, including renal clearance ($\mathrm{CL_r}$), hepatic clearance ($\mathrm{CL_h}$), and other contributing pathways. Renal excretion is the primary route for hydrophilic drugs and their metabolites, with elimination occurring primarily through the kidneys via glomerular filtration and reabsorption in the renal tubules. Reabsorption depends on the compound's permeability—lipophilic compounds are generally reabsorbed, while hydrophilic metabolites may not be as efficiently reabsorbed and are excreted in urine either passively or through active transport. Additionally, active transport mechanisms can aid in the reabsorption of some compounds. Hepatic clearance involves the metabolism of drugs in the liver, where lipophilic compounds are biotransformed into more hydrophilic metabolites. These metabolites can be excreted into bile and ultimately eliminated in the feces. Drug elimination from the body occurs via multiple routes: renal excretion typically removes drugs or their metabolites within hours, while biliary excretion through the liver may be slower, particularly if enterohepatic recirculation is involved.

Praziquantel Excretion

PZQ is predominantly excreted as hydroxylated, dihydroxylated, and other unidentified polar and conjugate metabolites via the kidneys. Less than 0.1% of the unchanged drug is recovered in the urine [41]. Approximately 80% of the administered dose is excreted in the urine within 4 days, with most of this occurring in the first 24 h post-dose. The remainder of the dose is likely excreted in the feces.

Praziquantel PK/PD

As discussed in the preceding sections, understanding the pharmacokinetics (PK) of a compound is essential to interpreting its pharmacodynamics (PD). The relationship between PK and PD—often referred to as the PK/PD relationship—explains how the concentration–time profile of a drug influences its therapeutic activity. When considering the PK/PD of PZQ, we must account for several factors that impact this relationship. First, the site of action of PZQ differs between the two main *Schistosoma* species. *S. mansoni* primarily infects the mesenteric and portal veins, while *S. haematobium* primarily infects the bladder. This difference impacts how we relate systemic concentrations to the drug's efficacy, as we must consider how drug levels in specific tissues, such as the mesenteric veins and bladder, contribute to therapeutic outcomes. Since the mesenteric and portal veins receive the drug from the intestines before it enters systemic circulation (Fig. 7.9), PZQ concentrations in these areas are largely determined by the rate of absorption. Blood from these veins then flows into the liver, where first-pass metabolism occurs, resulting in lower PZQ levels in systemic circulation than the portal vein and, consequently, in other organs. A study of praziquantel (PZQ) in an *S. mansoni* mouse model demonstrated 20-fold higher AUCs in the portal vein compared to systemic AUCs (Chap. 6) [29]. The study also found no association between systemic AUCs and the efficacy of PZQ against *S. mansoni*, as treatment with 1-aminobenzotriazole, an inhibitor of CYP450-mediated metabolism of PZQ, increased systemic AUCs without enhancing efficacy. This suggests that for *S. mansoni*, therapeutic effectiveness is not closely linked to systemic drug levels. However, this may not hold true for *S. haematobium*, for which systemic concentrations of PZQ could be more critical due to the parasite's primary location in the bladder.

This brings us to the second key consideration: PZQ is rapidly metabolized in the liver to form *trans*-4'-hydroxypraziquantel. The unbound concentrations of

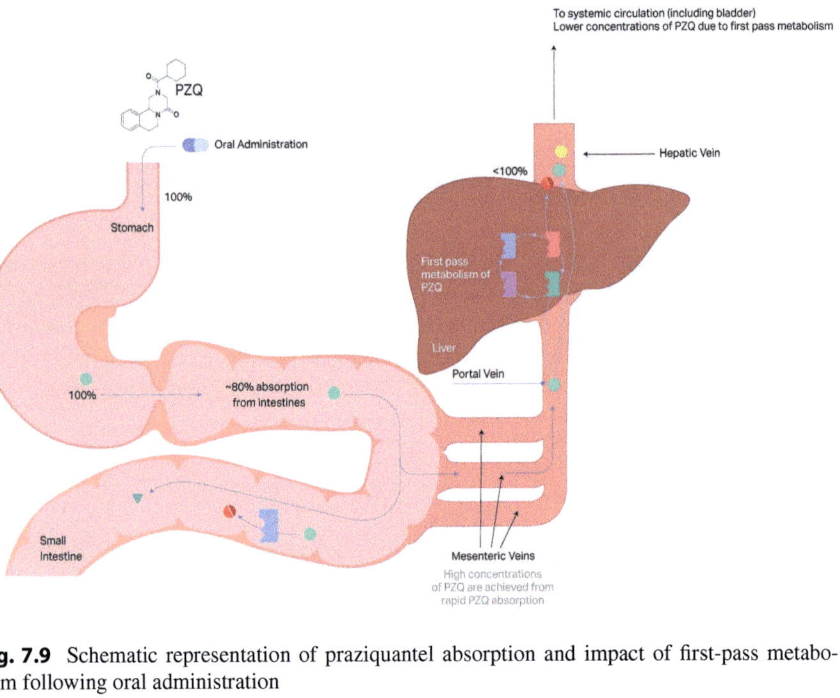

Fig. 7.9 Schematic representation of praziquantel absorption and impact of first-pass metabolism following oral administration

trans-4-hydroxypraziquantel can be up to 600 times higher than those of PZQ [35, 42] in humans. These elevated levels may compensate for the reduced potency of the metabolite, thereby contributing to its in vivo activity. This is particularly relevant for *S. haematobium*, where systemic drug levels are likely to be more critical.

Summary

The absorption, distribution, metabolism, and excretion (ADME) of a drug are crucial for defining in vivo drug levels at the active site and, by extension, its efficacy and/or toxicity. PZQ is efficiently absorbed, with up to 80% of the drug reaching systemic circulation. It is rapidly metabolized, with only 5% remaining in plasma 4 h after an oral dose. Despite rapid metabolism, efficacy is achieved both because of high drug levels in the portal vein—a primary site of infection—and the contribution of its metabolite, *trans*-4'-hydroxypraziquantel, at infections sites such as the bladder. PZQ is excreted primarily as hydroxylated and dehydroxylated metabolites, with other unidentified polar metabolites. The unique ADME of PZQ is therefore well matched to the active sites it needs to access for its pharmacological activity.

Competing Interests The authors have no conflicts of interest to declare that are relevant to the content of this chapter.

References

1. Kola I, Landis J (2004) Can the pharmaceutical industry reduce attrition rates? Nat Rev Drug Discov 3(8):711–716. https://doi.org/10.1038/nrd1470
2. Prentis R, Lis Y, Walker S (1988) Pharmaceutical innovation by the seven UK-owned pharmaceutical companies (1964–1985). Br J Clin Pharmacol 25(3):387–396. https://doi.org/10.1111/J.1365-2125.1988.TB03318.X
3. Ballard P, Brassil P, Bui KH, Dolgos H, Petersson C, Tunek A, Webborn PJH (2012) The right compound in the right assay at the right time: an integrated discovery DMPK strategy. Drug Metab Rev 44(3):224–252. https://doi.org/10.3109/03602532.2012.691099
4. Plowright AT, Johnstone C, Kihlberg J, Pettersson J, Robb G, Thompson RA (2012) Hypothesis driven drug design: improving quality and effectiveness of the design-make-test-analyse cycle. Drug Discov Today 17(1–2):56–62. https://doi.org/10.1016/J.DRUDIS.2011.09.012
5. Webborn PJH (2014) The role of pharmacokinetic studies in drug discovery: where are we now, how did we get here and where are we going? Future Med Chem 6(11):1233–1235. https://doi.org/10.4155/FMC.14.76
6. El M, Mandour M, El Turabi H, Homeida MMA, el Sadig T, Ali HM, Bennett JL, Leahey WJ, Hanon DWG (1990) Pharmacokinetics of praziquantel in healthy volunteers and patients with schistosomiasis. Trans R Soc Trop Med Hyg 84(3):389–393. https://doi.org/10.1016/0035-9203(90)90333-A
7. Saal C, Petereit AC (2012) Optimizing solubility: kinetic versus thermodynamic solubility temptations and risks. Eur J Pharm Sci 47(3):589–595. https://doi.org/10.1016/J.EJPS.2012.07.019

8. Clarysse S, Brouwers J, Tack J, Annaert P, Augustijns P (2011) Intestinal drug solubility estimation based on simulated intestinal fluids: comparison with solubility in human intestinal fluids. Eur J Pharm Sci 43(4):260–269. https://doi.org/10.1016/J.EJPS.2011.04.016
9. Kerns EH, Di L, Carter GT (2008) In vitro solubility assays in drug discovery. Curr Drug Metab 9(9):879–885. https://doi.org/10.2174/138920008786485100
10. Di L, Kerns EH (2006) Biological assay challenges from compound solubility: strategies for bioassay optimization. Drug Discov Today 11(9–10):446–451. https://doi.org/10.1016/J.DRUDIS.2006.03.004
11. Bergström CAS, Andersson SBE, Fagerberg JH, Ragnarsson G, Lindahl A (2014) Is the full potential of the biopharmaceutics classification system reached? Eur J Pharm Sci 57(1):224–231. https://doi.org/10.1016/J.EJPS.2013.09.010
12. Benet LZ, Broccatelli F, Oprea TI (2011) BDDCS applied to over 900 drugs. AAPS J 13(4):519–547. https://doi.org/10.1208/s12248-011-9290-9
13. El-Arini SK, Giron D, Leuenberger H (1998) Solubility properties of racemic praziquantel and its enantiomers. Pharm Dev Technol 3(4):557–564. https://doi.org/10.3109/10837459809028638
14. CDER/FDA (2000) Waiver of in vivo bioavailability and bioequivalence studies for immediate-release solid oral dosage forms based on a biopharmaceutics classification system guidance for industry. https://www.gmp-compliance.org/files/guidemgr/UCM070246.pdf. Accessed 20 May 2024
15. Lombardo FC, Perissutti B, Keiser J (2019) Activity and pharmacokinetics of a praziquantel crystalline polymorph in the Schistosoma mansoni mouse model. Eur J Pharm Biopharm 142:240–246. https://doi.org/10.1016/J.EJPB.2019.06.029
16. Špehar TK, Pocrnić M, Klarić D, Bertoša B, Čikoš A, Jug M, Padovan J, Dragojević S, Galić N (2021) Investigation of praziquantel/cyclodextrin inclusion complexation by NMR and LC-HRMS/MS: mechanism, solubility, chemical stability, and degradation products. Mol Pharm 18(11):4210–4223. https://doi.org/10.1021/acs.molpharmaceut.1c00716
17. Ei-Arini SK, Leuenberger H (1996) Dissolution properties of praziquantel-β-cyclodextrin systems. Pharm Dev Technol 1(3):307–315. https://doi.org/10.3109/10837459609022600
18. Cugovčan M, Jablan J, Lovrić J, Cinčić D, Galić N, Jug M (2017) Biopharmaceutical characterization of praziquantel cocrystals and cyclodextrin complexes prepared by grinding. J Pharm Biomed Anal 137:42–53. https://doi.org/10.1016/J.JPBA.2017.01.025
19. Di L, Artursson P, Avdeef A, Benet LZ, Houston JB, Kansy M, Kerns EH, Lennernäs H, Smith DA, Sugano K (2020) The critical role of passive permeability in designing successful drugs. ChemMedChem 15(20):1862–1874. https://doi.org/10.1002/CMDC.202000419
20. Lipinsk CA, Lombardo F, Dominy BW, Feeney PJ (2001) Experimental and computational approaches to estimate solubility and permeability in drug discovery and development settings. Adv Drug Deliv Rev 23(1–3):3–25. https://doi.org/10.1016/S0169-409X(96)00423-1
21. Smith D, Artursson P, Avdeef A, Di L, Ecker GF, Faller B, Houston JB, Kansy M, Kerns EH, Krämer SD, Lennernäs H, van de Waterbeemd H, Sugano K, Testa B (2014) Passive lipoidal diffusion and carrier-mediated cell uptake are both important mechanisms of membrane permeation in drug disposition. Mol Pharm 11(6):1727–1738. https://doi.org/10.1021/MP400713V
22. Sugano K, Kansy M, Artursson P, Avdeef A, Bendels S, Di L, Ecker GF, Faller B, Fischer H, Gerebtzoff G, Lennernaes H, Senner F (2010) Coexistence of passive and carrier-mediated processes in drug transport. Nat Rev Drug Discov 9(8):597–614. https://doi.org/10.1038/NRD3187
23. Artursson P, Palm K, Luthman K (2001) Caco-2 monolayers in experimental and theoretical predictions of drug transport. Adv Drug Deliv Rev 46(1–3):27–43. https://doi.org/10.1016/S0169-409X(00)00128-9
24. Amidon GL, Lennernäs H, Shah VP, Crison JR (1995) A theoretical basis for a biopharmaceutic drug classification: the correlation of in vitro drug product dissolution and in vivo bioavailability. Pharm Res 12(3):413–420. https://doi.org/10.1023/A:1016212804288

25. Lindenberg M, Kopp S, Dressman JB (2004) Classification of orally administered drugs on the World Health Organization model list of essential medicines according to the biopharmaceutics classification system. Eur J Pharm Biopharm 58(2):265–278. https://doi.org/10.1016/J.EJPB.2004.03.001
26. Bonate PL, Wang T, Passier P, Bagchus W, Burt H, Lüpfert C, Abla N, Kovac J, Keiser J (2018) Extrapolation of praziquantel pharmacokinetics to a pediatric population: a cautionary tale. J Pharmacokinet Pharmacodyn 45(5):747–762. https://doi.org/10.1007/S10928-018-9601-1
27. Castro N, Medina R, Sotelo J, Jung H (2000) Bioavailability of praziquantel increases with concomitant administration of food. Antimicrob Agents Chemother 44(10):2903–2904. https://doi.org/10.1128/AAC.44.10.2903-2904.2000
28. Koziolek M, Alcaro S, Augustijns P, Basit AW, Grimm M, Hens B, Hoad CL, Jedamzik P, Madla CM, Maliepaard M, Marciani L, Maruca A, Parrott N, Pávek P, Porter CJH, Reppas C, van Riet-Nales D, Rubbens J, Statelova M, Trevaskis NL, Valentová K, Vertzoni M, Čepo DV, Corsetti M (2019) The mechanisms of pharmacokinetic food-drug interactions – a perspective from the UNGAP group. Eur J Pharm Sci 134:31–59. https://doi.org/10.1016/J.EJPS.2019.04.003
29. Abla N, Keiser J, Vargas M, Reimers N, Haas H, Spangenberg T (2017) Evaluation of the pharmacokinetic-pharmacodynamic relationship of praziquantel in the Schistosoma mansoni mouse model. PLoS Negl Trop Dis. https://doi.org/10.1371/JOURNAL.PNTD.0005942
30. Murakami T, Yumoto R (2011) Role of phosphatidylserine binding in tissue distribution of amine-containing basic compounds. Expert Opin Drug Metab Toxicol 7(3):353–364. https://doi.org/10.1517/17425255.2011.548805
31. Treyer A, Mateus A, Wiśniewski JR, Boriss H, Matsson P, Artursson P (2018) Intracellular drug bioavailability: effect of neutral lipids and phospholipids. Mol Pharm 15(6):2224–2233. https://doi.org/10.1021/acs.molpharmaceut.8b00064
32. Steiner K, Garbe A (1976) The fate of praziquantel in the organism II. Distribution in rats. Eur J Drug Metab Pharmacokinet 1(2):97–106. https://doi.org/10.1007/BF03189263
33. Chiba M, Ishii Y, Sugiyama Y (2009) Prediction of hepatic clearance in human from in vitro data for successful drug development. AAPS J 11(2):262–276. https://doi.org/10.1208/S12248-009-9103-6
34. Bühring KU, Diekmann HW, Müller H, Garbe A, Nowak H (1978) Metabolism of praziquantel in man. Eur J Drug Metab Pharmacokinet 3(3):179–190. https://doi.org/10.1007/BF03189504
35. Westhoff F, Blaschke G (1992) High-performance liquid chromatographic determination of the stereoselective biotransformation of the chiral drug praziquantel. J Chromatogr B Biomed Sci Appl 578(2):265–271. https://doi.org/10.1016/0378-4347(92)80425-P
36. Nleya L, Thelingwani R, Li XQ, Cavallin E, Isin E, Nhachi C, Masimirembwa C (2019) The effect of ketoconazole on praziquantel pharmacokinetics and the role of CYP3A4 in the formation of X-OH-praziquantel and not 4-OH-praziquantel. Eur J Clin Pharmacol 75(8):1077–1087. https://doi.org/10.1007/S00228-019-02663-8
37. Vendrell-Navarro G, Scheible H, Lignet F, Burt H, Luepfert C, Marx A, Abla N, Swart P, Perrin D (2020) Insights into praziquantel metabolism and potential enantiomeric cytochrome P450-mediated drug-drug interaction. Drug Metab Dispos 48(6):481–490. https://doi.org/10.1124/DMD.119.089888
38. Gebreyesus TD, Makonnen E, Telele NF, Barry A, Mnkugwe RH, Gerba H, Dahl ML, Aklillu E (2024) CYP2C19 and CYP2J2 genotypes predict praziquantel plasma exposure among Ethiopian school-aged children. Sci Rep 14(1):1–11. https://doi.org/10.1038/s41598-024-62669-w
39. Mnkugwe RH, Minzi O, Kinung'hi S, Kamuhabwa A, Aklillu E (2021) Effect of pharmacogenetics variations on praziquantel plasma concentrations and schistosomiasis treatment outcomes among infected school-aged children in Tanzania. Front Pharmacol. https://doi.org/10.3389/fphar.2021.712084

40. Barry A, Kabatende J, Telele NF, Mnkugwe RH, Mugisha M, Ntirenganya L, Bienvenu E, Aklillu E (2023) Effect of pharmacogenetic variations on praziquantel plasma concentration and safety outcomes among school children in Rwanda. Sci Rep 13(1):1–11. https://doi.org/10.1038/s41598-023-28641-w
41. Patzschke K, Pütter J, Wegner LA, Horster FA, Diekmann HW (1979) Serum concentrations and renal excretion in humans after oral administration of praziquantel - results of three determination methods. Eur J Drug Metab Pharmacokinet 4(3):149–156. https://doi.org/10.1007/BF03189418
42. Friedrich L, Park S-K, Ballard P, Ho Baeurle TH, Maillard D, Bödding M, Keiser J, Marchant JS, Spangenberg T (2023) Metabolism of (R)-praziquantel versus the activation of a parasite transient receptor potential melastatin ion channel. ChemMedChem. https://doi.org/10.1002/CMDC.202300140

Nonclinical Safety

8

Claudia Demarta-Gatsi, Christian Brenneis, Philip Hewitt, Nicole Hübler, and Stefan Weigt

Abstract

Nonclinical safety investigations follow a strictly regulated, globally harmonized process and are essential for identifying potential adverse effects, thereby ensuring the safe use of new drugs. They are typically conducted in compliance with Good Laboratory Practice (GLP) quality standards and are required by regulatory authorities. For praziquantel (PZQ), a safe profile was consistently demonstrated in nonclinical safety pharmacology, general toxicology, genotoxicity, carcinogenicity, as well as developmental and reproductive toxicity studies. Although bioanalytical methods to establish exposure relationships and GLP quality standards were not established in the 1970s, the objectives, study designs, and key parameters investigated in in vivo studies largely resembled today's requirements. In conclusion, PZQ has demonstrated a notably favorable safety profile in nonclinical investigations, which was confirmed by its safe use following market approval.

Keywords

Nonclinical safety · Toxicology · Good Laboratory Practice · Toxicokinetics · Safety pharmacology · Genotoxicity · Carcinogenicity · Developmental and reproductive toxicity

---

Authors Claudia Demarta-Gatsi and Christian Brenneis have equally contributed to this chapter.

C. Demarta-Gatsi
Global Health R&D of Merck Healthcare, Ares Trading S.A., (a subsidiary of Merck KGaA, Darmstadt, Germany), Eysins, Switzerland

C. Brenneis · P. Hewitt · N. Hübler · S. Weigt (✉)
Chemical and Preclinical Safety, Merck Healthcare KGaA, Darmstadt, Germany
e-mail: stefan.weigt@merckgroup.com

T. Spangenberg (ed.), *Praziquantel*,
https://doi.org/10.1007/978-3-031-97397-0_8

Introduction

It is a scientific, ethical, and a regulatory requirement that before any potential new medicine can be administered, its safety profile must be investigated in nonclinical studies to determine safe doses for humans. In this context, nonclinical toxicology and safety pharmacology studies play a crucial role in identifying, understanding, and managing the potential adverse effects of pharmaceuticals.

Overall, the drug development process begins by evaluating the mode of action of the new drug candidate, the nature of its pharmacological target in non-disease organs/tissues, and the identification of animal species for nonclinical testing. To select the best drug candidate with the optimal safety profile, exploratory investigations using different in vitro systems are performed.

Before clinical testing of new drug candidates, regulatory guidelines described in the International Conference on Harmonization of Technical Requirements for Registration of Pharmaceuticals for Human Use (ICH) mandate the identification of toxicity levels and target organs, risk assessment, and analysis of mechanisms behind observed alterations, with a focus on mitigation. Specifically, for the nonclinical safety assessment of small molecules, the ICH guideline M3(R2) harmonizes the testing strategy to ensure that it is scientifically and ethically accepted worldwide [1]. Figure 8.1 illustrates the current practices in nonclinical safety testing from early drug discovery to New Drug Application (NDA) for New Chemical Entities (NCEs). It is important to note that this chapter does not aim to cover all study types currently required for the regulatory submission of a new drug entity, but rather to provide an overview of the most important ones.

For enabling an Investigational New Drug (IND) submission in the United States of America (USA) or an Investigational Medicinal Product Dossier (IMPD) submission in Europe or other regions that follow similar regulatory frameworks, and subsequent entry into clinical development, a nonclinical safety package consisting of Good Laboratory Practice (GLP) compliant safety pharmacology, sub-acute repeat-dose toxicity, and genotoxicity studies, is usually required. Concomitant with clinical Phases I–III, longer-duration (sub-chronic/chronic) repeat-dose toxicity studies, carcinogenicity, reproductive and developmental toxicity studies, or other toxicity studies may need to be performed—dependent on the indication and patient population. Except for Dose Range Finding (DRF) toxicity studies, which are designed to generate an orientation on tolerated doses and identify potential target organs, most of the toxicology and safety pharmacology studies are required to be conducted according to GLP regulations. The GLP principles describe the organizational process and the conditions under which nonclinical studies need to be planned, conducted, and monitored. They also describe how the experimental data must be documented, reported, and archived. The principles of GLP became effective in 1979, setting a precedent for quality and integrity in nonclinical laboratory studies. This initiative by the United States Food and Drug Administration (FDA) led to the establishment of international guidelines, including the Organization for Economic Co-operation and Development (OECD) GLP principles in 1981 [2, 3], with regulatory frameworks (e.g., ICH, OECD, FDA, or European Medicines Agency (EMA) guidelines) that provide detailed requirements for nonclinical safety study designs.

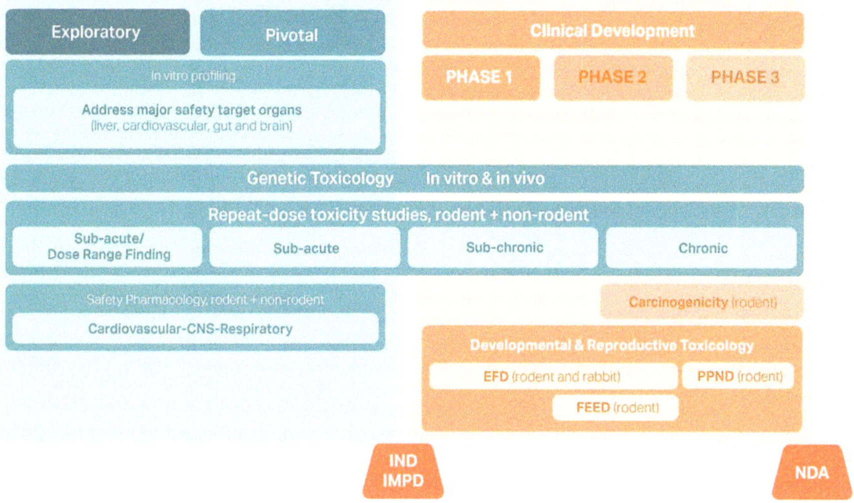

Fig. 8.1 Overview of current practices (2024) in nonclinical safety testing from early drug discovery to New Drug Application (NDA) for New Chemical Entities (NCEs). CNS Central Nervous System, FEED Fertility and Early Embryonic Development, EFD Embryo-Fetal Development, PPND Pre- and Postnatal Development, IND Investigational New Drug, IMPD European Investigational Medicinal Product Dossier

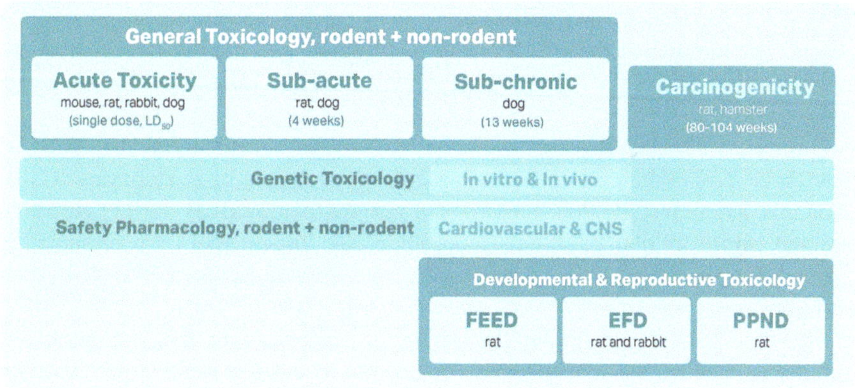

Fig. 8.2 Overview of the nonclinical safety studies conducted during the development of PZQ. LD_{50} dose at which 50% of the animals died, CNS Central Nervous System, FEED Fertility and Early Embryonic Development, EFD Embryo-Fetal Development, PPND Pre- and Postnatal Development

It is important to note that over the last 50 years, guideline content has evolved dynamically. The nonclinical development of praziquantel (PZQ), conducted in the 1970s, adhered to the guidelines of that time, when many in vitro investigations were not available and scientific and regulatory standards have not been defined or harmonized across countries. A summary of the nonclinical safety studies performed for PZQ are given in Fig. 8.2.

Nonclinical Safety Testing Strategy: In Vitro Testing

The current drug discovery process involves many safety-related in vitro models that are employed to select the best overall candidate(s) to move forward into in vivo general toxicity testing. The breadth of in vitro test systems used is company-specific, yet they tend to cover the major safety-related organ systems most affected clinically such as the cardiovascular system, liver, brain, kidney, and gastrointestinal systems. Other organs of interest are tested on a project/target specific basis. In addition, an early "Target Safety Assessment" is performed to understand the potential liabilities associated with hitting the intended clinical target. This may involve project (target)-specific assays being developed.

Most of these in vitro assays have not been qualified or recognized by regulatory agencies (e.g., FDA or EMA, see Chap. 11) and generally are in-house built cell systems that provide a general hazard assessment for a larger number of test molecules [4]. Importantly, the complexity of the in vitro models often increases as the potential drug candidate progresses along the drug development process.

In order to understand (and potentially reduce) off-target toxicity, in vitro safety pharmacology studies are an essential part of drug development and are employed to investigate interactions (e.g., agonistic and antagonistic effects) of novel drugs against specific, safety-related, molecular targets (e.g., specific ion channels, transporters, enzymes and receptors, as well as a battery of well-described protein kinases) distinct from the intended therapeutic target [5]. Such early hazard and off-target identification are also used to help predict potential clinical adverse effects. The number of these "off" targets being assessed usually increases as the drug candidate moves further into development, or if specific concerns arise during earlier testing.

During later stages of drug development, the final clinical candidates must be tested in several of these in vitro assays according to regulatory guidelines and GLP standards. These include the assessment of genetic toxicity (ICH S2(R1) guideline [6]) and cardiovascular liabilities (blockade of the human Kv11.1 potassium ion channel also known as the human Ether-à-go-go-Related Gene, *h*ERG) (ICH E14/S7B Q&A [7]).

Since the primary patient populations are in countries with high sunlight exposure, modern novel antischistosomal compounds would also need to be tested for their potential to cause/induce phototoxicity/photosensitivity. An assessment of the phototoxic potential of PZQ was not conducted, as it was not within the scope of drug development process at that time. ICH S10 guideline became effective in 2012 and provides a clear testing strategy, with a tiered approach to make this assessment [8].

For PZQ, no comprehensive in vitro testing battery was available (or required) at the time of development. Therefore, the only in vitro tests originally performed for PZQ were related to the assessment of genotoxicity (see "Genotoxicity" section).

More recently, additional off-target assays have been performed with PZQ. From a panel of 365 kinases, none were inhibited more than 50% at a nominal concentration of 1 μM. The only notable target (with 42% inhibition at 1 μM) was MST3-STK24. An additional in vitro panel of 44 ion channels and receptors was tested where only Kappa agonism was affected by PZQ—with a non-biologically relevant 30% inhibition observed at 10 μM. In addition, the cytotoxicity of PZQ and its two enantiomers, (*R*)-PZQ and (*S*)-PZQ, has been evaluated in eight cell lines using fluorescence microscopy for the morphology of apoptotic cells, and via lactate dehydrogenase assay for cytotoxicity [9]. The authors concluded that (*R*)-PZQ has lower cytotoxicity than the (*S*)-enantiomer and exhibits similar cytotoxicity to PZQ, which consists of (*R*)- and (*S*)-enantiomers in a 1:1 ratio.

Nonclinical Safety Pharmacology

According to the ICH S7A guideline [10], the objectives of safety pharmacology studies are: (1) to identify undesirable pharmacodynamic (PD) properties that may have relevance to human safety, (2) to evaluate adverse PD and/or pathophysiological effects of a substance observed in toxicology and/or clinical studies, and (3) to investigate the mechanism of the adverse PD effects observed. Identified liabilities during in vitro nonclinical safety investigations (e.g., off-target profiling, *h*ERG binding) can help to guide the selection of compound-specific safety pharmacology studies [11]. Some safety pharmacology endpoints can be integrated into the design of general toxicology studies. In other cases, these endpoints should be evaluated in dedicated nonclinical safety pharmacology studies.

In these studies, relevant physiological parameters (e.g., cardiovascular, respiratory, central nervous system (CNS)) are often continuously recorded over time using dedicated analysis systems (e.g., electrocardiogram (ECG), electroencephalogram (EEG), or plethysmography). The doses are typically based on expected therapeutic exposure levels and may include higher doses to evaluate exaggerated pharmacological effects, focusing on relevant exposure that reflects clinical use. The duration of exposure is generally short, focusing on acute effects that manifest within hours or days of administration.

When PZQ was initially developed no in vitro safety pharmacology studies such as *h*ERG assays were performed, therefore no compound-specific liabilities were known prior to the initiation of the animal studies. A variety of in vivo safety pharmacological studies were conducted in different animal species, which covered the effects on the cardiovascular system and the CNS. No respiratory system evaluations have been conducted for PZQ.

Overall, the oral administration of PZQ had no acute pharmacological safety effects in animals, which would suggest a risk for human patients. Safety pharmacology endpoints of cardiovascular, respiratory, and central nervous system function were also integrated into the repeat-dose toxicity studies in rats and dogs to some extent and no prohibitive findings were detected.

In Vivo Effects of PZQ on Cardiovascular Function

In anesthetized dogs, a dose of 10 mg/kg of PZQ was infused intravenously (i.v.) over a period of 10 min and hemodynamic parameters were investigated. Slightly increased cardiac output (the volume of blood pumped by the heart each minute), stroke volume (volume of blood pumped out of the heart's left ventricle during each systolic cardiac contraction), dP/dt (max) (the maximal rate of rise in ventricular pressure), and lowered peripheral resistance to flow was reported. The mean arterial pressure increased slightly, while the heart rate did not change significantly. An i.v. dose of 30 mg/kg of PZQ resulted in a slight decrease in heart rate, while all other hemodynamic changes were similar to those observed at 10 mg/kg, but more pronounced. An i.v. dose of 100 mg/kg of PZQ caused a major drop in peripheral resistance, resulting in a similarly significant drop in blood pressure.

In another study, conscious dogs were treated with oral doses of 10, 31.5, and 100 mg/kg. While no effect on blood pressure and heart rate was noted at doses of 10 and 31.5 mg/kg, there was a slight, temporary drop in blood pressure and increase in heart rate at the 100 mg/kg dose. The maximum effect occurred between 2- and 4-h following administration. These circulatory changes had completely subsided by 24 h.

Despite these weak effects at high doses, the cardiac safety of PZQ was supported by the absence of any treatment-related effects on the ECG or histopathology findings in the heart during oral, repeat-dose toxicity studies in dogs, up to the highest tested dose of 180 mg/kg/day.

In Vivo Effects of PZQ on Central Nervous System Function

The effects of PZQ on the CNS were evaluated in several species, including tests for orientation, motility, postural reflexes on a balance beam, traction reaction, anticonvulsant effect against seizures and catalepsy test.

In mice, orientation and motility inhibition were seen only at higher PZQ doses (ED_{50} = 884 mg/kg; "effective dose" producing 50% of an effect in 50% of the animals). In rabbits, intravenous doses of 25 mg/kg and higher induced frequent EEG peaks immediately following administration. These were particularly pronounced in the cortex EEG and occurred during a brief, convulsive phase. In addition, behavioral changes were seen at i.v. doses of 25–63 mg/kg.

Overall, the CNS symptoms that occurred at high i.v. doses of PZQ were considered an expression of nonspecific, cerebral intoxication at sublethal doses.

General Toxicology

General toxicity testing in animals provides essential data on the safety profile of a substance. It aims to identify dose–response relationships, determines the target organs affected, and characterizes the nature and severity of toxic effects.

Additionally, the studies enable establishing appropriate dose regimens for clinical studies, identify parameters for clinical monitoring for potential adverse effects, and provide essential data for regulatory submission.

Prior to starting general toxicity studies in animals, it is important to select the most relevant rodent (usually mouse or rat) and non-rodent species (usually dog, minipig, or nonhuman primate) in a data-driven approach. Importantly, cross reactivity of the development drug candidate toward its molecular target, such as binding and functional interference, needs to be demonstrated. Additionally, it is crucial that sequence homology and target distribution in the selected species aligns with that in humans. Furthermore, it should be ensured that the profile of drug metabolism resembles that observed in humans (Chap. 7). Finally, dedicated pharmacokinetic (PK) studies must indicate that the desired species can be sufficiently exposed.

In accordance with ICH M3(R2) guideline, the duration of the animal toxicity studies should be equal to or exceed the duration of the human clinical trials [1]. Evaluation for the potential of reversibility of toxicity (i.e., return to normal condition) should be provided when severe toxicity is observed with potential adverse clinical impact. The potential for reversibility can be evaluated during the repeat-dose toxicity studies that include a treatment-free recovery period or through a scientific assessment.

Typically, three dose levels are tested in male and female animals in repeat-dose toxicity studies. The low dose is selected to produce a PD effect, achieve the desired therapeutic effect, or ensure systemic exposure comparable to the anticipated clinical use. The highest dose is selected to identify target organ toxicity or is limited by factors such as the volume of administration or limit dose. The intermediate dose is the geometric mean between the high and low doses to demonstrate dose–response relationship.

Common measurements include clinical signs/symptoms (such as reduced activity or piloerection), body weight, food/water consumption, clinical chemistry, hematology, gross pathology and histopathology from all organs. Further, general toxicity studies typically include toxicokinetics (TK) evaluations to provide insights into the drug's Absorption, Distribution, Metabolism, and Excretion (ADME) profile (Chap. 7).

The endpoints in DRF and pivotal GLP-compliant repeat-dose toxicity studies are similar, but pivotal studies usually include additional ophthalmology and coagulation investigations and may incorporate specific safety biomarkers based on findings from DRF studies. The pivotal repeat-dose toxicity studies are crucial for defining the No Observed Adverse Effect Level (NOAEL) and/or No Observed Effect Level (NOEL), that may be used along with other information, such as the pharmacologically active dose, to determine the first dose in humans.

As described in Chap. 5, the pharmacological target of PZQ was identified only recently as an activator of a worm ion channel from the transient receptor potential melastatin family (TRPM) *Sm*.TRPM_{PZQ} [12]. Clearly, this channel is very specific to the trematode worms such as schistosomes causing schistosomiasis (Chap. 2). The most relevant human orthologue for *Sm*.TRPM_{PZQ} is *h*TRPM8 and has a role in cool temperature sensing and a quite restrictive expression in sensory neurons [13].

It has been subsequently shown that for PZQ, the activity against *h*TRPM8 is much lower than that against the trematode target. Since during the nonclinical development of PZQ in the 1970s, the molecular target of PZQ had not been identified, it could neither be considered for species selection nor for target-related toxicity liabilities. Instead, general toxicity studies used default species and study protocols that were common at that time.

Acute Toxicity

Historically, acute toxicity studies were performed with a lethality endpoint in at least two mammalian species. However, according to the ICH M3(R2) guideline [1], this information is now obtained through appropriately conducted dose-escalation or DRF studies that define a Maximum Tolerated Dose (MTD), the highest tested dose that does not result in unacceptable adverse effects or toxicity (Fig. 8.4).

For PZQ, acute (single-dose) toxicity studies have been conducted in mice, rats, rabbits, and dogs (Fig. 8.3). The mean lethal dose values indicate that the acute toxicity of PZQ is low after single oral, intramuscular, intraperitoneal, or subcutaneous administration. An oral single-dose toxicity study in juvenile rats showed no differences in the sensitivity compared to adult rats. In dogs, a lethal dose could not be established due to the emetic effect of PZQ in this species. Mean lethal dose values were > 2000 mg/kg in all other species tested, irrespective of the route of administration.

Sub-acute Repeat-Dose Toxicity

Sub-acute repeat-dose toxicity studies are typically initiated as DRF studies, designed to evaluate the potential adverse effects of a drug when administered repeatedly over a short period (up to 4 weeks) while determining the MTD and

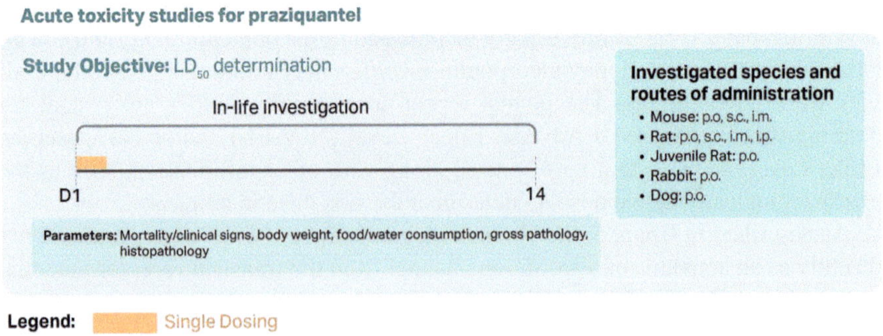

Fig. 8.3 Study design of acute toxicity studies done for PZQ in the 1970s. LD_{50} dose at which 50% of the animals died, D day, p.o. per os (oral), s.c. subcutaneous, i.m. intramuscular, i.p. interperitoneal

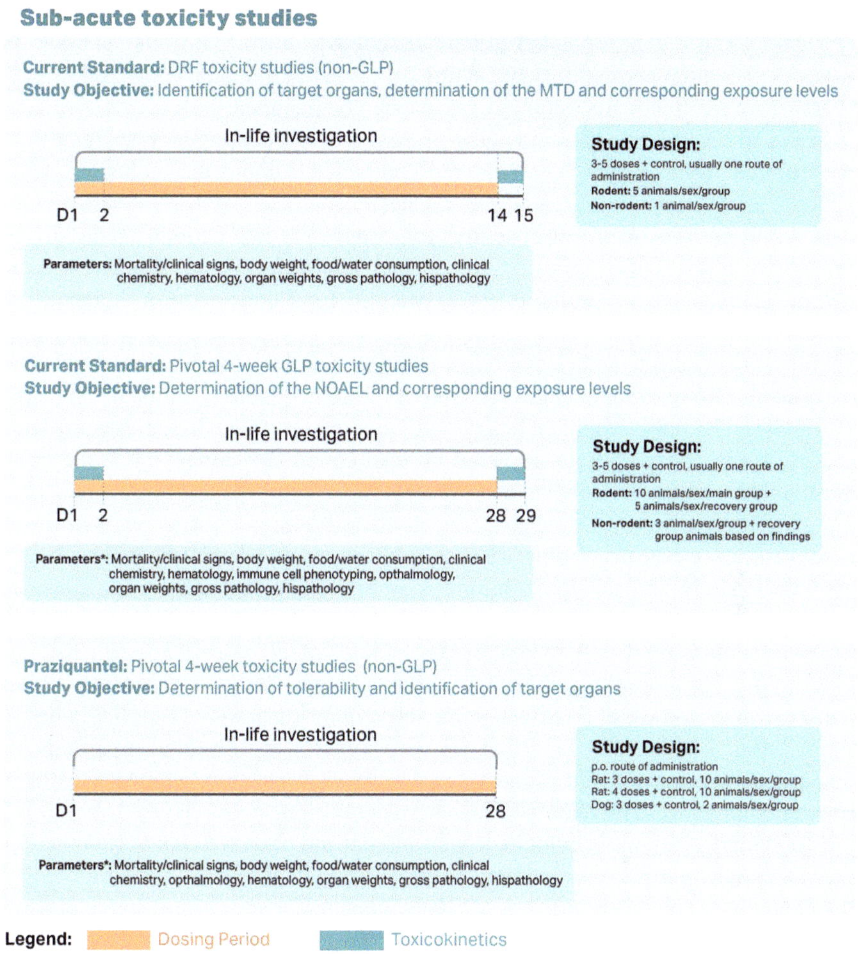

Fig. 8.4 Study design of sub-acute toxicity studies. The chart is divided into three sections: The upper and middle panels illustrate the study design of sub-acute non-pivotal Dose Range Finding (DRF) and pivotal 4-week GLP compliant repeat-dose toxicity studies according to current guidelines. The lower panel illustrates the design of the 4-week toxicity studies that were conducted for PZQ in the 1970s. GLP Good Laboratory Practice, MTD Maximum Tolerated Dose, NOAEL No Observed Adverse Effect Level, D day, p.o. per os (oral); *in non-rodents, safety pharmacology investigations (e.g., electrocardiography) can be included; this was also the case for PZQ

suitable dose levels for subsequent studies. Additionally, these studies aim to identify target organs that are significantly impacted by the drug, providing crucial information for understanding its safety profile (Fig. 8.4, upper panel).

Following the initial DRF studies, pivotal GLP-compliant repeat-dose toxicity studies are conducted to provide a more detailed evaluation using a higher number of animals. Typically, 10 (rodents) or 3 (non-rodents) animals per sex and group, along with additional animals for TK (rodents), are dosed daily over a 4-week period (Fig. 8.4, middle panel). However, the duration and dosing frequency may vary depending on the intended treatment schedule in the first clinical trial. To assess the potential reversibility of findings, a cohort of animals, which are investigated after a treatment-free recovery period, can be included.

In the 1970s, the nonclinical testing of PZQ was conducted under a framework that differed from today's standards and conducted without the inclusion of TK analysis, as illustrated in Fig. 8.4 (lower panel).

In a first rat 4-week repeat-dose toxicity study, PZQ was administered daily by oral gavage at doses of 0 (control), 33, 100, and 300 mg/kg. Parameters investigated included clinical signs, mortality, body weight, food consumption, hematology, clinical chemistry, urinalysis, gross pathology, organ weights, and histopathology. No mortality occurred, and clinical signs, food consumption, and body weight were not affected. No relevant changes were observed in hematology, clinical chemistry, or urinalysis. Kidney weights of rats of both sexes given 100 or 300 mg/kg/day were increased, but histopathology revealed no changes in the kidneys or in the other investigated organs. In conclusion, the NOEL in this study was considered to be 33 mg/kg/day.

Since the 4-week repeat-dose toxicity study in rats has been initiated without a prior DRF study and no dose-limiting toxicities have been observed, a second 4-week study with higher doses up to the limit dose of 1000 mg/kg was initiated. PZQ was administered daily by oral gavage at doses of 0 (control), 30, 100, 300, and 1000 mg/kg. No mortality occurred, and clinical signs, body weight, and food and water consumption were not affected. No relevant changes were observed in hematology, clinical chemistry, or urinalysis. Gross pathology revealed no significant changes. Liver and thyroid weights in rats of both sexes administered 300 or 1000 mg/kg/day were increased. In males, spleen weights were elevated at both 300 and 1000 mg/kg/day, along with increased adrenal and heart weights at 1000 mg/kg/day. In females, kidney and adrenal weights were increased at 1000 mg/kg/day. Histopathology showed no treatment-related changes in any organs. In conclusion, the NOEL in this study was considered to be 33 mg/kg/day.

To investigate sub-acute toxicity of PZQ in non-rodents, groups of two female and male Beagle dogs were treated for 4 weeks at dose levels of 0 (control), 20, 60, or 180 mg/kg/day. Parameters investigated were clinical signs, mortality, body weight, food consumption, body temperature, neurological examinations, ophthalmology, electrocardiography, hematology, clinical chemistry, urinalysis, gross pathology, organ weights, and histopathology. The only clinical sign—vomiting—occurred in one high dose dog and in clinical chemistry, alkaline phosphatase was increased in three of the four high dose animals. None of the other in-life or post-mortem parameters investigated were affected. In conclusion, the NOEL in this study was considered to be 60 mg/kg/day.

Sub-chronic Repeat-Dose Toxicity

Sub-chronic repeat-dose toxicity studies last for 13 weeks and are performed to support extended dosing in human clinical trials and to determine the doses suitable for subsequent chronic studies. Sub-chronic toxicity studies generally utilize 20 animals per sex per group for rodents, and 3–5 per sex for non-rodents (Fig. 8.5, upper panel), to ensure statistical validity and robustness of the data.

For PZQ, a sub-chronic study was only performed in dogs, but without the inclusion of TK analysis, as illustrated in Fig. 8.5 (lower panel). Groups of three male and three female dogs were treated for 13 weeks with 0 (control), 20, 60, or 180 mg/kg/day. The same parameters were investigated as in the 4-week study. The only clinical sign observed was vomiting in a few dogs during the first week of the study. Food consumption and body weight were temporarily depressed in high-dose dogs during the initial study phase but normalized over the course of the study. Liver weights were slightly higher in high dose dogs compared to the control while no microscopic changes were seen. Apart from that, none of the other in-life and postmortem parameters investigated

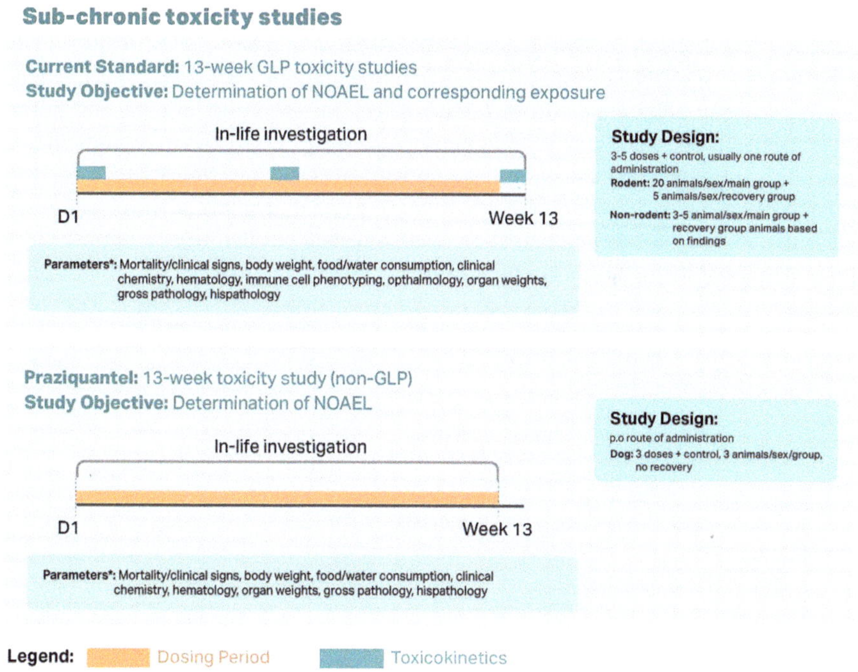

Fig. 8.5 Study design of sub-chronic toxicity studies. The chart is divided into two sections: The upper panel shows the study design for 13-week toxicity studies according to current guidelines. The lower panel illustrates the study design of the sub-chronic dog toxicity study that was conducted for PZQ in the 1970s; GLP Good Laboratory Practice, NOAEL No Observed Adverse Effect Level, D day, p.o. per os (oral); *in non-rodents, safety pharmacology investigations (e.g., electrocardiography) can be included; this was also the case for PZQ.

were affected. In conclusion, the NOAEL in this study was considered to be 60 mg/kg/day.

Chronic Repeat-Dose Toxicity

To support longer clinical trials and market authorization, chronic toxicity studies are conducted. According to the ICH S4 guideline [14], it is recommended to investigate chronic toxicity with a duration of 6 months in rodents and 9 months in non-rodents (Fig. 8.6). The endpoints/parameters investigated in chronic repeat-dose toxicity studies are in principle the same as the ones investigated in sub-chronic toxicity studies.

Considering that PZQ is administered as a single dose, repeat-dose toxicity studies with longer treatment duration were not deemed necessary; therefore, no chronic 6- or 9-month studies were performed for PZQ in any species.

Genotoxicity

Genetic toxicology focuses on assessing the potential of pharmaceuticals to damage genetic material (DNA) through various endogenous or exogenous mechanisms, leading to gene mutations, chromosomal alterations, heritable effects, and potentially contributing to cancer development. Screening for genotoxic compounds is crucial to mitigate human exposure and prevent associated health risks. Regulatory agencies worldwide have implemented measures to detect and identify genotoxic pharmaceuticals, aiming to control or eliminate their levels prior to human exposure. Genotoxicity testing is an integral part of safety evaluations for all new pharmaceuticals, with testing systems and protocol designs following the ICH S2(R1)

Chronic toxicity studies

Current Standard: Chronic GLP toxicity studies
Study Objective: Support clinical studies for chronic use

Study Design:
3 doses + control, usually one route of administration.

Rodent:
6-month duration, 20-25 animals/sex/main group + 5 animals/sex/recovery group.

Non-rodent:
9-month duration, 3 animals/sex/main group + 2 animals/sex/recovery group (only control + high dose).

Praziquantel: N.A.

Fig. 8.6 Study design overview of chronic toxicity studies according to current guidelines. GLP Good Laboratory Practice, N.A. not applicable

guideline [6]. Genotoxicity assays such as Ames Test (Bacterial Reverse Mutation Test), in vitro Mammalian Cell Gene Mutation Assay, in vitro Chromosomal Aberration Test, in vitro and in vivo Micronucleus Test, or in vivo Comet Assay are conducted according to current guidelines.

While some assays are standard in genotoxicity testing, not all are mandatory for every compound. The choice of assays is usually based on initial results, chemical structure, regulatory guidelines, and the drug's intended use.

A comprehensive genotoxicity package is available for PZQ, consisting of studies conducted in the 1970s as well as publications and assessments from the 1980s and 1990s, which includes additional assays beyond the standard requirements.

A large battery of in vitro and in vivo assays showed that PZQ lacks mutagenic potential. In vitro, genotoxicity investigations using *Salmonella typhimurium* and *Escherichia coli*, with and without metabolizing enzymes (i.e., liver S9 mix), yielded no indications of mutagenicity. Urine samples from treated mice, rats, and humans showed no mutagenic potential of PZQ metabolites in bacteria. In vivo, PZQ was not genotoxic in micronucleus tests in mice and rats or in a dominant lethal test, which is used to assess the potential for a substance to cause mutations that lead to dominant lethal effects in mice. The results of these original studies have also been published [15–17].

Investigations on the genotoxic potential of PZQ have been described in several publications. In one study, it tested negative in the plate incorporation test in five *Salmonella* strains at concentrations up to 1 mg/plate, with known mutagens used as positive controls [18]. Additionally, PZQ did not show genotoxicity in various test systems, including *Salmonella*, yeast, hamster and human cells, and *Drosophila melanogaster* [19]. PZQ also yielded negative results with and without S9 mix in the sister-chromatid exchange assay in human lymphocytes [20]. On the other hand, PZQ has been described as a weak co-mutagen, which increased the mutagenicity of several chemical mutagens and carcinogens in bacteria and animal cells [21].

Based on the available data, the genotoxicity and carcinogenicity of PZQ and other antischistosomal drugs were reviewed by the International Commission for Protection Against Environmental Mutagens and Carcinogens (ICPEMC) [22]. It was concluded that PZQ does not appear to present a genotoxic risk to humans exposed to the prescribed once-a-day treatment for different types of schistosomiasis. The Committee for Veterinary Medicinal Products (CVMP) of the European Agency for the Evaluation of Medicinal Products also assessed PZQ based on a comprehensive battery of in vitro and in vivo mutagenicity studies and concluded that PZQ is not a mutagenic substance [23].

Carcinogenicity

The main goal of carcinogenicity studies is to determine whether a substance has the potential to induce tumors or cancer. Carcinogenicity studies should only be performed for pharmaceuticals if the expected clinical use is continuous for at least 6 months (except when life expectancy in the indicated population is short) or there are any concerns derived from laboratory investigations, animal toxicology studies, and data in humans (ICH S1A guideline) [24]. ICH S1B(R1) guideline provides guidance on approaches for evaluating the carcinogenic potential of pharmaceuticals [25].

Certain other antischistosomal drugs (e.g., hycanthone) have been shown to exhibit a carcinogenic potential [26]. Therefore, comprehensive investigations of carcinogenicity to determine the long-term safety of PZQ were conducted in two species (rats and hamsters). Figure 8.7 illustrates the objective of carcinogenicity studies, provides basic information about the study design according to current guidelines, and outlines the carcinogenicity studies that have been conducted for PZQ. In the rat study, 50 male and 50 female Sprague-Dawley rats received 100 or 250 mg/kg/week of PZQ by oral gavage for 2 years (104 weeks), followed by a treatment-free recovery period up to week 130. The control group consisted of 100 male and 100 female rats receiving the vehicle only. Evaluated parameters included clinical signs, food and water consumption, body weight, hematology, clinical chemistry, and urinalysis. Gross pathological and histological examinations were performed on all animals that were moribund, found dead, or sacrificed at week 130. Treated animals had decreased survival time compared to controls, with statistical significance in high-dose males only. No treatment-related effects were observed in

Carcinogenicity studies

Study Objective: Incidence and timing (latency) of neoplasms including malignancy determined by histology

Current Standard:

3 doses + control, usually one route of administration
Rodent: 2-year study, 50-100 animals/sex/group
Transgenic mice: 6-month study, 25-30 animals/sex/group

Praziquantel:

Rat study:
p.o. route of administration, 2 doses + control
2-year treatment duration followed by a treatment free period until week 130, 50-100 animals/sex/group

Syrian golden hamster study:
p.o. route of administration, 2 doses + control
80-week treatment duration, 50-100 animals/sex/group

Fig. 8.7 Study design overview of carcinogenicity studies according to current guideline and carcinogenicity studies conducted for PZQ in the early 1980s; p.o. per os (oral)

hematology, clinical chemistry, or urinalysis, and there was no significant increase in tumor rates. In conclusion, PZQ did not show a carcinogenic potential in rats at doses up to 250 mg/kg/week.

In the hamster study, 50 male and 50 female Syrian golden hamsters received 100 or 250 mg/kg/week of PZQ orally by gavage for 80 weeks, with a control group of 100 male and 100 female hamsters receiving the vehicle only. Parameters evaluated were similar to those in the rat study. Body weights decreased in low-dose males up to week 19 and in high-dose females throughout the study, while food and water consumption increased in high-dose females. Survival time was not significantly different from controls. No treatment-related effects were observed in hematology, clinical chemistry, or urinalysis.

The number of benign neoplasms (an abnormal tissue mass) was higher in low-dose males but considered incidental due to lack of dose-dependency, as high-dose hamsters had fewer benign neoplasms than controls. The incidence of malignant neoplasms in high-dose groups was not statistically significant. In conclusion, PZQ did not show a carcinogenic potential in hamsters at doses up to 250 mg/kg/week.

The rat study for PZQ meets the typical guideline recommendation of a 2-year duration, while the hamster study falls short with a duration of only 80 weeks, and both studies utilized a once-weekly dosing regimen, which is less common than the standard daily dosing typically recommended for carcinogenicity studies. The rat carcinogenicity study for PZQ included a treatment-free recovery period, which is not specified in current guidelines. Moreover, a 6-month carcinogenicity study in transgenic mice is often used as a second carcinogenicity study in current practices.

Developmental and Reproductive Toxicity

The beginning of embryonal development is determined by fertilization, a fundamental process by which the fusion of the male spermatozoon with the female oocyte forms the zygote and results in the formation of a new organism. The entire process of embryogenesis is orchestrated by a complex regulation of gene expression as a result of cell division [27], three-dimensional movement, and cell differentiation in each individual embryo tissue type and organ. Any disruption or alteration of these well-coordinated processes will have a relevant impact on the developing embryo, potentially resulting in adverse events (including structural malformations, growth retardation, functional impairment, and/or death of an organism). Drugs that cross the placental barrier have the potential to affect fetal development during critical stages of embryogenesis. The degree of drug teratogenicity (i.e., the manifestation of developmental toxicology—ability to cause structural or functional abnormalities in the developing embryos) is determined by the timing and extent of exposure during development [28, 29]. Moreover, in response to the pregnancy, physiological changes (e.g., hormones levels, blood flow, etc.) in the mother may affect the PK of drugs that may reach and potentially harm the fetus [30–32].

For this reason, in addition to the many general toxicology and safety pharmacology studies described previously (which also include the pathological assessment of the female and male reproductive organs), nonclinical Developmental and Reproductive Toxicology (DART) testing is an essential part of drug development, since clinical studies rarely assess reproductive function or pregnancy in patients. In addition, DART studies are required to support inclusion of women of childbearing potential or pregnant women in clinical trials. Therefore, as per the latest regulatory standards (ICH S5(R3) guideline) [33], nonclinical studies for assessing reproductive toxicity for new drug candidates are crucial to support the conduct of clinical trials and market authorization.

The DART package encompasses animals in various life stages, ranging from conception to sexual maturity, including mature adults, to identify developmental toxicity across different segments of the life cycle (Fig. 8.8). To further elucidate these stages and their correlated events, the reproductive life cycle has been subdivided into six periods: from mating to fertilization (A), from conception/fertilization to implantation (B), from implantation to hard palate closure (embryonic period, C), from hard palate closure to birth (fetal period, D), from birth to weaning (E), and from weaning to sexual maturation/production of gametes (F).

Fertility and Early Embryonic Development

The intent of the Fertility and Early Embryonic Development (FEED) studies is to detect potential toxicity resulting from treatment before mating (>2 weeks prior to cohabitation), and through mating and implantation (gestational day (GD) 6) (i.e., stages A and B; Fig. 8.8) (Fig. 8.9, upper panel). In particular, the study assesses adverse effects of the test compound on maturation of gametes, mating performance, fertilization, preimplantation and implantation stages of the embryo, none of which are captured in general repeat-dose toxicity studies. In females, adverse effects can be detected on the estrus cycle, tubal transport, implantation, and embryo development during the preimplantation stages [34]. While, in males, possible adverse functional effects on libido and/or epididymal sperm maturation (viability, morphology, and motility) are evaluated [35, 36]. It is crucial to carefully plan these studies based on existing knowledge (e.g., previous general toxicity or/and Embryo-Fetal Development (EFD) toxicity studies) to prevent requests for additional tests later.

Parental toxicity, such as the animal's behavior, clinical signs, mortality and general physical condition, food consumption, and body weight are examined during the study. After mid-pregnancy (usually GD13–15 as per ICH S5(R3) guideline), mated females are euthanized and uterine contents examined (i.e., corpora lutea counts, number of implantation sites, live and dead embryos) [33]. In contrast, the male's euthanasia occurs after completion of the uterine dissections of the female animals. Optionally, sperm samples can be collected from a cauda epididymis for spermatological examinations (i.e., evaluating percent motionless and sperm concentration). In both groups, reproductive organs and tissues with gross lesions are examined for histological changes.

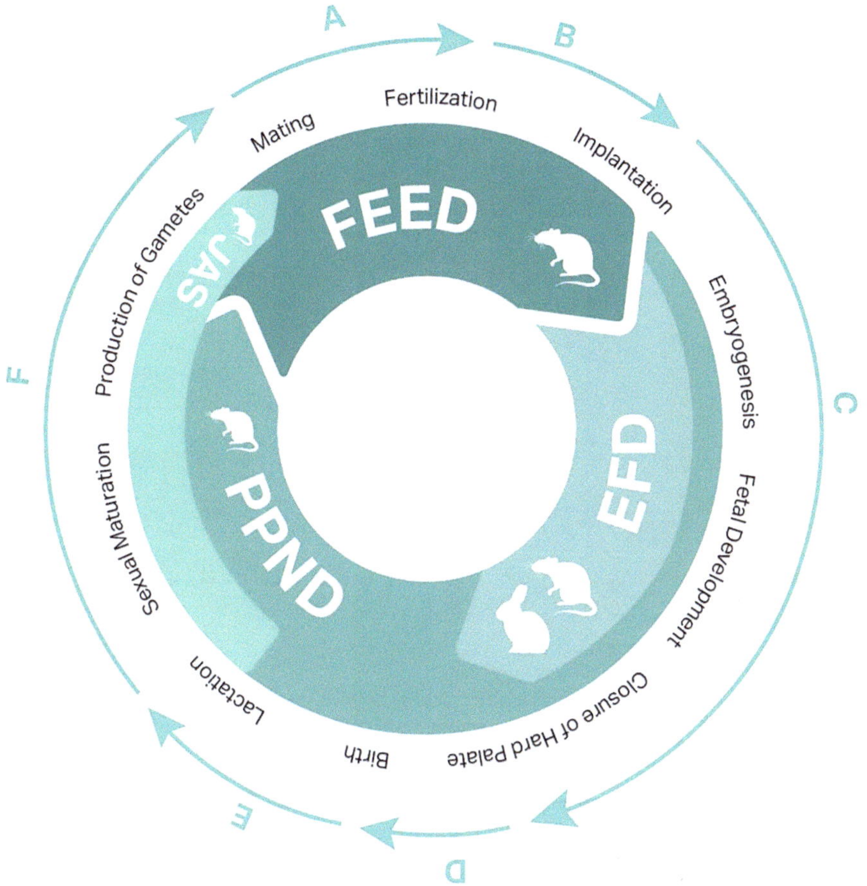

Fig. 8.8 Life cycle segments for reproduction and development studies. (**A**) mating to fertilization, (**B**) fertilization to implantation, (**C**) implantation to closure of the hard palate (end of major organogenesis), (**D**) closure of hard palate to birth, (**E**) birth to weaning, (**F**) weaning to sexual maturation/production of gametes. Standard DART testing typically includes Fertility and Early Embryonic Development (FEED) studies (Segment 1, **A-B**), Embryo-Fetal Development (EFD) toxicity studies (Segment 2, **C**), and Pre- and Postnatal Development (PPND) toxicity studies (Segment 3, **C-F**). Additionally, Juvenile Animal Studies (JAS) (**E** and/or **F**) and general toxicity (**F**)

Potential adverse effects of PZQ on gamete maturation, mating, fertilization, and early embryonic development and late effects in second filial (F2) offspring were evaluated in rats. The study assessed tolerability and fertility in the parent (filial 0 (F0)) animals, as well as early embryonic development and general reproductive performance in the first filial (F1) generation (Fig. 8.9, lower panel). PZQ was administered orally by gavage to groups of 24 male and 60 female Wistar rats at doses of 0 (control), 30, 100 and 300 mg/kg/day. Males were dosed for 10 weeks before mating and throughout the mating period, which lasted up to a maximum of

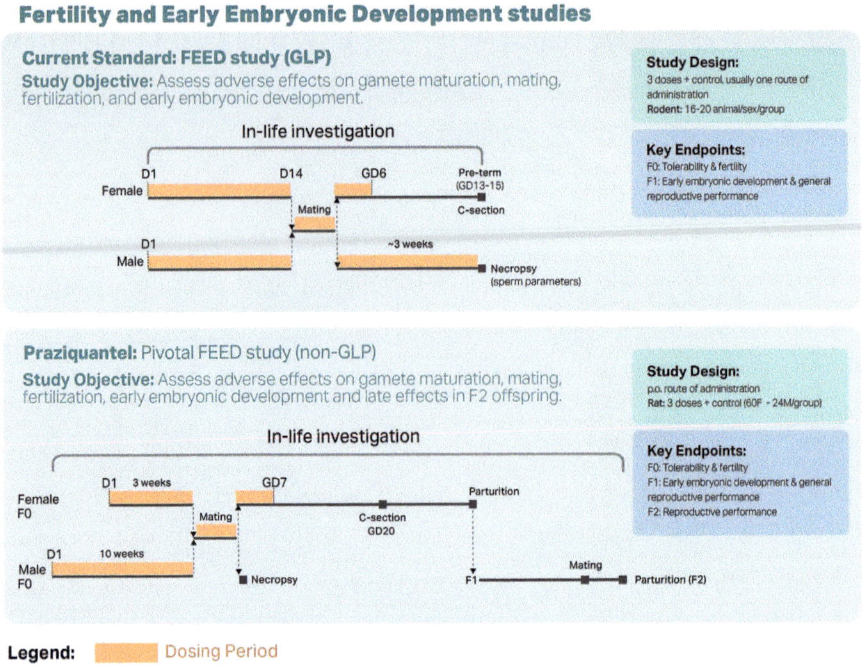

Fig. 8.9 Study design of FEED studies. The chart is divided into two sections: The upper panel shows a FEED study design according to current ICH S5(R3) guideline. The lower panel depicts the study design used to investigate fertility, early embryonic development, and overall reproductive performance for PZQ in the 1970s. FEED Fertility and Early Embryonic Development, GLP Good Laboratory Practice, F0 filial 0 (gestating female), F1 first filial generation, F2 second filial generation, F females, M males, D day, GD gestational day, p.o. per os (oral), C-section caesarean section

4 weeks, while females were dosed for 3 weeks before mating and throughout the mating period until gestational day 7 (GD7).

Half of the dams underwent caesarean section on GD20 and were sacrificed, while the other half were allowed to litter and rear their pups during lactation. One male and one female from each litter of the F1 offspring generation in the control and high-dose groups were raised to sexual maturity and mated. Parameters evaluated for the F0 generation included appearance and behavior, mortality, and body weight. After mating, male rats underwent macroscopic examination and determination of testicular weights.

Dams sacrificed on GD20 were examined for the number of corpora lutea (a temporary collection of cells that forms on the ovary after ovulation), implantations (fertilized eggs (blastocysts) attached to the uterine wall), live embryos or fetuses, sex of all fetuses, litter weight, mean fetal weight per litter, mean placental weight per litter, and external, visceral, and skeletal malformations. Animals allowed to deliver spontaneously were examined for body weight development, general tolerability (appearance and behavior), mortality, duration of gestation, course of delivery, lactation performance, and at the end of the weaning for the number of implantations and anatomic–pathological effects.

Investigations of the F1 generation included body weight, number at birth, sexual behavior, appearance, inspection for external deformities, general behavior, and developmental and functional tests. One male and one female each from the control and high-dose groups were paired and allowed to give birth to their litters. Reproductive performance was determined by insemination and impregnation rates, duration of gestation, number of living and dead pups, mean weight of pups at birth, and the appearance and behavior of pups.

In the F0 generation, salivation occurred in one female in the mid-dose group and in one female and one male in the high-dose group, while diarrhea was observed in individual males in the high-dose group. The NOEL was considered 30 mg/kg/day.

No treatment-related effects were observed in the F1 generation obtained by caesarean section or natural birth. Additionally, no effects on reproductive performance were seen in the F1 generation, and no late effects on the F2 generation were identified.

In conclusion, under the defined experimental conditions of this study, oral doses of up to 300 mg/kg/day in rats did not impair fertility or general reproductive performance and caused no late effects in F1 and F2 offspring. Oral administration of 100 and 300 mg/kg/day PZQ led to mild clinical symptoms in the F0 generation. However, male and female fertility and general reproductive performance of the F0 and F1 generations were not affected by treatment, and no late effects in F1 and F2 offspring were observed up to 300 mg/kg/day.

Embryo Fetal Development Toxicity

Today, pharmaceutical companies must use in vivo studies to assess the Embryo-Fetal Development (EFD) toxicity of drug candidates before administering them to women of childbearing potential. The need for such stringent testing protocols arose from the thalidomide tragedy in the late 1950s and early 1960s [37], which raised awareness about the risks of birth malformations caused by certain drugs. Before this tragedy, drug testing practices were less rigorous, often relying on testing in only one species, typically rodents. However, the thalidomide case highlighted the limitations of this approach, as rodents proved less sensitive to the drug compared to non-rodent species (e.g., rabbits and/or nonhuman primates). Consequently, there was a paradigm shift toward mandatory testing of drug candidates in both rodent (usually rats) and non-rodent (usually rabbits) species to ensure a better safety assessment [38, 39].

Assessing the developmental toxicity of drugs and chemicals in laboratory animals enables the evaluation of potential adverse effects from the implantation of the blastocyst (early-stage embryo, GD5–6) to the closure of the hard palate (GD16–19) on pregnant females, embryos, and fetus (i.e., stage C and D; Fig. 8.8). In the EFD toxicity study, the drug candidate is administered daily by oral gavage (although other routes may be used if justified) to at least 16 presumed pregnant females

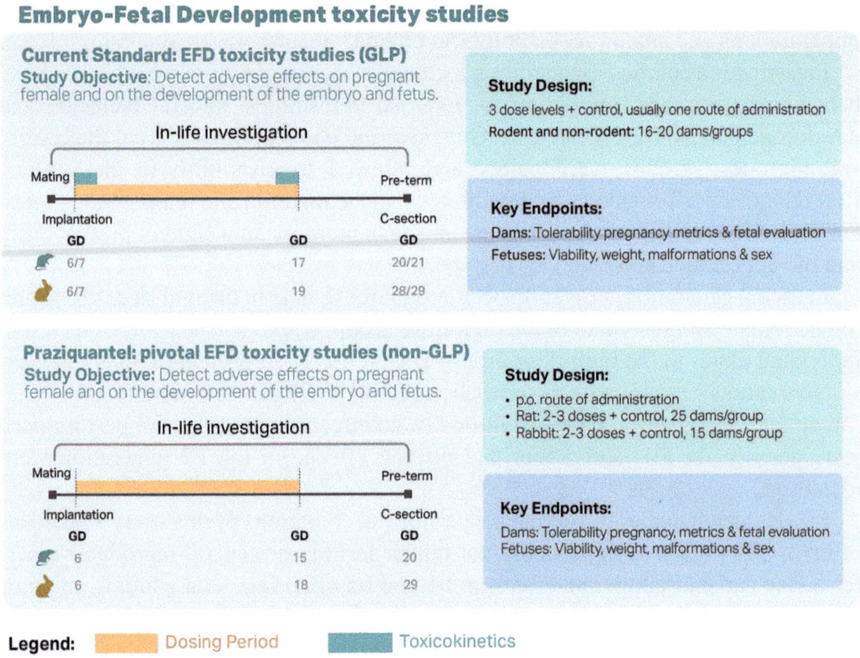

Fig. 8.10 Study design of EFD toxicity studies. The chart is divided into two sections: The upper panel shows the EFD study design according to current ICH S5(R3) guideline. The lower panel illustrates the design of the EFD studies conducted for PZQ in the 1970s. EFD Embryo-Fetal Development, GLP Good Laboratory Practice, GD gestational day, p.o. per os (oral), C-section caesarean section

during the organogenesis period (GD6/7–17 for rats and GD6/7–19 for rabbits) (Fig. 8.10, upper panel). Ideally, three geometrically spaced dose levels are tested, from a low clinically relevant dose to a high dose with minimal maternal toxicity [40, 41]. The low dose provides maternal plasma exposure just above that expected in pregnant patients, while the high dose is either the MTD or 25 times the human exposure at the recommended maximum concentration [33]. To effectively support the EFD risk assessment, it is important to collect relevant data from general toxicology studies to inform dose selection and study design. For EFD studies in rodents, existing general toxicity data (e.g., exposure vs. toxicity) may allow for direct progression to pivotal EFD studies. Nevertheless, a preliminary EFD (pEFD) approach may be appropriate if developmental or maternal toxicity concerns exist [42–44]. Regarding the inclusion of rabbits, due to the lack of prior studies, a PK/tolerability DRF study in nonpregnant or pregnant animals will aid in informing dose selection.

Maternal toxicity (clinical signs, mortality, physical condition, food consumption, and body weight) is carefully monitored in EFD studies, as it can lead to adverse reproductive outcomes, including increased resorptions, fetal deaths, and litter losses. Distinguishing between malformations and other fetal losses is difficult, underscoring the importance of tracking maternal stress indicators and selecting doses to minimize maternal toxicity [33, 45, 46]. Before expected parturition, dams are euthanized (GD20/21 for rats and GD28/29 for rabbits), and a caesarean section is performed to assess fetal abnormalities. The study primarily evaluates fetal survival and structure through external examinations, followed by comprehensive visceral and skeletal examinations for possible malformations [47, 48]. Additionally, macroscopic evaluations of the dam's viscera, including implantation sites and placenta appearance at the maternal–fetal interface, are performed.

Maternal systemic drug exposure is typically evaluated using a separate cohort of pregnant TK animals at both the start and end of the dosing period. Additionally, another vital aspect of these studies is assessing fetal drug concentrations at the end of the dosing period to assess transplacental passage and provide evidence of fetal exposure during organogenesis [33, 49].

The potential adverse effects of PZQ on the pregnant female, as well as on embryonic and fetal development, were assessed in rats and rabbits (Fig. 8.10, lower panel) without prior pEFD or DRF studies or TK analysis.

A series of three EFD toxicity studies were conducted in Wistar rats. In each study, groups of 25 rats received oral doses of PZQ suspended in different vehicles: 0 (control), 33, 100 and 300 mg/kg/day for the first study; 0, 30, 100 and 300 mg/kg/day for the second study; and 0, 100 and 300 mg/kg/day for the third study. Dosing occurred from GD6 to 15, and parameters evaluated during the in-life phase included appearance, behavior, mortality, body weight, and food consumption.

In all studies, no deaths occurred, and clinical signs, food consumption, and body weight development were similar between treated and control groups. On GD20, fetuses were delivered via caesarean section, and the uteri were examined for reproductive outcomes, including the numbers of corpora lutea, implantations, live and dead fetuses, early and late resorptions, and abortions. Fetal weights were measured, and fetuses were assessed for gross external, visceral, and skeletal malformations.

The results showed no treatment-related effects on reproductive parameters in any of the studies. Mean fetal weights were unaffected, and no dead fetuses were found. While some skeletal malformations were observed in the first study, they were deemed spontaneous and not related to treatment. Although these skeletal abnormalities are common spontaneous occurrences in this strain of rats, the study was repeated twice more using also a second vehicle to rule out any potential treatment effects. The second and third studies reported no malformations in any fetuses.

Overall, all three studies concluded that oral administration of PZQ at doses up to 300 mg/kg/day did not result in maternal toxicity, embryofetal toxicity, or teratogenicity. The dose of 300 mg/kg/day is considered the NOEL for maternal and embryofetal toxicity and teratogenicity across all studies.

Two EFD studies were conducted on pregnant female Himalayan rabbits. In both studies, PZQ was administered daily by oral gavage from GD6 to 18 to groups of 15 mated female rabbits at doses of 0 (control), 30, 100 and 300 mg/kg in the first study, and 0, 100 and 300 mg/kg in the second study, using the same vehicles as in the rat studies.

In both studies, in-life parameters evaluated included appearance, behavior, body weight, and food consumption. The rabbits were sacrificed on GD29, and their uteri were examined for the number and distribution of corpora lutea, implantation sites, and uterine contents. Fetuses were weighed and assessed for gross external, visceral, and skeletal malformations, and the sex of each fetus was determined.

Throughout both studies, none of the animals exhibited clinical abnormalities. However, in the first study, mean food consumption was slightly reduced, and body weight gains stagnated, with high-dose rabbits experiencing transient weight loss. In the second study, mean food consumption and body weights were reduced in the 100 and 300 mg/kg/day groups. There were no treatment-related effects on any other parameters investigated.

Overall, both studies concluded that oral administration of PZQ induced slight maternal toxicity (evidenced by decreased food consumption and body weight) but was not toxic nor teratogenic for the embryo. The NOEL for embryofetal toxicity and teratogenicity was determined to be 300 mg/kg/day across both studies.

Pre- and Postnatal Development Toxicity

Pre- and Postnatal Development (PPND) toxicity studies are conducted to evaluate the effects of a drug on both the gestating female (F0) and the developing offspring (F1) from the time of implantation through birth and early development (i.e., stage C and F; Fig. 8.8). Based on existing knowledge from the previous EFD studies, where maternal toxicity and toxicokinetic exposure during pregnancy were assessed, PPND doses are selected. Furthermore, dose levels causing more than mild maternal toxicity should be avoided, with exposures ideally exceeding the maximum recommended human dose by 3–10 times [50]. In the PPND study, the drug candidate is administered daily by oral gavage to presumed pregnant females from early pregnancy (GD6/7) through weaning, approximately postnatal day (PND) 20 (Fig. 8.11, upper panel).

Usually conducted after FEED and EFD studies, PPND studies in rats provide additional information on functional reproductive endpoints, such as parturition (the process of giving birth) and lactation and assess the growth and behavior of the offspring. Due to natural delivery of pups, the evaluation of the placenta and accurate counting of stillborn offspring, particularly if cannibalism occurs, can be complicated. Therefore, the number of viable offspring is recorded on the delivery day. Dams are necropsied at weaning to assess uterine implantation sites and collect visible abnormalities, with paired control tissues for comparison. The litter is the unit of measure for endpoints evaluated in pregnancy studies. Pup evaluations start at birth, recording the number of live pups and their sex. Pups are monitored daily for clinical signs, mortality, preweaning developmental milestones (e.g., pinna unfolding), and reflex ontogeny (e.g., response to light) [33]. They are also weighed twice weekly during lactation to track growth. Around PND4, the litter size is reduced to four to five pups per sex to ensure uniformity across litters and consistency with historical data, as litter size affects pup growth rates. The offspring

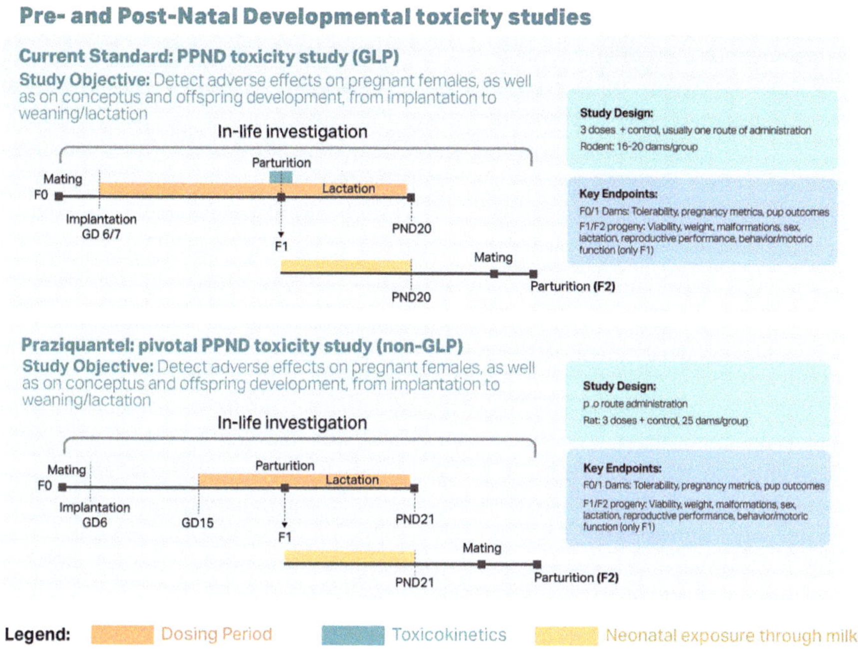

Fig. 8.11 Study design of PPND toxicity studies. The chart is divided into two sections: The upper panel shows the PPND study design according to current ICH S5(R3) guideline. The lower panel illustrates the PPND study conducted for PZQ in the 1970s. PPND Pre- and Postnatal Development, GLP Good Laboratory Practice, GD gestational day, PND postnatal day, F0 filial 0 (gestating female), F1 first filial generation, F2 second filial generation, p.o. per os (oral)

remain with the biological mother, who continues to be exposed to the drug until weaning (PND21), and their survival and growth depend in part on successful lactation. Important to consider is that the transfer of the drug through breast milk is possible, resulting in indirect exposure of the pups. After weaning of the F1 (PND21), at least one male and one female from each litter are selected for posweaning behavior tests and mating. F1 females are evaluated for estrous cyclicity and paired (avoiding sibling pairings) for a maximum of 14 days for the assessment of reproductive capacity.

Maternal systemic exposure is measured on GD20. Additionally, a quantitative aliquot of milk from F0 dams can be collected during the lactation period to determine the extent of neonatal exposure through maternal milk until weaning. Exposure levels in fetuses and pups can also be measured early in the postnatal period (GD20, PND4 and 20) to assess drug transfer across the placenta and from the dams. Although this is not routinely performed, it contributes to the overall study interpretation. However, even when pup exposure is confirmed, it can be challenging to determine the extent of gestational placental versus lactational transfer, the impact of the immature postnatal gastrointestinal system, and the translational relevance to humans.

The potential adverse effects of PZQ on pregnant and lactating females, as well as on conceptus and offspring development, from GD15 to wearing were assessed in Wistar rats. Groups of 25 presumed pregnant rats (F0 generation) received doses of 0 (control), 30, 100 and 300 mg/kg/day PZQ orally by gavage from GD15 to PND21 (see Fig. 8.11, lower panel). The F0 dams were allowed to deliver spontaneously and nurse their progeny up to PND28. Parameters investigated for dams included mortality, clinical findings, body weight, food consumption, number of pregnant animals, duration of pregnancy, and living and dead pups at birth. Maternal animals were sacrificed on PND28 and examined macroscopically and for the number of implantation sites.

The F1 progeny was examined for live and dead newborns, sex distribution, body weight, external abnormalities, viability and lactation, clinical findings, and motoric function. After weaning on PND28, male and female F1 animals of each litter were separated. Ten males and ten females were randomly selected from each group and reared until the age of 12 weeks. In weeks 15 and 16, one male and one female were paired. Males were sacrificed at the end of week 16 and underwent necropsy, while females were allowed to deliver spontaneously and nurse their F2 progeny up to PND28. Parameters investigated for F1 dams included mortality, clinical findings, body weight, food consumption, number of pregnant animals, duration of pregnancy, and living and dead pups at birth. Maternal animals were sacrificed on PND28 and examined macroscopically and for the number of implantation sites.

The F2 progeny was examined for live and dead newborns, sex distribution, body weight, external abnormalities, viability and lactation, and clinical findings. On PND28, F2 pups were sacrificed, underwent necropsy, and were examined for skeletal and visceral malformations.

None of the parameters investigated in the F0, F1, and F2 generations showed a treatment-related effect. The NOEL for pre- and postnatal development can be considered to be 300 mg/kg/day.

Juvenile Toxicity

Developing drugs for children requires addressing their unique physiological and developmental differences compared to adults [51]. Juvenile Animal Studies (JAS) bridge the gap between fetal exposure in utero or neonatal exposure through milk, and the onset of exposure in adult toxicity studies [52]. Regulatory guidelines such as the FDA Guidance for Industry on the Nonclinical Safety Evaluation of Pediatric Drug Products [53] provide a framework for evaluating pediatric medicines and emphasize the importance of JAS to assess safety and detect adverse effects on postnatal growth and development. Unlike routine toxicity evaluations, JAS are tailored on a case-by-case basis [54], focusing on the therapeutic target, toxicological profile, and developmental stage of the pediatric population. These studies help identify developmental toxicities or other adverse effects that cannot be adequately assessed in standard evaluations or safely measured in pediatric clinical trials, especially when previous nonclinical toxicity data are insufficient to support pediatric trials.

For PZQ, no dedicated JAS have been performed. However, the results obtained from toxicity studies with PZQ, including a PPND study in rats and a single-dose toxicity study in juvenile rats, showed no differences in sensitivity and target organ toxicity relevant to the pediatric population.

At the time PZQ was developed, there were generally fewer regulatory requirements for conducting dedicated juvenile toxicity studies compared to current standards. Toxicity testing often emphasized adult animals, and the specific requirements for juvenile studies were not as stringent or well-defined as they are today.

Summary

In conclusion, the nonclinical safety investigations of PZQ have demonstrated a notably favorable safety profile. PZQ had no acute pharmacological safety effects in animals, which would suggest a risk for human patients. Despite the absence of in vitro screening studies during its early development, recent investigations have confirmed the lack of significant off-target effects. A comprehensive battery of in vitro and in vivo genotoxicity assays showed that PZQ lacks mutagenic potential. Comprehensive studies across various species, including rodents (rats, hamsters, mice) and non-rodents (rabbits, dogs), consistently demonstrated in general toxicity and carcinogenicity studies, as well as DART studies (FEED, EFD, and PPND) that PZQ had no toxicologically relevant findings. The study designs included endpoints, group sizes, and durations that were already close to the approaches described in current guidelines. However, the pivotal nonclinical safety studies were not conducted according to GLP, which was not an established requirement at that time. The lack of global harmonization and clear guidelines, which were later established by entities such as the ICH, led to the testing of more species and animals than would be necessary today. Additionally, due to the lack of analytical methods, the TK of PZQ were not determined, and therefore the relationship between exposure and safety-related effects was not investigated. Nevertheless, more than 40 years of clinical experience from both the development phase and post-market approval confirm the safe use of PZQ, underscoring its reliability as a therapeutic agent against schistosomiasis (Chap. 10).

Acknowledgments The authors thank Laura Comotto for her valuable review of the "Developmental and Reproductive Toxicity" section.

Conflict of Interest All authors are employees of Merck Healthcare KGaA.

References

1. International Conference on Harmonisation of Technical Requirements for Registration of Pharmaceuticals for Human Use (2009) Guidance on nonclinical safety studies for the conduct of human clinical trials and marketing authorization for pharmaceuticals M3(R2). https://database.ich.org/sites/default/files/M3_R2__Guideline.pdf Accessed 27 Jan 2025
2. Pharma.Tips (2025) https://www.pharma.tips/the-history-of-good-laboratory-practices-from-concept-to-regulation/. Accessed 27 Jan 2025
3. AAΔE Independent Authority for Public Revenue. https://www.aade.gr/en/chemical-laboratories/accreditation-according-principles-good-laboratory-practice-glp. Accessed 27 Jan 2025
4. Pognan F, Beilmann M, Boonen HCM, Czich A, Dear G, Hewitt P, Mow T, Oinonen T, Roth A, Steger-Hartmann T, Valentin JP, Van Goethem F, Weaver RJ, Newham P (2023) The evolving role of investigative toxicology in the pharmaceutical industry. Nat Rev Drug Discov 22(4):317–335. https://doi.org/10.1038/s41573-022-00633-x
5. Green JR, Mahalingaiah PKS, Gopalakrishnan SM, Liguori MJ, Mittelstadt SW, Blomme EAG, Vleet TR (2023) Off-target pharmacological activity at various kinases: potential functional and pathological side effects. J Pharmacol Toxicol Methods 123:107468. https://doi.org/10.1016/j.vascn.2023.107468
6. International Conference on Harmonisation of Technical Requirements for Registration of Pharmaceuticals for Human Use (2011) Guidance on genotoxicity testing and data interpretation for pharmaceuticals intended for human use S2(R1). https://database.ich.org/sites/default/files/S2%28R1%29%20Guideline.pdf. Accessed 27 Jan 2025
7. International Conference on Harmonisation of Technical Requirements for Registration of Pharmaceuticals for Human Use (2022) ICH E14/S7B Implementation Working Group, Clinical and Nonclinical Evaluation of QT/QTc Interval Prolongation and Proarrhythmic. Potential, Questions and Answers E14/S7B Q&As. https://database.ich.org/sites/default/files/E14-S7B_QAs_Step4_2022_0221.pdf. Accessed 27 Jan 2025
8. International Conference on Harmonisation of Technical Requirements for Registration of Pharmaceuticals for Human Use (2013) Photosafety Evaluation of Pharmaceuticals S10. https://database.ich.org/sites/default/files/S10_Guideline.pdf. Accessed 27 Jan 2025
9. Sun Q, Mao R, Wang D, Changyan H, Zheng Y, Dequn S (2016) The cytotoxicity study of praziquantel enantiomers. Drug Des Devel Ther 10:2061–2068. https://doi.org/10.2147/DDDT.S98096
10. International Conference on Harmonisation of Technical Requirements for Registration of Pharmaceuticals for Human Use (2000) Safety pharmacology studies for human pharmaceuticals S7A. https://database.ich.org/sites/default/files/S7A_Guideline.pdf. Accessed 27 Jan 2025
11. Wakefield ID, Pollard C, Redfern WS, Hammond TG, Valentin JP (2002) The application of *in vitro* methods to safety pharmacology. Fundam Clin Pharmacol 16(3):209–218. https://doi.org/10.1046/j.1472-8206.2002.00099.x
12. Park SK, Sprague DJ, Rohr CM, Chulkov EG, Petrow I, Kumar S, Marchant JS (2024) The anthelmintic meclonazepam activates a schistosome transient receptor potential channel. J Biol Chem 300:105528. https://doi.org/10.1016/j.jbc.2023.105528
13. Bautista DM, Siemens J, Glazer JM, Tsuruda PR, Basbaum AI, Stucky CL, Jordt SE, Julius D (2007) The menthol receptor TRPM8 is the principal detector of environmental cold. Nature 448(7150):204–208. https://doi.org/10.1038/nature05910
14. International Conference on Harmonisation of Technical Requirements for Registration of Pharmaceuticals for Human Use (1998) Duration of chronic toxicity testing in animals (Rodent and non rodent toxicity testing) S4. https://database.ich.org/sites/default/files/S4_Guideline.pdf. Accessed 27 Jan 2025
15. Machemer L, Lorke D (1978) Mutagenicity studies with praziquantel, a new anthelmintic drug, in mammalian systems. Arch Toxicol 39(3):187–197. https://doi.org/10.1007/BF00368227

16. Frohberg H (1984) Results of toxicological studies on praziquantel. Arzneimittelforschung 34(9B):1137–1144
17. Obermeier J, Frohberg H (1977) Mutagenicity studies with praziquantel, a new anthelmintic drug: tissue-, host-, and urine-mediated mutagenicity assays. Arch Toxicol 38:149–161. https://doi.org/10.1007/BF00293649
18. Yi-chang N (1981) Mutagenicity and teratogenicity of pyquiton (praziquantel). Personal communication
19. Bartsch H, Kuroki T, Malaveille C, Loprieno N, Barale R, Abbondandolo A, Bonatti S, Rainaldi G, Vogel E, Davis A (1978) Absence of mutagenicity of praziquantel, a new, effective, anti-schistosomal drug, in bacteria, yeasts, insects and mammalian cells. Mutat Res 58(2–3):133–142. https://doi.org/10.1016/0165-1218(78)90002-2
20. Zhao S-Y, Li C-B, Ma Z-R, Ding F (1981) *In vitro* induction of SCE in human lymphocytes by several paracitisides with and without s-9. Acta Biol Exp Sin 14:123
21. Billings PC, Heidelberger C (1982) Effects of praziquantel, a new antischistosomal drug, on the mutation and transformation of mammalian cells. Cancer Res 42(7):2692–2696
22. Kramers PG, Gentile JM, Gryseels BJ, Jordan P, Katz N, Mott KE, Mulvihill JJ, Seed JL, Frohberg H (1991) International commission for protection against environmental mutagens and carcinogens. Icpemc publication no. 18. Review of the genotoxicity and carcinogenicity of antischistosomal drugs; is there a case for a study of mutation epidemiology? Report of a task group on mutagenic antischistosomals. Mutat Res 257(1):49–89. https://doi.org/10.1016/0165-1110(91)90019-r
23. European Medicines Agency (1996) Committee for veterinary medicinal products. Praziquantel. Summary report (1). https://www.ema.europa.eu/en/documents/mrl-report/praziquantel-summary-report-1-committee-veterinary-medicinal-products_en.pdf. Accessed 27 Jan 2025
24. International Conference on Harmonisation of Technical Requirements for Registration of Pharmaceuticals for Human Use (1995) Guideline on the need for carcinogenicity studies of pharmaceuticals S1A. https://database.ich.org/sites/default/files/S1A%20Guideline.pdf. Accessed 27 Jan 2025
25. International Conference on Harmonisation of Technical Requirements for Registration of Pharmaceuticals for Human Use (2022) Testing for carcinogenicity of pharmaceuticals S1B(R1). https://database.ich.org/sites/default/files/S1B-R1_FinalGuideline_2022_0719.pdf. Accessed 27 Jan 2025
26. Bulay O, Urman H, Patil K, Clayson DB, Shubik D (1979) Carcinogenic potential of hycanthone in mice and hamsters. Int J Cancer 23(1):97–104. https://doi.org/10.1002/ijc.2910230117
27. Siggia ED, Warmflash A (2018) Modeling mammalian gastrulation with embryonic stem cells. Curr Top Dev Biol 129:1–23. https://doi.org/10.1016/bs.ctdb.2018.03.001
28. Deltsidou A, Lykeridou K, Plessas ST (2000) Congenital anomalies and aetiology. Rev Clin Pharmacol Pharmacokinetics Int Ed 14(3):111–115
29. Lammens M, van Vugt JMG, Willemsen M, van der Voorn P, van Bokhoven H, ten Donkelaar HJ (2014) Clinical Neuroembryology: development and developmental disorders of the human central nervous system, 2nd edn. Springer, Berlin/Heidelberg, pp 105–164
30. Sachdeva P, Patel BG, Patel BK (2009) Drug use in pregnancy; a point to ponder! Indian J Pharm Sci 71(1):1–7. https://doi.org/10.4103/0250-474X.51941
31. Buhimschi CS, Weiner CP (2009) Medications in pregnancy and lactation: part 1. Teratology. Obstet Gynecol 113(1):166–188. https://doi.org/10.1097/AOG.0b013e31818d6788
32. Eke AC (2022) An update on the physiologic changes during pregnancy and their impact on drug pharmacokinetics and pharmacogenomics. J Basic Clin Physiol Pharmacol 33(5):581–598. https://doi.org/10.1515/jbcpp-2021-0312
33. International Conference on Harmonisation of Technical Requirements for Registration of Pharmaceuticals for Human Use (2020) Detection of reproductive and developmental toxicity for human pharmaceuticals S5(R3). https://database.ich.org/sites/default/files/S5-R3_Step4_Guideline_2020_0218_1.pdf. Accessed 27 Jan 2025

34. Vidal JD, Colman K, Bhaskaran M, de Rijk E, Fegley D, Halpern W, Jacob B, Kandori H, Manickam B, McKeag S, Parker GA, Regan KS, Sefing B, Thibodeau M, Vemireddi V, Werner J, Zalewska A (2021) Scientific and regulatory policy committee best practices: documentation of sexual maturity by microscopic evaluation in nonclinical safety studies. Toxicol Pathol 49(5):977–989. https://doi.org/10.1177/0192623321990631
35. Creasy DM, Chapin RE (2018) Male reproductive system. In: Fundamentals of toxicologic pathology, 3rd edn. Academic, London, pp 459–516
36. Takakura I, Creasy DM, Yokoi R, Terashima Y, Onozato T, Maruyama Y, Chino T, Tahara T, Tamura T, Kuroda J, Kusama H (2014) Effects of male sexual maturity of reproductive endpoints relevant to DART studies in Wistar Hannover rats. J Toxicol Sci 39(2):269–279. https://doi.org/10.2131/jts.39.269
37. Vargesson N (2015) Thalidomide-induced teratogenesis: history and mechanisms. Birth Defects Res C Embryo Today 105(2):140–156. https://doi.org/10.1002/bdrc.21096
38. Swaters D, van Veen A, van Meurs W, Turner JE, Ritskes-Hoitinga M (2022) A history of regulatory animal testing: what can we learn? Altern Lab Anim 50(5):322–329. https://doi.org/10.1177/02611929221118001
39. Ridings JE (2013) The thalidomide disaster, lessons from the past. Methods Mol Biol 947:575–586. https://doi.org/10.1007/978-1-62703-131-8_36
40. Andrews PA, Blanset D, Costa PL, Green M, Green ML, Jacobs A, Kadaba R, Lebron JA, Mattson B, McNerney ME, Minck D, de Castro Oliveira L, Theunissen RT, DeGeorge JJ (2019) Analysis of exposure margins in developmental toxicity studies for detection of human teratogens. Regul Toxicol Pharmacol 105:62–68. https://doi.org/10.1016/j.yrtph.2019.04.005
41. Andrews PA, McNerney ME, DeGeorge JJ (2019) Exposure assessments in reproductive and develop-mental toxicity testing: an IQ-DruSafe industry survey on current practices and experiences in support of exposure-based high dose selection. Regul Toxicol Pharmacol 107:104413. https://doi.org/10.1016/j.yrtph.2019.104413
42. Barrow P (2023) An assessment of the reliability of 52 enhanced preliminary embryofetal development studies to detect developmental toxicity. Birth Defects Res 115(2):218–223. https://doi.org/10.1002/bdr2.2108
43. Barrow P, Clemann N (2021) Review of embryo-fetal developmental toxicity studies performed for pharmaceuticals approved by FDA in 2018 and 2019. Reprod Toxicol 99:144–151. https://doi.org/10.1016/j.reprotox.2020.06.013
44. Barrow P (2018) Review of embryo-fetal developmental toxicity studies performed for pharmaceuticals approved by FDA in 2016 and 2017. Reprod Toxicol 80:117–125. https://doi.org/10.1016/j.reprotox.2018.04.008
45. Tyl RW (2012) Commentary on the role of maternal toxicity on developmental toxicity. Birth Defects Res B Dev Reprod Toxicol 95(3):262–266. https://doi.org/10.1002/bdrb.21015
46. Beyer BK, Chernoff N, Danielsson BR, Davis-Bruno K, Harrouk W, Hood RD, Janer G, Liminga UW, Kim JH, Rocca M, Rogers J, Scialli AR (2011) ILSI/HESI maternal toxicity workshop summary: maternal toxicity and its impact on study design and data interpretation. Birth Defects Res B Dev Reprod Toxicol 92(1):36–51. https://doi.org/10.1002/bdrb.20281
47. Makris SL, Solomon HM, Clark R, Shiota K, Barbellion S, Buschmann J, Ema M, Fujiwara M, Grote K, Hazelden KP, Wah Hew K, Horimoto M, Ooshima Y, Parkinson M, Wise LD (2009) Terminology of developmental abnormalities in common laboratory mammals (version 2). Reprod Toxicol 28(3):371–434. https://doi.org/10.1016/j.reprotox.2009.06.010
48. Barrow PC (ed) (2013) Teratogenicity testing, vol 947: methods in molecular biology. Humana Press, Springer/Heidelberg, p 599
49. International Conference on Harmonisation of Technical Requirements for Registration of Pharmaceuticals for Human Use (1994) Note for guidance on toxicokinetics: the assessment of systemic exposure in toxicity studies S3A. https://database.ich.org/sites/default/files/S3A_Guideline.pdf. Accessed 27 Jan 2025
50. Halpern W (2023) General design considerations in reproductive and developmental toxicity studies. In: Jagadeesh G, Balakumar P, Senatore F (eds) The quintessence of basic and clinical research and scientific publishing. Springer, Singapore

51. Kearns GL, Abdel-Rahman SM, Alander SW, Blowey DL, Leeder JS, Kauffman RE (2003) Developmental pharmacology-drug disposition, action, and therapy in infants and children. N Engl J Med 349(12):1157–1167. https://doi.org/10.1056/NEJMra035092
52. Rao DB, Hoberman AM, Brown PC, Varela A, Bolon B (2021) Regulatory perspectives on juvenile animal toxicologic pathology. Toxicol Pathol 49(8):1393–1404. https://doi.org/10.1177/01926233211046869
53. US Food and Drug Administration (2006) Guidance for Industry Nonclinical Safety Evaluation of Pediatric Drug Products. https://www.fda.gov/media/119658/download
54. Cappon GD, Bailey GP, Buschmann J, Feuston MH, Fisher JE, Hew KW, Hoberman AM, Ooshima Y, Stump DG, Hurtt ME (2009) Juvenile animal toxicity study designs to support pediatric drug development. Birth Defects Res B Dev Reprod Toxicol 86(6):463–469. https://doi.org/10.1002/bdrb.20220

Controlled Human *Schistosoma mansoni* Infection Model and Biomarkers

9

Marie-Astrid Hoogerwerf and Meta Roestenberg

Abstract

Controlled human infection models have been in use for decades, supporting vaccine development. For schistosomiasis, no licensed vaccine is available, despite urgent need. A controlled human infection model for schistosomiasis has been developed at the Leiden University Medical Center. This model uses a single-sex schistosome infection in order to prevent egg production and has now been tested with both male-only and female-only infections. Although symptoms of the acute schistosomiasis syndrome occur in a dose-dependent fashion, the model is generally well-tolerated. Circulating anodic antigen has been successfully used to detect the infection, resulting in an attack rate of nine out of eleven participants who received 20 male cercariae and six out of ten participants receiving 20 female cercariae. Participants were treated with praziquantel; however, treatment was not successful in four out of ten participants in the female infection group, although most self-cleared during follow-up. The male-only model is now sufficiently developed to serve as a challenge model for testing the efficacy of novel vaccines and drugs.

Keywords

Controlled human infection model · CAA · Acute schistosomiasis symptoms · Vaccine- and drug efficacy

---

M.-A. Hoogerwerf (✉) · M. Roestenberg
Leiden University Center for Infectious Diseases, Leiden University Medical Center, Leiden, The Netherlands
e-mail: m.hoogerwerf@lumc.nl

T. Spangenberg (ed.), *Praziquantel*,
https://doi.org/10.1007/978-3-031-97397-0_9

Controlled Human Infection Models

Controlled human infection models (CHIMs) have been in use since the eighteenth century, when Jenner's cowpox inoculation laid the foundation for the concept of vaccination (Chap. 1). Since then, the concept of experimentally infecting healthy individuals has been implemented in several disease areas including infections with malaria tropica, caused by the parasite *Plasmodium falciparum*, as a treatment for syphilis, and experimental hepatitis A infections in care facilities [1–3]. Despite their contentious history, human challenge studies have led to important discoveries in disease pathogenicity by confirming Koch's postulates (the criteria for establishing a causal relation between a pathogen and a disease) and uncovering important transmission routes of infectious agents. This was demonstrated as early as the late nineteenth century, when human experiments confirmed mosquitoes as the vector for yellow fever [4]. Controlled infection models have also provided valuable insights into the relation between blood group and pathogenicity of norovirus [5], immunological responses to malaria infection [2] and transmission of respiratory viruses [6].

Today, CHIMs are increasingly applied as an effective and quick method to assess vaccine and drug efficacy with the aim of reducing the overall costs associated with the development of new preventive and therapeutic agents. As new CHIMs are developed, the number of pathogens for which models are available is ever increasing [7]. The most often performed CHIMs include influenza and rhinovirus, but *Plasmodium falciparum* is the third most studied pathogen to date (Fig. 9.1) [7].

The value of CHIMs lies in the well-characterized, hence the reference to "controlled," circumstances under which infections are performed. There is precise knowledge of the timing of infection, route of infection, and infectious dose, allowing for a careful dissection of immune responses across a variety of different parameters including the heterogeneity of CHIM study subjects. Once carefully

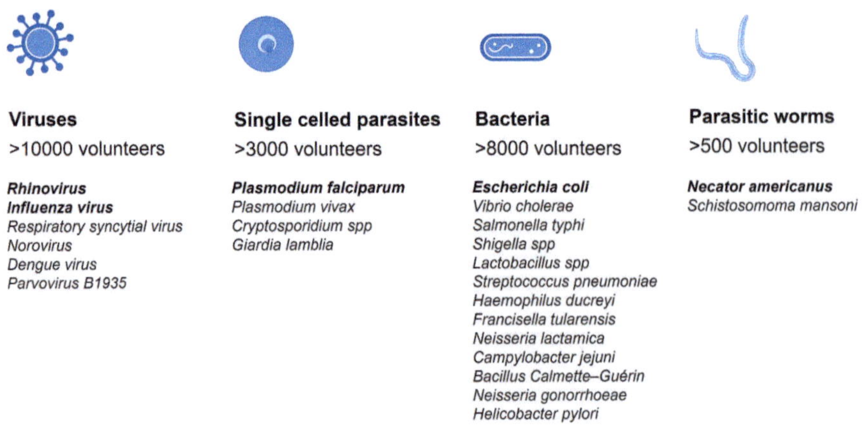

Fig. 9.1 Overview of pathogens used in CHIMs

established, the understanding of attack rates following a controlled exposure allows for the use of much smaller sample sizes to evaluate vaccine and drug efficacy compared with studies involving naturally exposed populations in real-world settings. In addition, through restrictive inclusion and exclusion criteria, the heterogeneity of the study population can be drastically reduced. This provides a rapid and cost-effective way to show an early proof-of-efficacy for therapeutics without having to invest in large-scale field studies. Once a compound demonstrates efficacy in a CHIM, it can justify further investment in larger-scale studies. Although CHIMs have considerable limitations—such as limited comparability with field infections and unproven predictive value—they tend to overestimate therapeutic effects due to their controlled conditions and use of healthy participants. For this reason, CHIMs are often seen as gatekeepers: products ineffective in a CHIM are generally unlikely to work in the field and should instead be rerouted for improved design or formulation, allowing to save resources for other investments.

In vaccine-CHIM studies, participants receive either the investigational vaccine or a placebo, after which they are experimentally infected with the pathogen of interest and rigorously monitored to ensure early detection of infection. A significant difference in attack rates between the immunized group versus the placebo group is a signal of vaccine efficacy. This approach has led to several seminal studies in vaccine development. Most notably, the initial proof-of-concept for the now-licensed malaria vaccine RTS,S was generated using a controlled human infection model in the 1990s [8]. The oral cholera vaccine [9] is another example where a CHIM study supported licensure for travelers without the need for field studies, as efficacy in the intended population was sufficiently demonstrated. Similarly, controlled infection studies with a *Salmonella typhi* model have generated important insights into which antibodies can be used as correlates of protection, that serotype influences the development of protective immunity and as proof-of-concept for vaccine efficacy studies [10].

Although CHIM studies have traditionally been most valuable for evaluating preventive strategies, they can be used to test the effect of therapeutics as well. In such a setup, participants are exposed to the pathogen of interest and subsequently treated with the investigational drug, a placebo or a comparator drug at a predetermined interval. Measuring clearance of infection, decrease of symptoms, or infection intensity can then inform on the effectiveness of the drug, depending on the pathogen and intended outcome of the treatment. For example, several antimalarial drugs have been evaluated in this way, with both positive and negative outcomes [2]. This is also true for several drugs against respiratory viruses [11, 12]. Similar to vaccines, CHIMs therapeutic studies allow for rapid and efficient demonstration of a treatment effect, allowing a firmer base for larger scale studies in patients. In addition, the prespecified interval between infection and treatment allows for a more precise delineation of when specific treatments could be effective. For example, the effectiveness of the antiviral drug oseltamivir against early influenza symptoms was demonstrated using a controlled infection model [13]. Similarly, the exact timing of treatment is of particular interest for novel therapeutics for schistosomiasis, as sensitivity of the schistosome to a compound can vary across its developmental stages, as seen with praziquantel (PZQ). Given the important consequence of reduced susceptibility to treatment regimens and mass drug administration (MDA) programs, CHIM studies, through their inherent synchronized infection nature, could be critical tools to investigate therapeutics active at a specific life cycle stage (Chap. 3).

Schistosomiasis Controlled Human Infection Model

Given the global burden of schistosomiasis, the pressing need for novel interventions and the limited budget available for research and development (R&D) in neglected diseases, the use of a CHIM model to prioritize candidate products in the pipeline is self-evident (Chap. 3). In the development of any CHIM, utmost care must be taken to ensure the safety of participants, researchers, and bystanders. To this end, CHIM should adhere to four main principles. First, controlled, experimental infections should not cause irreversible pathology in participants and should be either treatable or self-limiting. Second, the infection needs to be detectable both as the primary outcome of the study and to follow-up the effectiveness of treatment, ideally using a quantitative marker to allow measurement of infection intensity. Third, any infectious agent should be produced in a controlled manner, adhering wherever possible to Good Manufacturing Practice (GMP) guidelines. Finally, any bystander risks should be minimized.

Several years ago, a controlled human infection model was established for schistosomiasis at the Leiden University Medical Center (LUMC), the Netherlands. The primary goal of the model was to aid in the development of an effective antischistosomal vaccine and to better understand the complexity of host–pathogen interactions in schistosomiasis. The CHIM uses cercariae from *Schistosoma mansoni* (*S. mansoni*) to infect healthy, adult volunteers. All participants are screened prior to inclusion and are excluded if there is evidence of previous *Schistosoma* exposure or plans to travel to an endemic area during the study period.

S. mansoni, together with *S. haematobium*, are the species responsible for the majority of the global schistosomiasis disease burden. As such, *S. mansoni* is a main target for vaccine development. Its life cycle is somewhat easier to maintain in laboratory conditions compared to *S. haematobium*, making it more amenable for controlled production (Chap. 6). Both life cycles had been sustained in a laboratory setting at LUMC for several decades. The LUMC *S. mansoni* strain originates from Puerto Rico.

To establish the schistosomiasis CHIM, preventing egg production by paired male and female adult worms in infected participants was considered key to avoid pathology (Chap. 2). Schistosomal pathology is largely driven by granulomatous inflammation resulting from the deposition of eggs, which can lead to fibrosis and scarring when present in large quantities. Therefore, infections in the CHIM are performed using either male of female single-sex cercariae. By performing single-sex infections, the burden of infection and potential pathology is reduced, but clearly this also limits the use of the CHIM model for egg-related targets, e.g., anti-fecundity vaccines or therapeutics. Moreover, eggs are thought to be key manipulators of the human host immune system. These mechanisms consequently cannot be studied in the CHIM.

To ensure that cercariae are single-sex, a real-time polymerase chain reaction (PCR) to amplify parasite DNA was developed to test for the sex of cercariae as part of the quality control (QC) during cercarial production. The challenge agent, infectious cercariae, is produced in a contained environment following GMP guidelines, although not in a clean room. Consequently, they are formally non-GMP, as only processes performed within the clean room at the LUMC facility fall under its GMP license. Production starts with selecting *Biomphalaria glabrata* snails as intermediate hosts (Fig. 9.2). Miracidiae are obtained by light-induced hatching of eggs obtained from livers of infected hamsters. *B. glabrata* snails are then infected with a single miracidium and housed individually. After 5 weeks, snails shed cercariae and the sex of these cercariae is determined using a multiplex real-time PCR that targets the *Schistosoma*-specific Internal Transcribed Spacer 2 (ITS2) sequence and the *S. mansoni* W1-repeat. This W1-repeat is much lower in copy number in male worms compared to females of the Puerto Rico strain and can therefore be used to identify male from non-male sex. The ITS2 cycle threshold (Ct, cutoff >30) is used as reference. The sex of cercariae is determined by calculating the delta (ΔCt) between the ITS2 Ct and the WT1 cycle threshold. Male cercariae exhibit a ΔCt>10, whereas females show a ΔCt<4. After identifying the sex of the cercariae infecting the snail, the latter is then shed again to generate the clinical trial material. To further ensure safety of the challenge material, the water used to keep the snails is tested for bioburden, and snails are kept in water containing antibiotics before production of cercariae for human infection [14]. The challenge material is characterized as an auxiliary medicinal product under the EU regulation 536/2014. To produce cercariae, a product dossier is created detailing all production steps, quality controls, and materials used. All data on production are monitored and analyzed by a QC officer. The challenge agent is subsequently released for use by the qualified person (QP) [14]. The excipient of the challenge material is ordinary (mineral) water. Cercariae are individually picked from the water and checked for viability under microscopy. They are then transferred to a well-plate containing water, where individual doses are prepared. Doses are applied to the participant within 2 h of

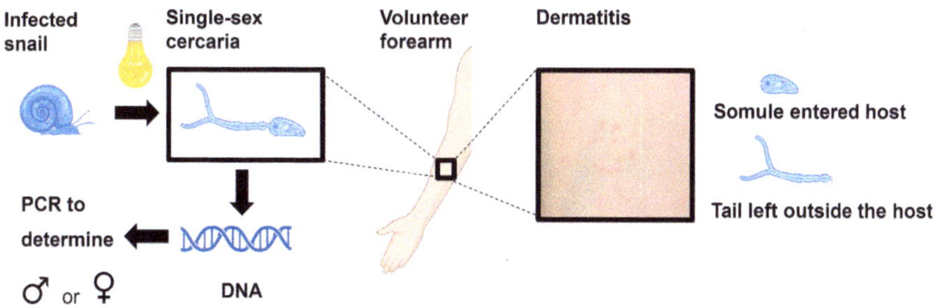

Fig. 9.2 Preparation of single-sex cercaria until skin penetration (dermatitis). (Made in part with BioRender®)

preparation as the viability of cercariae in freshwater is short (±80% remain viable up to 4 h). The infectious dose of water containing the cercariae is pipetted into an open-ended tube that is applied to the skin of the participant and left in place for 30 min. After this time, the water is pipetted off the skin and returned to the well. Contents of the well are checked for cercarial tails, indicating entry of the head into the skin, or remaining complete cercariae [14].

After infection, participants are monitored weekly or bi-weekly at the study center. At each visit, adverse events (AEs) are recorded, and at prespecified time points, blood is collected for safety lab, eosinophil (a variety of white blood cells) counts, and worm proximal biomarker measures, i.e., the circulating anodic antigen (CAA). Urine samples are taken to detect CAA, and fecal samples are analyzed by PCR to ensure no eggs are being produced.

Use of Circulating Anodic Antigen (CAA) as Biomarker

As egg production is avoided in the CHIM, egg excretion in feces cannot be used as a measure of infection, i.e., as a distal biomarker. As an alternative, the serum level of CAA is used as a proximal biomarker. CAA is produced in the gut of the schistosome, from where it is regurgitated into the bloodstream of the host and eventually excreted in urine. CAA's polysaccharide structure renders it very stable, and it has not been identified in organisms other than schistosomes (Fig. 9.3). CAA carries a strong negative charge and has a molecular weight of at least 10 kDa, although this can vary. The concentration of CAA in the bloodstream is believed to correlate with the infection burden of schistosomes in the host. As early as the 1960s, circulating antigens were detected in hamsters and mice infected with *S. mansoni* [15]. These antigens were later identified as the CAA and circulating cathodic antigen (CCA), and their levels were found to be highly sensitive and specific for diagnosing infection [16]. Currently, an upconverting phosphor lateral flow assay (UCP-LF) is used for the measurement of CAA in urine and serum, with a sensitivity as low as 1 pg/mL [17]. In the controlled schistosomiasis infection model, CAA levels were initially measured in both serum and urine. From the earliest infection studies, serum CAA was used as the primary endpoint due to its greater stability and sensitivity compared to the urine assay. A serum CAA level above 1 pg/mL indicates a detectable infection, while levels below 0.5 pg/mL are considered negative. Measurements between 0.5 and 1 pg/mL are classified as inconclusive and are retested over time. In the CHIM studies, infection levels are significantly lower than those seen in endemic areas, where CAA levels as high as 1000 pg/mL are commonly recorded. As a result, point-of-care (POC) diagnostic tools, such as the POC-CCA urine dipstick—which detects circulating cathodic antigen (CCA, an antigen similar to CAA)—lack the sensitivity needed to detect these low-level infections common in CHIM studies, despite being effective in high-endemic settings [14, 18].

The main advantage of using CAA as a measure of infection is its quantitative nature, which allows for detection of fluctuations in antigen levels over time. Although CAA production per worm has shown to be variable and dependent on its

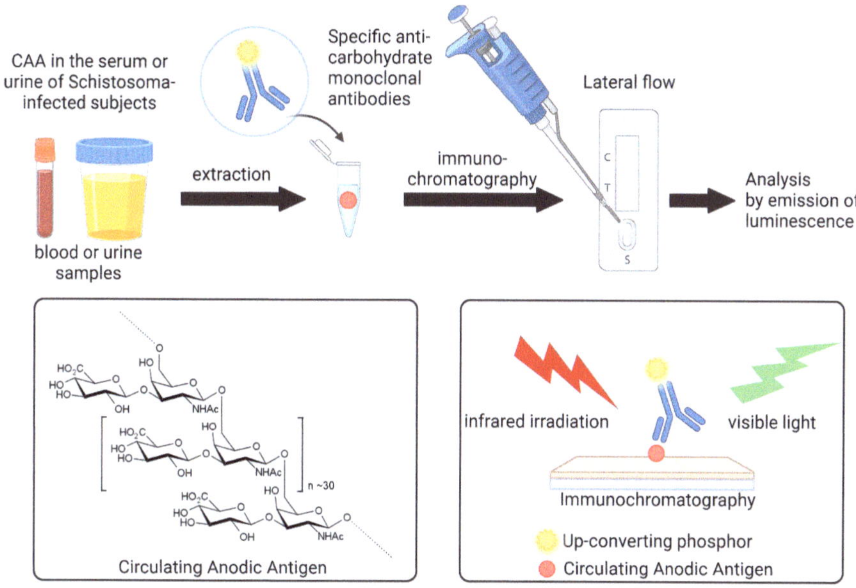

Fig. 9.3 Circulating anodic antigen schematic structure and UCP-CAA assay [19]. (Made in part with BioRender®)

metabolic rate and maturation stage, monitoring these levels over time provides a measure of worm fitness. Moreover, CAA can be used as a reliable biomarker of infection or cure upon treatment. This differs from the use of assays such as antibody detection, which generally stay positive for years after treatment.

Safety and Detectability

In the initial dose-finding CHIM, 10, 20, and 30 male cercariae were applied to the forearms of study participants. Sixteen out of 17 participants developed mild local cercarial dermatitis, which lasted longer in higher dose groups. Itching was present initially and characterized as mild. The dose of 10 cercariae resulted in detectable CAA levels in two out of three participants and was well tolerated. The dose was then escalated to 30 cercariae to improve detection. At this dose, all three participants had detectable CAA levels for 3 weeks after exposure. However, all three participants experienced severe adverse events (AE) between 2.5 and 5 weeks post-infection. Notably, one participant experienced severe fatigue and fevers that lasted for weeks—and preferred not to receive any immunosuppressive treatment. Overall, reported AEs during this timeframe were fever, fatigue, night sweats, and headaches, which responded to treatment with nonsteroid anti-inflammatory drugs such as ibuprofen, diclofenac, or acetaminophen. One participant developed periorbital edema around the left eye, characterized by swelling of the connective tissue, which was successfully treated with antihistamines. No other causative infection was detected and an increase in eosinophiles (eosinophilia) occurred in most participants 1–2 weeks after symptom onset. Consequently, these AEs were considered to be evidence of an acute schistosomiasis syndrome, which has also been referred to as Katayama fever. In contrast to textbook descriptions, the array of AEs was quite heterogeneous and predominantly mild to moderate in nature. Before these CHIM studies, the underlying mechanism of acute schistosomiasis was believed to stem from inflammatory responses to egg deposition. However, given the lack of egg production in the CHIM, these studies unequivocally demonstrated that acute schistosomiasis syndrome can occur independently of egg production [20]. Although symptoms were manageable with nonsteroid anti-inflammatory drugs, the high incidence of severe AEs caused the infectious dose to be lowered to 20 cercariae in the subsequent cohort. This resulted in detectable infection in 9 out of 11 participants (82%). Two out of 11 participants reported one severe AE each (fever, headache, sweats), whereas four participants experienced mild to moderate symptoms of acute schistosomiasis (i.e., flu-like symptoms or malaise). As the 20 cercariae dose gave sufficiently high attack rates with a better safety profile, this dose was taken forward for further studies [21].

Since animal studies have previously suggested that immune responses to male and female worms may differ, and antigenic targets for vaccines may also differ between males and females, a female-only model was subsequently developed. As there are rare reports of single female worms producing eggs in vitro, the infection time in the female CHIM was reduced to decrease chances of egg production. Similar to the male model, three participants were exposed to 10 cercariae, followed by 10 participants who were exposed to 20 cercariae. AEs profiles were comparable to the male model. However, of the participants exposed to 10 female cercariae, none had detectable CAA levels. Of those exposed to 20 cercariae, 6 out of 10 (60%) became positive 3 weeks post-infection. CAA levels reached in this group were similar to the male-only group. No eggs were detected in any of the fecal samples obtained from study participants. It had been hypothesized that unpaired female worms may remain immature, resulting in lower CAA production in vitro and potentially eliciting stronger regulatory immune responses in the human host. However, the female-only study showed no evidence of this hypothesis, as the AEs profiles and CAA outcomes were comparable [22].

Immunological Aspects

In both the male and female single-sex models, a subset of participants exhibited an increase in circulating eosinophils between weeks 2 and 8, followed by a subsequent decline (Fig. 9.4). However, this pattern was not uniform across all participants. In the male model, 11 out of 17 participants (65%) developed noticeable eosinophilia, while in the female model, this figure was 6 out of 11 participants (55%). Eosinophil counts were not related to cercarial dose, CAA levels, or

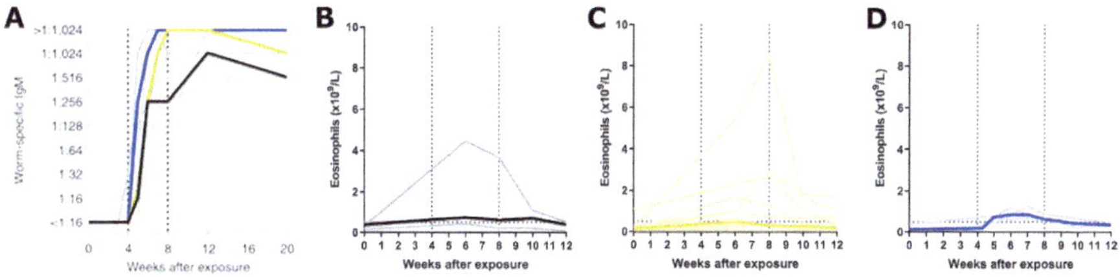

Fig. 9.4 Summary of key immunological features upon *Schistosoma* infection. (**a**) IgM against adult worm antigens over time. The gray, yellow, and blue lines represent data of individual volunteers exposed to 10 ($n = 3$), 20 ($n = 10$) or 30 ($n = 3$) cercariae, respectively, and the darker lines represent the median of each group. (**b–d**) Eosinophil counts ($\times 10^9$/L) per volunteer and the median per group. The thin green, red, and blue lines represent data of individual volunteers infected with 10 ($n = 3$), 20 ($n = 11$), or 30 ($n = 3$) cercariae, respectively, while thick lines represent the median of each group [21]

AEs. These findings may have important implications for the diagnosis of acute schistosomiasis syndrome in clinical practice, as eosinophilia was thought to be the hallmark of acute schistosomiasis [21, 22].

All participants exposed to either male or female cercariae developed specific antibodies in the blood serum for adult-worm Immunoglobulin M (IgM) (seroconversion). Levels were detected by immunofluorescence assay between 4 and 7 weeks after exposure. Furthermore, Immunoglobulin G (IgG) levels against adult-worm antigen were detectable from week 8, with seroconversion observed in all participants at week 16. This was predominantly an IgG1 response [21, 22]. As the most prevalent immunoglobulin subclass in humans, IgG1 plays a crucial role in promoting opsonization and complement activation in response to pathogen proteins, similar to IgG3. In contrast, IgG2 is most strongly linked to antibody responses targeting bacterial polysaccharide components. IgG4, on the other hand, is associated with allergic reactions and chronic parasitic infections, and it does not exhibit complement binding activity [23]. At follow-up, a subset of participants also showed detectable IgG levels against soluble egg antigen (SEA), ranging from 4 to 52 weeks post-infection. Although this could suggest that exposure to eggs triggered the immune response, PCR confirmed that no eggs were present in any of the participants. Consequently, the presence of IgG against SEA is likely attributable to cross-reactivity between the antigens of the eggs and those of the adult worm. Additionally, no changes in IgG4 or Immunoglobulin E (IgE) levels against crude adult-worm antigens were observed in either the male or female models.

Although the lack of egg development represents an important difference between CHIM and natural infections, it appeared that the responses to acute, early infection can be studied in detail in a CHIM setting. In-depth analysis of the cellular immunologic responses of the participants of both male and female CHIM showed that a mixed T-helper 1 (Th1) and T-helper 2 (Th2) immune response was associated with the reported AEs. Th1 responses are characterized by the production of IgG1, macrophage activation, and induction of cellular cytotoxicity, and are typically associated with pro-inflammatory responses. Th2 responses stimulate IgE and eosinophil synthesis and are associated with immunity against parasitic infections and the development of asthma and allergic diseases. In individuals with severe acute schistosomiasis syndromes, a significantly pronounced Th1 inflammatory response is observed at week 4 postinfection, coinciding with the peak of symptoms. By week 8 post-infection, the immunological profiles of participants with and without severe symptoms converged toward a more regulatory and Th2-driven response, resembling the responses observed in chronic, natural infections. However, extensive immunological assessments of both CHIM samples and natural infections are necessary to accurately delineate their similarities and differences [24].

The lack of differences between the male and female models, along with the early development of a Th2 response at just 8 weeks post-infection, contradicts previously established dogmas. These earlier beliefs suggested that the Th2 response found in natural infections was primarily derived from eggs, and that male-only infections were more pro-inflammatory than female-only infections [25, 26].

CHIM and Praziquantel

The development of PZQ as a therapeutic drug for schistosomiasis long predates the use of CHIM, and therefore CHIMs were not part of its clinical development. However, the availability of a known and effective therapeutic has aided the establishment of the CHIM for schistosomiasis, whereby PZQ was used as rescue medication at the end of the experiment. Moreover, the longitudinal close follow-up of study participants after clearance of the infection has revealed interesting insights into PZQ effects. At a prespecified timepoint post-infection, all participants in the controlled schistosomiasis infection studies were treated with PZQ to clear infection with adult worms. In the male model, initial treatment consisted of single oral dose of 40 mg/kg at week 12 post-infection, which is the World Health Organization (WHO) recommended dose. Doses were split in two, according to guidelines, to be taken on the same day, and consumed with a fatty meal. Eight out of 14 positive participants had CAA levels dropping below detection threshold after the first treatment but the six remaining participants showed persistent infection. In these participants CAA levels became undetectable after a second treatment with a single dose of 40 mg/kg PZQ [21]. Interestingly, despite decades of research and an excellent Cochrane review summarizing and interpreting the results of medical research on PZQ efficacy [27], the latter remains much debated as data from endemic areas are always clouded by asynchronous infections. The WHO-recommended regimen already incorporates a re-treatment after several weeks to target previously immature worms. However, initial data from CHIM studies suggest that PZQ efficacy is much lower than previously estimated based on data from endemic settings, indicating that re-treatment is necessary regardless of infection synchronicity.

In the female infection model, participants were treated with a 60 mg/kg single oral dose of PZQ at weeks 8 and 12 post-infection. The dose was higher than in the male model, as treatment with a 40 mg/kg single dose was not efficacious in almost half of participants, and animal data suggests that the effective concentration for treating single-sex female infections is higher than that required for mixed or male-only infections (Chap. 6). An additional dose was administered at week 12 to target immature worms that may not have been effectively treated at week 8 [28]. Despite this aggressive treatment regimen, 4 out of 10 participants still had detectable CAA levels after PZQ administration. One participant self-cured before receiving the additional dose. Even a repeated PZQ treatment at 60 mg/kg for 3 consecutive days failed to achieve a cure, as did a full course of Coartem® (Artemether-Lumefantrine). However, two participants self-cured during the follow-up period, while one remained CAA-positive (Fig. 9.5).

The difference in susceptibility to treatment between male and female worms is poorly understood. It has been hypothesized that female worms remain immature in the absence of males, rendering them less susceptible to PZQ. Interestingly, the CAA results and the AEs reported by participants do not align with compromised development of female worms, as the AEs were similar, CAA levels were comparable and immune responses were alike. Additionally, although artemisinin is known to be more effective against immature stages, it did not result in a cure in this particular case [29]. An alternative hypothesis could be that immature female worms may linger in the liver venules rather than the mesenteric vessels, resulting in exposure to lower drug concentrations due to the first-pass metabolism.

The observed reduced susceptibility of female worms to PZQ may have implication for deworming efforts in the field. While it is unclear how many female worms remain single during natural infections, it is reasonable to assume that such situations may occur. Additionally, it is unclear whether the reduced susceptibility phenotype is unique to single-sex female worms, or whether it applies to female worms in general. If the latter is true, this may explain the rapid resurgence of patent egg production after PZQ treatment, through pairing with maturing male worms. In animal models, female worms have shown greater resistance to PZQ treatment through repulsion and regeneration of their surface when coated with complement, faster regeneration after PZQ administration, and upregulation of gene expression associated with tegument maintenance after incubation with serum (Chap. 6) [30].

Because one of the key criteria for an acceptable CHIM is the availability of either effective treatment or self-limiting infections, the reduced susceptibility to PZQ compromises the use of the female model as a CHIM. Although participants can clearly self-cure, this process may take an extended period of time. Therefore, the development of novel compounds that are broadly active and effectively kill female worms may facilitate further advancement of the female CHIM model. For now, only the male CHIM model will continue to be used for vaccine and therapeutic drug testing.

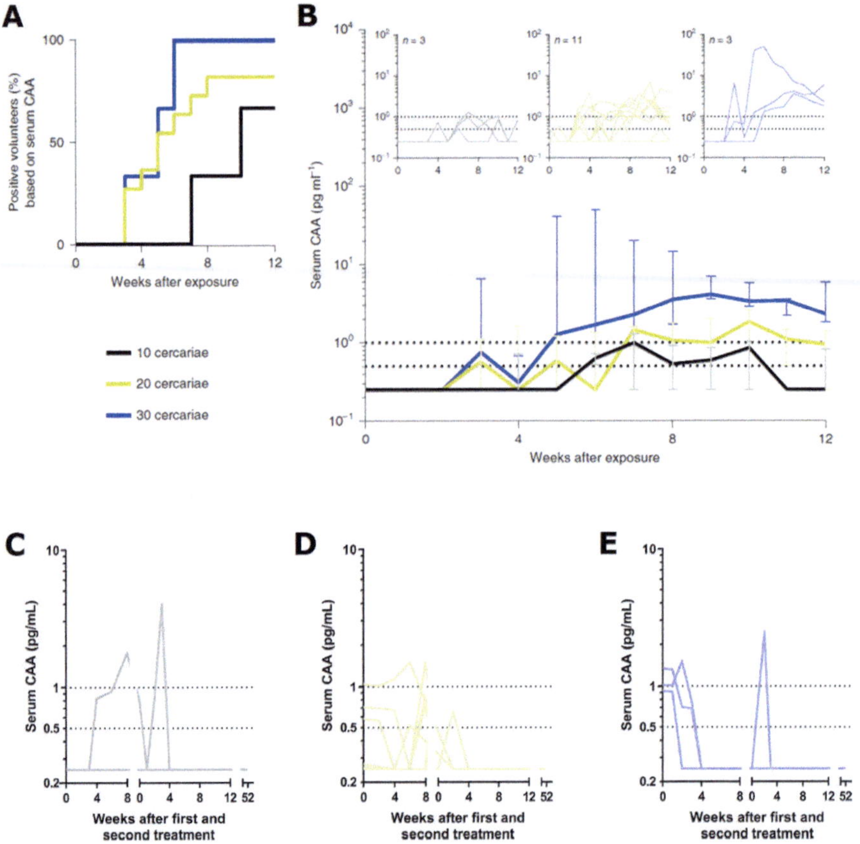

Fig. 9.5 CAA levels in single-sex male-infected participants before and after PZQ treatment. Prepatent period and serum CAA levels. (**a**) The time to patency (serum CAA level > 1 pg/mL) in weeks per dose group (log-rank test, $P = 0.21$). (**b**) The serum CAA levels for individual volunteers in pg/mL (top) and the median serum CAA levels and interquartile range for groups infected with 10, 20, or 30 cercariae (bottom). The gray, yellow, and blue lines represent data of individual volunteers or groups exposed to 10 ($n = 3$), 20 ($n = 11$), or 30 ($n = 3$) cercariae, respectively. Serum CAA levels after treatment. (**c–e**) Serum CAA levels in pg/mL after the first treatment, second treatment, and at week 52. All values below the detection threshold of 0.5 pg/mL are plotted at 0.25 pg/mL. The gray, yellow, and blue lines represent data of individual volunteers infected with 10 ($n = 3$), 20 ($n = 11$), or 30 ($n = 3$) cercariae, respectively [21]

Stage-Specific Anthelminthic Treatment

The reduced efficacy of PZQ against immature worms leaves a critical gap in the treatment of schistosomiasis. The development of novel compounds thus focusses on identifying therapeutics that act also against juvenile schistosomes to aid eradication efforts. Because of the synchronized nature of CHIM for schistosomes, this model is particularly relevant for testing the stage specificity of novel compounds.

By dosing at specific intervals post-infection, the activity of compounds can be tested across an array of maturation stages. In addition, the availability of both male and female infection models would offer the opportunity to assess the selectivity of novel compounds against one sex or the other, provided that the issue of rescue treatment is adequately addressed.

Further Applications of the *Schistosoma* Model

Now that the controlled human *Schistosoma* infection model has been established, researchers can now utilize it to evaluate the efficacy of vaccines and drugs. Additionally, the interesting finding of acute schistosomiasis syndrome in a subset of study participants enables a more thorough examination of immune responses—specifically, assessing whether these responses are protective and identifying which host factors may influence susceptibility to these symptoms. A new study has been initiated to explore the effects of repeated short-term exposure to *Schistosoma* infection, abrogated by treatment at week 8 after each exposure [31]. This chemo-attenuation approach has already proven successful in establishing protective immunity against malaria challenges [32], and more recently, against hookworm [33]. By investigating the effects of short-term exposure, specific immune responses to the early stages of infection can be examined, which can help identify novel antigenic targets for vaccine development. Furthermore, studying early-stage infections may enhance our understanding of protective responses and tolerance to schistosomes. Implementing the schistosome CHIM in endemic areas is particularly important, as all studies conducted to date have involved Schistosoma-naïve, healthy European participants, who do not accurately represent the populations for which vaccines and drugs are primarily intended (Chap. 3). Research on other pathogens, such as *Plasmodium falciparum*, has demonstrated that efficacy of vaccines is generally lower in preexposed individuals than in naïve participants [34, 35]. Establishing the schistosome CHIM in areas with high levels of preexposure is therefore highly relevant, as it allows for the exploration of attack rates and immune responses. Efforts are currently underway to initiate a study on *S. mansoni* in Uganda [36], which will investigate immune responses, AEs, and worm development as measured by CAA in individuals who are preexposed. It is anticipated that preexposure may reduce the likelihood of acute schistosomiasis syndromes, potentially enhancing the tolerability of the schistosome CHIM. However, in case of preexisting immunity, higher cercarial doses may be necessary to induce an attack rate comparable to that of unexposed populations. Additional complexities arise from the need to manage the release of the cercarial challenge agent in Uganda, as the Puerto Rican strain of *S. mansoni* is not endemic in sub-Saharan Africa. As a precaution, participants in the schistosome CHIM will be instructed to avoid freshwater sources to prevent natural infections. Additionally, using potentially non-endemic *B. glabrata* snails to culture these cercariae poses a biosafety risk. Nevertheless, because the novel treatments and vaccines will primarily target preexposed populations, establishing a Ugandan schistosome CHIM is likely to significantly enhance our understanding of schistosomiasis in endemic populations, including their host responses, and provide an excellent platform for vaccine and drug testing.

Summary

With the development of a CHIM for schistosomiasis, a valuable new tool for drug- and vaccine testing has become available and may provide an early proof of efficacy before engaging substantial development costs. This model has been enabled by advances in PCR techniques for determining cercarial sex, the use of CAA as method of detection, and the availability of PZQ as a rescue medication. Although the schistosome CHIM is constrained by its single-sex nature, and the limited sensitivity of female worms to PZQ treatment, this model is an important step forward with many potential novel applications in drug, vaccine, and antigen discovery.

Competing Interests M.R. has received funding for the conduct of the female cercariae infection study from the European Union's Horizon 2020 research and innovation program under grant agreement No. 815643. Views and opinions expressed are, however, those of the authors only and do not necessarily reflect those of the European Union or the Horizon 2020 research and innovation program. Neither the European Union nor the granting authority can be held responsible for them.

References

1. Nicol WD (1946) Treatment of neurosyphilis; a comparison between malaria plus tryparsamide and malaria therapy. Br J Vener Dis 22(3):112–121. https://doi.org/10.1136/sti.22.3.112
2. Stanisic DI, McCarthy JS, Good MF (2017) Controlled human malaria infection: applications, advances, and challenges. Infect Immun. https://doi.org/10.1128/iai.00479-17
3. Krugman S (1986) The Willowbrook hepatitis studies revisited: ethical aspects. Rev Infect Dis 8(1):157–162. https://doi.org/10.1093/clinids/8.1.157
4. Reed W, Carroll J, Agramonte A, Lazear JW (1900) The etiology of yellow fever - a preliminary note. Public health Pap Rep 26:37–53
5. Lindesmith L, Moe C, Marionneau S, Ruvoen N, Jiang X, Lindblad L, Stewart P, LePendu J, Baric R (2003) Human susceptibility and resistance to Norwalk virus infection. Nat Med 9(5):548–553. https://doi.org/10.1038/nm860
6. Dayananda P, Chiu C, Openshaw P (2022) Controlled human infection challenge studies with RSV. In: Current topics in microbiology and immunology. Springer, Berlin, Heidelberg. https://doi.org/10.1007/82_2022_257
7. Roestenberg M, Hoogerwerf M-A, Ferreire DM, Mordmüller B, Yazdanbakhsh M (2018) Experimental infection of human volunteers. Lancet Infect Dis 18(10):312–322. https://doi.org/10.1016/S1473-3099(18)30177-4
8. Gordon DM, McGovern TW, Krzych U, Cohen JC, Schneider I, LaChance R, Heppner DG, Yaun G, Hollingdale M, Slaoui M, Hauser P, Voet P, Sadoff JC, Ballou WR (1995) Safety, immunogenicity, and efficacy of a recombinantly produced Plasmodium falciparum circumsporozoite protein-hepatitis B surface antigen subunit vaccine. J Infect Dis 171(6):1576–1585. https://doi.org/10.1093/infdis/171.6.1576

9. Chen WH, Cohen MB, Kirkpatrick BD, Brady RC, Galloway D, Gurwith M, Hall RH, Kessler RA, Lock M, Haney D, Lyon CE, Pasetti MF, Simon JK, Szabo F, Tennant S, Levine MM (2016) Single-dose live oral cholera vaccine CVD 103-HgR protects against human experimental infection with vibrio cholerae O1 El tor. Clin Infect Dis 62(11):1329–1335. https://doi.org/10.1093/cid/ciw145
10. Sztein MB, Booth JS (2022) Controlled human infectious models, a path forward in uncovering immunological correlates of protection: lessons from enteric fevers studies. Front Microbiol 13. https://doi.org/10.3389/fmicb.2022.983403
11. Balasingam S, Wilder-Smith A (2016) Randomized controlled trials for influenza drugs and vaccines: a review of controlled human infection studies. Int J Infect Dis 49:18–29. https://doi.org/10.1016/j.ijid.2016.05.013
12. Lambkin-Williams R, Noulin N, Mann A, Catchpole A, Gilbert AS (2018) The human viral challenge model: accelerating the evaluation of respiratory antivirals, vaccines and novel diagnostics. Respir Res 19:123. https://doi.org/10.1186/s12931-018-0784-1
13. Sherman AC, Mehta A, Dickert NW, Anderson EJ, Rouphael N (2019) The future of flu: a review of the human challenge model and systems biology for advancement of influenza vaccinology. Front Cell Infect Microbiol 9. https://doi.org/10.3389/fcimb.2019.00107
14. Janse JJ, Langenberg MCC, Oosterhoud JK-V, Ozir-Fazalalikhan A, Brienen EAT, Winkel BMF, Erkens MAA, van der Beek MT, van Lieshout L, Smits HH, Webster BL, Zandvliet ML, Verbeek R, Westra IM, Meij P, Visser LG, van Diepen A, Hokke CH, Yazdanbakhsh M, Roestenberg M (2018) Establishing the production of male *Schistosoma mansoni* cercariae for a controlled human infection model. J Infect Dis 218(7):1142–1146. https://doi.org/10.1093/infdis/jiy275
15. Deelder AM, Kornelis D, van Marck EA, Eveleigh PC, van Egmond JG (1980) *Schistosoma mansoni*: characterization of two circulating polysaccharide antigens and the immunological response to these antigens in mouse, hamster, and human infections. Exp Parasitol 50(1):16–32. https://doi.org/10.1016/0014-4894(80)90004-1
16. van Lieshout L, Polderman AM, de Vlas SJ, de Caluwé P, Krijger FW, Gryseels B, Deelder AM (1995) Analysis of worm burden variation in human *Schistosoma mansoni* infections by determination of serum levels of circulating anodic antigen and circulating cathodic antigen. J Infect Dis 172(5):1336–1342. https://doi.org/10.1093/infdis/172.5.1336
17. Corstjens PLAM, DeDood CJ, Kornelis D, Tjon Kon Fat EM, Alan Wilson R, Kariuki TM, Nyakundi RK, Loverde PT, Abrams WR, Tanke HJ, van Lieshout L, Deelder AM, van Dam GJ (2014) Tools for diagnosis, monitoring and screening of *Schistosoma* infections utilizing lateral-flow based assays and upconverting phosphor labels. Parasitology 141(14):1841–1855. https://doi.org/10.1017/S0031182014000626
18. Hoekstra PT, Madinga J, Lutumba P, van Grootveld R, Brienen EAT, Corstjens PLAM, van Dam GJ, Polman K, van Lieshout L (2022) Diagnosis of schistosomiasis without a microscope: evaluating Circulating Antigen (CCA, CAA) and DNA detection methods on banked samples of a community-based survey from DR Congo. Trop Med Infect Dis 7. https://doi.org/10.3390/tropicalmed7100315
19. Vermeer HJ, van Dam GJ, Halkes KM, Kamerling JP, Vliegenthart JF, Hokke CH, Deelder AM (2003) Immunodiagnostically applicable monoclonal antibodies to the circulating anodic antigen of *Schistosoma mansoni* bind to small, defined oligosaccharide epitopes. Parasitol Res 90(4):330–336. https://doi.org/10.1007/s00436-003-0860-3
20. Langenberg MCC, Hoogerwerf M-A, Janse JJ, van Lieshout L, Corstjens PLAM, Roestenberg M (2119) Katayama syndrome without *Schistosoma mansoni* eggs. Ann Intern Med 170(10):732–733. https://doi.org/10.7326/L18-0438
21. Langenberg MCC, Hoogerwerf M-A, Koopman JPR, Janse JJ, Kos-van Oosterhoud J, Feijt C, Jochems SP, de Dood CJ, van Schijlenburg R, Ozir-Fazalalikhan A, Manurung MD, Sartono E, van der Beek MT, Winkel BMF, Verbeek-Menken PH, Stam KA, van Leeuwen FWB, Meij P, van Diepen A, van Lieshout L, van Dam GJ, Corstjens PLAM, Hokke CH, Yazdanbakhsh M, Visser LG, Roestenberg M (2020) A controlled human *Schistosoma mansoni* infection model to advance novel drugs, vaccines and diagnostics. Nat Med 26(3):326–332. https://doi.org/10.1038/s41591-020-0759-x

22. Koopman JPR, Houlder EL, Janse JJ, Casacuberta-Partal M, Lamers OAC, Sijtsma JC, de Dood C, Hilt ST, Ozir-Fazalalikhan A, Kuiper VP, Roozen GVT, de Bes-Roeleveld LM, Kruize YCM, Wammes LJ, Smits HH, van Lieshout L, van Dam GJ, van Amerongen-Westra IM, Meij P, Corstjens PLAM, Jochems SP, van Diepen A, Yazdanbakhsh M, Hokke CH, Roestenberg M (2023) Safety and infectivity of female cercariae in *Schistosoma*-naïve, healthy participants: a controlled human *Schistosoma mansoni* infection study. Lancet 97:104832. https://doi.org/10.1016/j.ebiom.2023.104832
23. Vidarsson G, Dekkers G, Rispens T (2014) IgG subclasses and allotypes: from structure to effector functions. Front Immunol 5. https://doi.org/10.3389/fimmu.2014.00520
24. El H, Stam KA, Koopman JPR, König MH, Langenberg MCC, Hoogerwerf M-A, Niewold P, Sonnet F, Janse JJ, Casacuberta Partal M, Sijtsma JC, de Bes-Roeleveld LHM, Kruize YCM, Yazdanbakhsh M, Roestenberg M (2024) Early symptom-associated inflammatory responses shift to type 2 responses in controlled human schistosome infection. Sci Immunol 9:eadl1965. https://doi.org/10.1126/sciimmunol.adl1965
25. Boissier J, Chlichlia K, Digon Y, Ruppel A, Moné H (2003) Preliminary study on sex-related inflammatory reactions in mice infected with *Schistosoma mansoni*. Parasitol Res 91(2):144–150. https://doi.org/10.1007/s00436-003-0943-1
26. Sombetzki M, Koslowski N, Rabes A, Seneberg S, Winkelmann F, Fritzsche C, Loebermann M, Reisinger EC (2018) Host defense versus immunosuppression: unisexual infection with male or female *Schistosoma mansoni* differentially impacts the immune response against invading cercariae. Front Immunol 9. https://doi.org/10.3389/fimmu.2018.00861
27. Danso-Appiah A, Olliaro PL, Donegan S, Sinclair D, Utzinger J (2013) Drugs for treating *Schistosoma mansoni* infection. Cochrane Database Syst Rev 2019:CD000528. https://doi.org/10.1002/14651858.CD000528.pub2
28. Pica-Mattoccia L, Cioli D (2004) Sex- and stage-related sensitivity of *Schistosoma mansoni* to in vivo and in vitro praziquantel treatment. Int J Parasitol 34(4):527–533. https://doi.org/10.1016/j.ijpara.2003.12.003
29. Bergquist R, Elmorshedy H (2018) Artemether and praziquantel: origin, mode of action, impact, and suggested application for effective control of human schistosomiasis. Trop Med Infect Dis 3. https://doi.org/10.3390/tropicalmed3040125
30. Winkelmann F, Frank M, Rabes A, Koslowski N, Schulz C, Bischofsberger M, Reisinger EC, Sombetzki M (2021) Human serum activates the tegument of female schistosomes and supports recovery from praziquantel. Parasitol Res 120(1):209–221. https://doi.org/10.1007/s00436-020-06968-x
31. Roestenberg M. Repeated controlled human *Schistosoma mansoni* infection. ClinicalTrials.gov identifier: NCT05085470. Updated January 1, 2023. https://clinicaltrials.gov/study/NCT05085470. Accessed 23 Oct 2024.
32. Roestenberg M, McCall M, Hopman J, Wiersma J, Luty AJ, van Gemert GJ, van de Vegte-Bolmer M, van Schaijk B, Teelen K, Arens T, Spaarman L, de Mast Q, Roeffen W, Snounou G, Rénia L, van der Ven A, Hermsen CC, Sauerwein R (2009) Protection against a malaria challenge by sporozoite inoculation. N Engl J Med 361(5):468–477. https://doi.org/10.1056/nejmoa0805832
33. Hoogerwerf M-A, Janse JJ, Kuiper VP, van Schuijlenburg R, Kruize YCM, Sijtsma JC, Nosoh BA, Koopman J-PR, Verbeek-Menken PH, Westra IM, Meij P, at Brienen E, Visser LG, van Leishout L, Jochems SP, Yazdanbakhsh M, Roestenberg M (2023) Protective efficacy of short-term infection with Necator americanus hookworm larvae in healthy volunteers in the Netherlands: a single-centre, placebo-controlled, randomised, controlled, phase 1 trial. https://doi.org/10.1016/S2666-5247(23)00218-5
34. de Jong SE, van Unen V, Manurung MD, Stam KA, Goeman JJ, Jochems SP, Höllt T, Pezzotti N, Mouwenda YD, Eunice Betouke Ongwe M, Lorenz F-R, Kruize YCM, Azimi S, König MH, Vilanova A, Eisemann E, Lelieveldt BPF, Roestenberg M, Lee Sim BK, Reinders MJT, Fendel R, Hoffman SL, Kremsner PG, Koning F, Mordmüller B, Lell B, Yazdanbakhsh M (2021) Systems analysis and controlled malaria infection in Europeans and Africans elucidate naturally acquired immunity. Nat Immunol 22(5):653–665. https://doi.org/10.1038/s41590-021-00911-7

35. van Dorst MMAR, Pyuza JJ, Nkurunungi G, Kullaya VL, Smits HH, Hogendoorn PCW, Wammes LJ, Everts B, Elliott AM, Jochems SP, Yazdanbakhsh M (2024) Immunological factors linked to geographical variation in vaccine responses. Nat Rev Immunol 24(4):250–262. https://doi.org/10.1038/s41577-023-00941-2
36. Driciru E, Koopman JPR, Cose S, Siddiqui AA, Yazdanbakhsh M, Elliott AM, Roestenberg M (2021) Immunological considerations for *Schistosoma* vaccine development: transitioning to endemic settings. Front Immunol 12. https://doi.org/10.3389/fimmu.2021.635985

Clinical Efficacy of Praziquantel on *Schistosoma* Infections

10

Wilhelmina M. Bagchus

Abstract

This chapter summarizes the clinical development of praziquantel (PZQ) and its active enantiomer (*R*)-PZQ focusing on the clinical studies that were performed to support their registration for the treatment and/or prophylaxis of infections due to different *Schistosoma* species. It begins with a general description of the three different clinical trial phases (I, II, and III) is given. Thereafter, the phase I and the phase II/III trials done for registering PZQ for patients suffering from infections with *Schistosoma mansoni*, *S. haematobium*, and *S. japonicum* in the 1980s are summarized. This is followed by a summary of the more recent trials performed for registering the child-friendly (*R*)-PZQ dispersible tablet formulation consisting of a phase I trial in healthy volunteers followed by a phase II and a phase III trials in infected children from 3 months to 6 years of age.

Keywords

Praziquantel · (*R*)-PZQ (Arpraziquantel) · Clinical pharmacology · Clinical development · Phase I/II/III · Safety · Efficacy · Dose-finding

Clinical Trial Phases

It takes a lot of research and investment to develop a new drug treatment (Chap. 1). Clinical trials in humans and subsequent registration (Chap. 11) are the last steps of the development path. Clinical trials are conducted in sequential steps, also called "phases" with each phase serving a unique purpose. These phases are designed to answer certain questions about the experimental drug while aiming to keep the

W. M. Bagchus (✉)
Retired from the Merck Institute for Pharmacometrics, Ares Trading S.A., (an affiliate of Merck KGaA, Darmstadt, Germany), Lausanne, Switzerland

T. Spangenberg (ed.), *Praziquantel*,
https://doi.org/10.1007/978-3-031-97397-0_10

participants in each trial as safe as possible and the number of participants as low as possible. Below is a general summary of the different clinical phases, followed by a description of the clinical development of praziquantel (PZQ) and of its most active enantiomer (Chap. 4), (*R*)-praziquantel or (*R*)-PZQ, also known by its international nonproprietary name Arpraziquantel (ArPZQ), which is compared with a standard development pathway.

General and Preclinical

Researchers investigate experimental drugs in a laboratory setting, generating in vitro data in cell lines and/or in vivo data in animals, before studying them in humans. These are so-called "preclinical" or "nonclinical" studies, aimed at gathering essential safety and pharmacological data on the experimental drugs. Preclinical studies are required by regulatory authorities such as the United States Food and Drug Administration (FDA) and the European Medicines Agency (EMA) before the experimental drug can be evaluated in humans. Following this phase, researchers must seek permission from government bodies and independent groups, such as Ethics Committees, to start clinical trials in humans. Clinical trials are often registered on the ClinicalTrials.gov database [1], especially when investigators plan to publish their findings. It is a US government website maintained by the National Library of Medicine (NLM), part of the National Institutes of Health (NIH). The NIH does not review or approve clinical trials but instead publishes key elements of the trial protocol and outcome. This includes the study's purpose, design, intended population, main eligibility criteria, medications used, status (e.g., recruiting), main outcome measurements, and details on where (country, clinic, contact information) and by whom (principal investigator, sponsor) the trial is conducted. Upon completion of the trial, results are also summarized and published, with all data provided by the study's sponsor (Chaps. 6, 7, and 8).

Phase I

Phase I clinical trials represent the first step after preclinical studies and typically mark the first time a new experimental drug, referred to as an investigational medicinal product (IMP), is evaluated in humans. These trials are commonly known as First-in-Human (FIH) studies. A phase I clinical trial evaluates the drug in a small group of participants, typically involving from 15 to 100 subjects, administering different doses in a sequential dose-escalating design and (if possible) enrolling healthy volunteers. The subjects participating in clinical studies must sign an Informed Consent Form (ICF) summarizing the details of the study and of the IMP. In addition, the subjects need to fulfill a set of trial-specific inclusion criteria and must not meet any predefined exclusion criteria. To correctly assess the tolerability and safety of the IMP, subjects receiving placebo are included at each dose level in a double-blind setting (i.e., neither subject nor investigator is aware of which subject is receiving IMP or placebo). Researchers conduct phase I clinical trials to:

- Evaluate the tolerability and safety of the IMP at different doses, including single and, if necessary, multiple doses.
- Investigate the human pharmacology of the IMP, if possible, following different dosing regimens.
- Assess how the effects observed in nonclinical studies (toxicological and pharmacological) translate into humans (Chap. 8).
- Analyze the relationship between the dose of IMP given and the concentration observed in blood as a proxy for efficacy and tolerability (Chap. 7).

Currently, incorporating pharmacological aspects in FIH studies is encouraged, such as including biomarkers indicative of efficacy (if available) or utilizing a challenge arm in healthy volunteers (Chap. 9). This was not employed during the development of PZQ. At that time, no human biomarkers had been identified, as the target—though unknown—was present in the infectious agent (schistosome) rather than in humans (Chap. 5). Additionally, challenge models for schistosomiasis had not yet been developed (Chap. 9).

A new guideline for conducting FIH studies was adopted in 2017. Since then, before starting a FIH study, toxicologists, pharmacologists, clinical pharmacologists, and clinicians define three key parameters using all available in vitro and in vivo data from animals and/or pharmacological models:

- *Minimal Anticipated Biological Effect Level (MABEL)*: The lowest dose of the experimental drug anticipated to elicit a biological effect in humans, which is crucial for determining a safe starting dose.
- *Pharmacologically Active Dose (PAD)*: The dose of the experimental drug anticipated to produce a pharmacological effect in the body.
- *No Observed Adverse Effect Level (NOAEL)*: The highest dose at which no adverse effects were observed in the toxicological studies, providing a safety margin to ensure that participants are not exposed to harmful levels of the experimental drug [2].

In general, the starting dose is set at a level expected to achieve the exposure necessary for a minimal biological effect, thereby avoiding the administration of a dose that lacks any relevant effect. Subsequent doses may fall within the range of anticipated pharmacological activity. If feasible, biomarkers may be employed to confirm pharmacological activity, although this is not always achievable.

The highest dose to be evaluated in a FIH study is a topic of debate. Factors to consider include the type of toxicity associated with the compound, the expected and observed safety in subjects, as well as the predicted and actual exposures. In general the dose should remain below the NOAEL. However, a thorough evaluation by a dedicated team of experts is essential. They must assess all available safety, tolerability, exposure, and biological/pharmacological data after a specific dose has been administered to humans before advancing to a higher dose in the FIH study or starting multiple dosing.

Later in the development process, additional phase I studies—commonly referred to as clinical pharmacology studies—are (mostly) conducted in healthy volunteers once the FIH study (or the phase II/III studies) has demonstrated a favorable benefit–risk profile for the IMP. These studies may include food-effect studies, drug–drug interaction studies, ADME (absorption, distribution, metabolism, and excretion) studies, bioavailability/bioequivalence studies, and investigations in specific patient populations, such as individuals with liver impairment or pediatric patients.

These studies are essential for obtaining posology descriptions, such as understanding how the compound should be administered (e.g., with or without food). They also help determine whether dose adjustments are required for specific populations (e.g., reduced doses for individuals with kidney or liver impairment) and particular circumstances (e.g., modifying doses when coadministering other medications).

Phase II

Phase II clinical trials are conducted to assess the effectiveness of the experimental drug in individuals with a specific disease or condition. Typically, these trials involve 100–500 participants. Depending on the required number of subjects and the prevalence of the disease, one or more clinical sites may be utilized. Researchers design phase II clinical trials to:

- Determine if the experimental drug is effective in individuals with a specific disease or condition.
- Explore the optimal dose and dosing regimen of the experimental drug.
- Evaluate the tolerability of the experimental drug at doses that demonstrate therapeutic effects.

Phase II clinical trials are often done in very well-defined, homogeneous adult patient populations (e.g., participants from a specific ethnic group, or those with either a severe or a mild form of the disease) to effectively evaluate the efficacy and tolerability of various doses of the IMP. An escalating sequential dose design may be used when there are uncertainties regarding the expected tolerability in the patient population or when the exposure data from healthy volunteers in the phase I study does not adequately represent those in the patient population. Preferably, several different dose levels—typically lower than the maximum doses or exposures assessed in phase I—should be administered simultaneously to different adult patient cohorts using a randomized double-blind design. These cohorts should include participants receiving either a placebo or a single dose of a comparator molecule. Randomizing subjects to receive either the IMP or a placebo/comparator enables an unbiased comparison of the IMP's effects against those of the placebo/comparator, as well as the effects of different dose administrations. The double-blind design ensures that neither the investigator nor the subjects know which treatment is being administered, thereby enhancing the objectivity of the safety, tolerability, and efficacy results.

Phase III

If the results of phase I and phase II clinical trials are positive—indicating a favorable benefit/risk ratio—the experimental drug is assessed in a larger patient population with broader inclusion and exclusion criteria. These studies often involve multiple clinics or institutions across various countries, allowing for the enrollment of participants from diverse age groups, ethnic backgrounds, disease severities as well as participants with mild liver or kidney impairments. Typically, these studies involve 1000–5000 participants, though the size can vary based on the indication. Researchers design phase III clinical trials to:

- Confirm the safety profile and efficacy findings observed in earlier phases for the dose intended for registration.
- Evaluate the effectiveness of the experimental drug in comparison to other treatments, determining whether it shows comparable (non-inferiority) or greater efficacy (superiority).

During or after phase III, additional studies may be conducted to evaluate the IMP in specific populations, such as children, the elderly, or patients with severe renal or liver impairments. Further clinical pharmacology studies are often performed in healthy volunteers to assess factors such as food effects, drug interactions, ADME parameters, and both absolute and relative bioavailability – especially if the marketed formulation differs from the IMP used in the phase III trials.

Praziquantel (PZQ)

Developed in the early 1970s, PZQ did not follow the current consensus and guidelines established by the International Council for Harmonisation (ICH) and FIH protocols for designing development studies and conducting dose-finding studies, as these guidelines were introduced later (Chap. 11). The following section summarizes the development process of PZQ, highlighting its use in veterinary medicine before being developed for human treatment (Chap. 1).

Preclinical/Nonclinical

Animal efficacy studies demonstrated that the effective dose of PZQ needed to be adjusted according to body weight (kg). In addition, both single-dose and three-times a-day (ter in die, t.i.d.) dosing every 4 h were evaluated in preparation for the clinical efficacy studies, given PZQ's relatively short half-life (Chap. 7) [3]. Based on animal studies the expected therapeutic dose for PZQ was estimated to range from 1 × 5 mg/kg up to a maximum of 3 × 25 mg/kg body weight (Chap. 6).

Phase I: Tolerability of PZQ in Healthy Volunteers

Preliminary safety and pharmacokinetic (PK) data for PZQ were collected from 12 out of 18 adult male volunteers at doses of 1 × 5 mg/kg and 1 × 10 mg/kg to determine optimal times for measuring individual PK parameters. Standard tolerance assessments, which involved questioning subjects about their experiences, did not reveal any adverse reactions, allowing for the evaluation of higher dosages.

In the next study, a total of 36 male volunteers participated in a double-blind study consisting of two parts to assess the drug's PK and tolerance, compared to a placebo. The active substance was administered orally at: 1 × 20 mg/kg, 3 × 10 mg/kg, 1 × 50 mg/kg, and 3 × 25 mg/kg. In each treatment group, some volunteers received the same placebo dosage (once or thrice) as those in the active group to eliminate any psychological factors associated with drug intake. It remains unclear why the highest single dose evaluated in this study was limited to 50 mg/kg. In retrospect, evaluating a higher dose could have been advantageous for the development of the more active enantiomer, (*R*)-PZQ, as it might have provided valuable insights into clinically relevant changes observed in laboratory parameters or during clinico-physiological examinations. PZQ was administered in the form of 200 mg scored tablets during a standard breakfast consisting of 1 bread roll, 10 g of butter, 1 teaspoon of jam, 2 cups of tea without sugar, and 1 glass of orange juice [3].

Respiratory frequency, pulse rate, blood pressure, and electrocardiogram (ECG) readings were monitored prior to drug administration. Blood chemistry parameters, including sodium and potassium concentrations and hepatic enzymes, were also assessed. Furthermore, psychological tests were conducted to evaluate any potential psychopharmacological effects of the drug, given that PZQ was initially developed for antipsychotic indications. In some participants, an electroencephalogram (EEG) was performed. All volunteers underwent a thorough physical examination, including an assessment of cerebral functions that could potentially be impaired by the drug. These assessments were monitored consistently during and after administration, with some tests conducted up to 7 days post-administration.

In summary, no clinically relevant changes were observed in any laboratory parameters, nor in the medical, neurological, or clinico-physiological examinations of healthy volunteers. Only the 3 × 25 mg/kg dosage produced a slight, transient disturbance in general well-being. It is unclear whether this disturbance contributed to the restriction of the maximum dose in later phases to 3 × 20 mg/kg.

Phase I: PK of PZQ in Healthy Volunteers

In the cited studies, blood samples were collected to measure PZQ concentrations. The data showed that PZQ is rapidly absorbed following oral administration, and undergoes significant first-pass metabolism (oxidation by liver cytochrome P450 enzymes). Its metabolites are primarily excreted in the urine. In healthy volunteers, maximum serum concentrations (T_{max}) were reached within 1–2 h max.

In another study using a radioactive ^{14}C-labeled PZQ, it was demonstrated that the elimination half-life of PZQ from serum in humans was 1–1.5 h, while that of PZQ plus its metabolites was 4–5 h. The renal elimination half-life for PZQ plus metabolites ranged from 4 to 6 h, with cumulative renal excretion exceeding 80% of the dose within 4 days, 90% of which occurred on the first day [5]. Only total PZQ concentrations were measured, and the differences between the (*R*)- and (*S*)-enantiomers and their metabolites were not identified. This lack of enantiomeric data significantly complicated the clinical development of (*R*)-PZQ as systemic levels of the (*R*)- and (*S*)-enantiomers were later observed to differ markedly after PZQ administration for reasons that remain unknown.

Phase II/III

For PZQ, a nonstandard development approach was taken to expedite the process, combining elements of both phase II (e.g., dose- and regimen-finding, placebo comparisons) and phase III development. The studies were conducted with various *Schistosoma* species in heterogeneous populations, including children—characteristics more typical of phase III studies or often done after the IMP is registered in the adult population. Close professional cooperation between informed representatives from the drug manufacturers (Bayer and E. Merck, now Merck KGaA, Darmstadt) and the World Health Organization (WHO) enabled the use of a standardized clinical trial design and agreed-upon technical protocols across all three studies. Parasitological methods for therapeutic assessment, however, differed depending on the infecting *Schistosoma* species. The design and outcomes of these studies are described in the following sections.

S. haematobium Clinical Study in Zambia

The trial began with a double-blind assessment comparing placebo and PZQ oral doses of 1 × 20, 2 × 20, or 3 × 20 mg/kg to evaluate tolerance and efficacy in patients with a minimum schistosome egg count of 50 per 10 mL of urine (Chap. 2). This was followed by a single-blind trial, in which only the participants did not know whether they were receiving the treatment or a placebo. The trial assessed the efficacy of three oral doses of 20 mg/kg given at 4-h intervals and a single dose of 50 mg/kg. A total of 80 African participants, aged 7–22 years (76 under 16 years), were enrolled in the trial, including 50 males and 30 females. Inclusion was based on a geometric mean egg excretion rate of 50 eggs per 10 mL of urine, calculated from at least two pretreatment urine samples.

Parasitological examinations were conducted on urine samples collected over 3 consecutive days for all treated patients. Quantitative hatching tests were performed monthly for up to 6 months and subsequently at 9, 12, 15, 18, and 24 months. To confirm egg viability, at least one positive pretreatment hatching test was required during the 3 consecutive days of sample collection [6].

These initial clinical studies demonstrated the following:

1. *Patient tolerance*: Excellent tolerance was observed in a predominantly school-age Zambian cohort. However, this was assessed through a symptom questionnaire, and many of the treated children struggled to understand the symptoms. As a result, they could only describe major disruptions in sensation or function.
2. *Therapeutic efficacy*: There was a high therapeutic effect against light or moderate *S. haematobium* infections at all tested dose levels. Six months after treatment, only one case of parasitological failure was detected among the 73 patients who were followed up. After 1 year, 66 of the 79 patients (83.5%) with *S. haematobium* infection were followed up, and two cases of parasitological failure (2.5%) were identified. Two years posttreatment, 45 of the 79 patients (57%) showed negative urine tests, 7 (9%) had positive hatching tests, and 27 patients (34%) were absent.
3. *Lack of adverse effects on health parameters*: There were no adverse effects on hematological functions, liver tests, ECG, or clinical parasitology measurements, even in patients with sickle cell trait, glucose-6-phosphate dehydrogenase (G6PD) deficiency, or multiple parasitic infections.

After treatment, the mean changes in liver enzymes (including total serum bilirubin, alanine aminotransferase, and aspartate aminotransferase), random blood sugar, hemoglobin, and white blood cell count were not clinically significant. The observed values reflected the inherent variability of each technique. Drug treatment did not consistently influence biochemical and hematological functions, except for eosinophilia, which is defined as an elevated level of eosinophils, a type of white blood cell. In this clinical study, eosinophilia was specifically defined as a doubling of the pretreatment eosinophil count posttreatment. According to this definition, 10 out of 24 (42%) drug-treated patients and 6 of 20 (30%) placebo-treated individuals showed eosinophilia.

S. mansoni Clinical Study in Brazil

All patients participating in the trial were male, aged between 20 and 48 years. The first part of the study focused on tolerance as the primary endpoint, assessed in a double-blind design. Eight subjects received a dosage of 1 × 20 mg/kg, eight received 2 × 20 mg/kg, and 13 received 3 × 20 mg/kg, with a 4-h interval between each administration. The number of subjects receiving placebo was comparable. The proportion of subjects without side effects were 87.5%, 86%, and 62% respectively, while in the placebo group this was 100%, 86%, and 92%, indicating good tolerability, with higher frequency of (mild) side effects at the highest dose (e.g., abdominal discomfort and headache).

In the subsequent phase, efficacy was evaluated as a primary endpoint using a single-blind design, involving 30 subjects who received 3 × 20 mg/kg and 31 subjects who received 50 mg/kg. The inclusion criterion for *S. mansoni* infections required a geometric mean excretion rate of 100 eggs per gram of stool, indicating medium infection intensity, calculated from two immediate pretreatment stool samples [7]. The therapeutic activity of the drug was assessed through three consecutive daily stool examinations using the Kato-Katz technique [8], with two slides prepared from each stool sample, as well as three hatching tests—a procedure used to detect *S. mansoni* eggs—on each daily stool sample [9]. Follow-up examinations were performed at 1, 2, 3, 6, and 12 months posttreatment. Patients were considered cured when *S. mansoni* eggs or miracidia were not detected in feces at any time during the follow-up period. All patients given the placebo were still excreting viable *S. mansoni* eggs in feces at 6 months, while doses of PZQ at 3 × 20 mg/kg and 1 × 50 mg/kg produced cure rates of 96% and 92%, respectively, at that time. At 1 year, 28 patients from each of the dose groups were followed up. In each group, 27 out of 28 patients (96.4%) remained cured. At these dosage levels, the mean number of eggs per gram of stool in patients who were not parasitologically cured, measured at 6 months and 1 year, was reduced to 2% of the pretreatment egg count. The frequency of side effects increased with higher dosages, with reported symptoms including nausea, epigastric pain, headache, dizziness, and drowsiness. These side effects were mild to moderate in severity and resolved within 48 h. The results of laboratory tests (e.g., hematology, liver tests) showed no significant changes resulting from treatment.

S. japonicum Clinical Studies in Japan and Philippines

The first clinical trials of PZQ in subjects with *S. japonicum* infections in Japan were planned to assess the drug's tolerance. Three double-blind studies compared to a placebo, involving a total of 51 patients, were conducted with dosages of PZQ at 1 × 20 mg/kg, 2 × 20 mg/kg, and 3 × 20 mg/kg administered in a single day. Their ages ranged from 24 to 72 years, and there were 38 male and 13 female patients. For *S. japonicum* infections, a geometric mean excretion rate of 100 eggs per gram of stool, calculated from two immediate pretreatment stool samples, was used as the inclusion criterion. The occurrence of adverse effects was greater in patients receiving PZQ at a dose of 3 × 20 mg/kg compared to those receiving lower doses or placebo. Typically, the side effects—including drowsiness, headache, lumbago, abdominal fullness, and epigastric discomfort—lasted for several hours but eventually resolved. Laboratory test results indicated no significant changes associated with the treatment. The overall assessment indicated that all patients treated with PZQ at the lower dose levels exhibited either excellent or good tolerance.

Additionally, PZQ was evaluated for tolerance and efficacy compared to a placebo in two double-blind clinical trials conducted in the Philippines, primarily involving male patients aged 6 years and above. The drug was administered orally at a dose of 3 × 20 mg/kg at 4-h intervals to a total of 82 patients, some without advanced disease and others with hepatosplenic involvement. A total of 43 patients received a placebo. In a separate single-blind trial, 42 patients were given a single oral dose of 50 mg/kg. Undesirable side effects were reported in 53% of patients receiving 3 × 20 mg/kg and in 70% of those given a dose of 50 mg/kg. The side effects primarily included abdominal discomfort, fever, sweating, and occasional dizziness, but were generally transient and mild. At 6 months posttreatment, 60 out of 75 patients treated with 3 × 20 mg/kg and 29 out of 41 patients treated with 1 × 50 mg/kg were completely free of *S. japonicum* eggs. At 12 months posttreatment, 25 of the 33 patients and 14 of the 26 patients in the respective groups were cured, indicating that the divided dosing regimen produced a superior therapeutic outcome [10].

PZQ was approved in Germany and the United States in the early 1980s at a recommended dose of 3 × 20 mg/kg for the single-day treatment of schistosomiasis. It remains unclear whether only the studies mentioned above were submitted or if additional efficacy data were included in the registration files (Chap. 11). Since then, other generic PZQ products have become available worldwide; for example, Cipla Ltd. received approval for its 600 mg PZQ tablet in 2015 through the WHO's prequalification process.

Post-approval Studies

Already in 1984, Wegner [11] documented the safe and effective treatment of over 25,000 patients in Africa, Asia, and South America using single-dose or single-day treatment regimens against all five pathogenic *Schistosoma* species (*S. haematobium*, *S. mansoni*, *S. japonicum*, *S. intercalatum*, and *S. mekongi*). Through testing on various trematode species across different host animals, PZQ demonstrated high efficacy regardless of the geographic origin of the parasite species. These studies also included randomized trials comparing the efficacy of PZQ with that of older, less expensive treatments under diverse field conditions. For instance, studies in Tanzania [12] and Brazil [13] showed that oxamniquine was as effective as PZQ in eliminating *S. mansoni* infections in the gastrointestinal tract, while metrifonate was as effective as PZQ for treating *S. haematobium* infections in the urinary tract [14, 15]. These studies helped establish effective dosage guidelines for PZQ treatment across various *Schistosoma* species. For *S. mansoni*, *S. haematobium*, and *S. intercalatum*, a single dose of 40 mg/kg body weight was found effective, while for *S. japonicum* and *S. mekongi*, a total dose of 60 mg/kg administered in three divided doses over 1 day was recommended. In cases of heavy infection with *S. mansoni* or *S. haematobium*, some experts suggested increasing the dosage to 50–60 mg/kg, split into two doses, to ensure full efficacy. In addition to its established efficacy against schistosomiasis, PZQ's versatility as an antiparasitic agent led to its use in treating a broader range of parasitic infections between 1983 and 1988. During this time, it became a treatment option for infections caused by intestinal, tissue, and lung flukes, as well as intestinal and tissue cestodes, including neurocysticercosis. These expanded indications underscored PZQ's effectiveness across various parasitic diseases and were later included in the official labeling and package leaflet.

A 2009 Cochrane review assessed the efficacy and safety of PZQ specifically for treating urinary schistosomiasis. This systematic review included 24 randomized controlled trials (RCTs) involving 6315 participants. The trials compared single-agent and combination regimens of PZQ, metrifonate, and artemisinin derivatives. The review concluded that the standard recommended single oral dose of 40 mg/kg PZQ was highly effective, with its oral administration route, low toxicity, and strong efficacy supporting its use in regional and national control programs [16]. A 2014 Cochrane review further assessed the efficacy and safety of PZQ, incorporating 30 RCTs with a total of 8165 participants. Of these, 24 trials were conducted among children in sub-Saharan Africa, while 21 included participants over 20 years old. On average, a single 40 mg/kg dose of PZQ reduced egg excretion by approximately 60% compared to placebo within 1–2 months posttreatment. Cure rates varied widely across studies, from 22.5% to 83.3%, but exceeded 60% in five out of seven trials. Additionally, the urinary egg count was reduced by over 95% in five of six trials, indicating a strong impact on worm burden. Splitting the 40 mg/kg dose into two doses over 12 h did not improve efficacy and, in fact, increased side effects such as vomiting and dizziness in a trial with 220 participants [17]. These findings support the efficacy of a single-dose regimen, which was associated with fewer side effects compared to split dosing. The summary of product characteristics (SPC) for PZQ is regularly updated with findings from Cochrane reviews and other independent studies to provide healthcare providers with current safety and efficacy data for informed treatment decisions. In 2012, at the launch of the Pediatric PZQ Program, it was noted that safety data for children under 7 years of age were insufficient. However, by 2019, the indication was expanded to include children aged 1–7, supported by additional reviews and studies confirming the drug's safety and efficacy in this younger age group.

(*R*)-Praziquantel

PZQ is a 1:1 racemic mixture of two enantiomers: (*R*)-praziquantel, also known as Arpraziquantel, and (*S*)-praziquantel. The (*R*)-enantiomer has been identified as the more active form, while the (*S*)-enantiomer shows reduced activity and is responsible for the drug's unpleasant taste (Chaps. 4, 5, and 6) [18]. The original 600 mg pill of PZQ was too large for children to swallow and needed to be broken into smaller segments (quartered), which could lead to imprecise dosing. In 2012, the Pediatric PZQ Consortium launched a plan to develop (*R*)-PZQ as a single entity for children under 7 years of age, using a more child-friendly formulation (less bitterness) with the expectation that a lower dose would be required (since only (*R*)-PZQ would be administered) and smaller tablets would be needed. To address potential development risks, two pediatric dispersable tablet (DT) formulations were persued in parallel up to phase III. The first contained only (*R*)-PZQ, while the second contained racemic PZQ as a backup in case (*R*)-PZQ development failed. Both formulations targeted the pediatric population aged 3 months to 6 years.

Preclinical/Nonclinical

Although the toxicological profile of (*R*)-PZQ had already been evaluated as part of overall PZQ toxicology, it was still prudent to verify it independently. Therefore, the toxicological profile of (*R*)-PZQ as a single agent was confirmed through a 4-week repeat-dose toxicity study in rats. (*R*)-PZQ was tested against PZQ, and the results showed a comparable profile (Chap. 8). At that time, the target through which PZQ exerted its effects remained unknown (Chap. 5). The pharmacological activity of (*R*)-PZQ was demonstrated through both in vitro and in vivo experiments (Chap. 6). In vitro studies showed that the activity of PZQ on schistosome worms was primarily driven by (*R*)-PZQ, with minimal contribution from (*S*)-PZQ (eudysmic ratio ~ 300, indicating a much higher potency of (*R*)-PZQ) [19]. In 2017, Zwang and Olliaro conducted a review to investigate whether available clinical data indicated a difference in the effective dose of PZQ between preschool-age and school-age children infected with *S. mansoni* and *S. haematobium*, assessing both the efficacy and safety of treatment in these age groups [20]. Forty-seven studies, including both comparative and non-comparative, were identified, enrolling a total of 15,549 children. Of these, 14,340 children (92%) were assessed 3–8 weeks post-treatment with 40 mg/kg PZQ (the WHO-recommended dose; n = 8380, 56%) or with comparators (n = 5960, 44%). The median age of participants was 10 years (range 1–19), with 11% (n = 1694) classified as preschool-age children. Based on the findings from the reviewed studies, it was concluded that PZQ 40 mg/kg was effective in reducing infection intensity in all *Schistosoma* species, with no significant differences between preschool-age (ages 1–5) and school-age children. However, preschool-age children accounted for only 11% of the total sample, which may limit the generalizability of the findings for this age group [20].

A nonstandard, adapted clinical development program for (*R*)-PZQ was initiated to develop the compound for children under 7 years of age, as it was not considered a new active substance. The program was designed in accordance with existing pediatric development guidelines, including the 2014 FDA guidance and the EU Guideline on the Investigation of Chiral Active Substances (April 1994). Regular discussions were held with regulatory authorities, such as the FDA, EMA, and the

German Federal Institute for Drugs and Medical Devices (BfArM), to refine the clinical studies. These studies aimed to demonstrate the efficacy and safety of a single dose of the new dispersable tablet (DT) in the targeted pediatric age group (3 months to 6 years). Since the safety of (*R*)-PZQ had already been established in humans (as part of the racemic formulation), a classic FIH study was not required. Instead, a bridging approach was adopted to utilize existing safety and efficacy data from the PZQ formulation, with the goal of accelerating the development process for this pediatric population. First, a clinical pharmacology study in healthy volunteers was conducted to assess the safety and bioavailability of (*R*)-PZQ DT compared to PZQ. However, the bioavailability of (*R*)-PZQ was much lower than predicted and the dose–response relationship was nonlinear (Chap. 7). As bridging was not possible, a phase II dose-finding study was required in pediatric patients infected with *S. mansoni*. This study needed to explore doses that would possibly lead to high levels of (*R*)-PZQ in the blood, which had not been previously encountered with PZQ, and the safety of which was unknown. Therefore, the 60 mg/kg dose of (*R*)-PZQ was included only after assessing the safety profile of the lower 30 and 45 mg/kg doses in a subset of patients. For safety reasons, only children above 2 years of age were included in Part 1 of the dose-finding study, while younger children were included in Part 2 after selecting the most effective dose. Following phase II, a phase III study was carried out in Cote d'Ivoire and Kenya. While phase II focused solely on *S. mansoni* infections, phase III was expanded to include children infected with *S. haematobium*, facilitating the compound's registration for both indications. Initially, only an (*R*)-PZQ arm was planned, but the EMA recommended including a PZQ arm. This arm was not meant to serve as a direct comparison to (*R*)-PZQ, but rather as a contingency measure if (*R*)-PZQ efficacy did not meet expectations. Consequently, the number of subjects in the PZQ group was limited, as its inclusion was intended solely as a backup [21].

Phase I

The study was an open-label, randomized, five-period crossover study involving 36 male participants, who received various treatments in a predetermined sequence over five periods. In the first two periods, (*R*)-PZQ DT was given at a dose of 20 mg/kg and compared with the current PZQ formulation at 40 mg/kg (containing 20 mg/kg (*R*)-PZQ) in a crossover design, with single oral doses taken with food, measuring PZQ, (*R*)-PZQ and (*S*)-PZQ concentrations. In the third, fourth, and fifth periods, doses of 10 and 30 mg/kg of (*R*)-PZQ DT were administered, with one 20 mg/kg (*R*)-PZQ DT dose administered without food. Data from these periods were compared both with each other and with those from the first and second periods.

(*R*)-PZQ DT, administered as multiple 150 mg tablets dispersed in water to achieve a target dose of 20 mg/kg, showed a systemic presence of only (*R*)-PZQ and no (*S*)-PZQ (Fig. 10.1a left). PZQ was administered as several 500 mg Cysticide® tablets to reach a dose of 40 mg/kg. After PZQ administration, both (*R*)- and (*S*)-PZQ were measured and higher systemic concentrations of (*S*)-PZQ (Fig. 10.1a

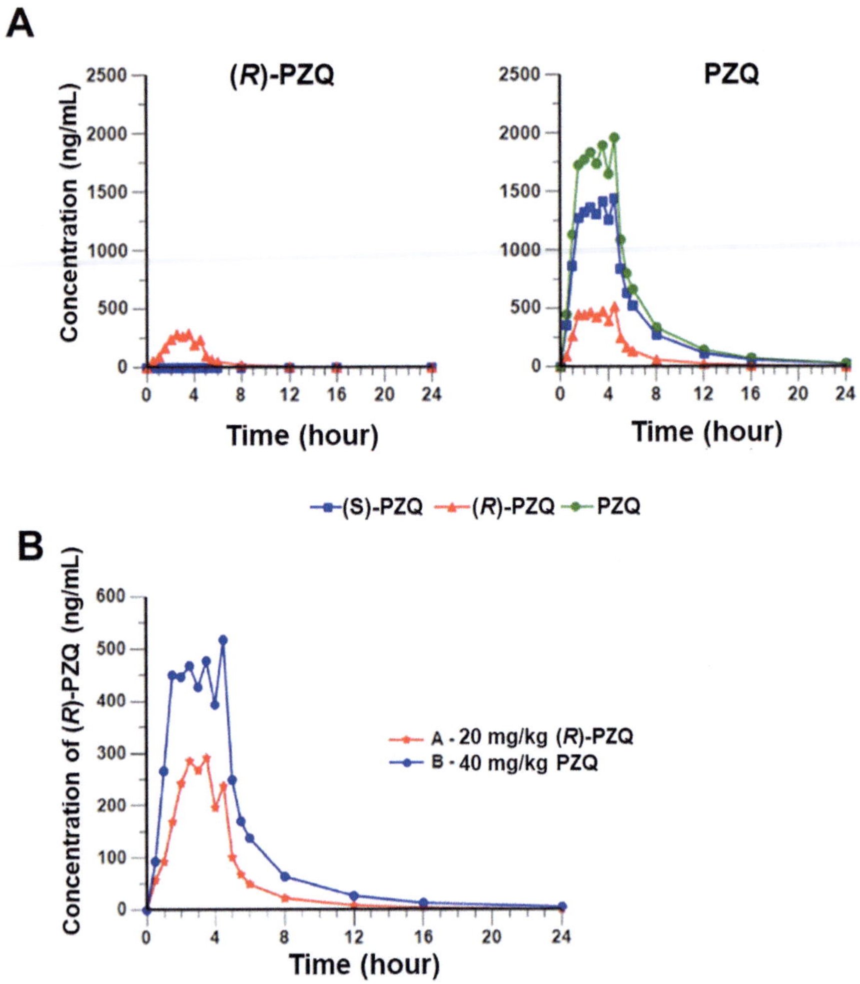

Fig. 10.1 Panel **a** (left) shows the concentration–time profile of (*R*)-PZQ (red) and (*S*)-PZQ (blue) after dosing with 20 mg/kg (*R*)-PZQ, confirming absence of conversion of (*R*) to (*S*)-PZQ. Panel **a** (right) shows the concentration time profile of PZQ (green), (*S*)-PZQ (blue) and (*R*)-PZQ (red) after dosing with 40 mg/kg PZQ. Although PZQ contains both enantiomers in a 1:1 ratio, systemic concentrations of the (*R*)-enantiomer are much lower than of the (*S*)-enantiomer. Panel **b** compares the concentration–time profile of (*R*)-PZQ (red) after administration of (*R*)-PZQ DT at 20 mg/kg with that of (*R*)-PZQ (blue) after administration of PZQ at 40 mg/kg. Although the administered dose of the (*R*)-enantiomer is the same, systemic concentration are much lower after administration of (*R*)-PZQ DT

right, blue) than those of (*R*)-PZQ (Fig. 10.1a right, red) were observed. Consistent with previous in vitro and in vivo findings, this effect is not due to the conversion of (*R*)-PZQ to (*S*)-PZQ, as no (*S*)-PZQ was detected after administration of (*R*)-PZQ DT alone. Moreover, although both formulations contained equivalent amounts of

(*R*)-PZQ (20 mg/kg), the expected equal exposure was not achieved (exemplified in Fig. 10.1b).

Calculation of the AUC showed that the relative bioavailability of (*R*)-PZQ administered as (*R*)-PZQ DT at 20 mg/kg was only 40% of the bioavailability of (*R*)-PZQ achieved after administration of PZQ at 40 mg/kg. The reason for this lower bioavailability remains unknown.

After adjusting for administered dose, (*R*)-PZQ showed a nonlinear, supra-proportional (quadratic) increase in both C_{max} and AUC when administered at 10, 20, and 30 mg/kg. This indicates that at higher doses, the systemic exposure of the drug was greater than expected if the relationship was linear. This could be due to changes in the way the body absorbs, distributes, metabolizes, or excretes the drug at different dose levels [21]. This non-linear increase makes predicting bioavailability at higher doses of (*R*)-PZQ DT uncertain.

Food consisting of 16.4% protein, 9.9% fat, and 73.7% carbohydrates (658.6 calories) significantly increased both the peak concentration (C_{max}) of (*R*)-PZQ by 2.9-fold and the overall exposure (AUC) by 1.6-fold, with a greater impact on C_{max} than AUC. This food effect on (*R*)-PZQ DT was less pronounced than in the study done by Castro [4], which investigated the effect of meals on the bioavailability of PZQ. In that study, a high-carbohydrate meal (15.3% protein, 10.5% fat, 74.2% carbohydrates; 674.5 calories) and a high-fat meal (19.6% protein, 32.5% fat, 47.9% carbohydrates; 656 calories) both resulted in a much higher increase in bioavailability, with C_{max} increasing by 6.2-fold for the high-carbohydrate meal and 3.4-fold for the high-fat meal, and AUC increasing by 3.7-fold and 2.8-fold, respectively, compared to when PZQ was taken without food (Chap. 7). The half-life and T_{max} of (*R*)-PZQ were not affected by food, which is consistent with findings for PZQ in the Castro study [4].

(*R*)-PZQ DT at doses of 10, 20, and 30 mg/kg was safe and well-tolerated. However, the relative bioavailability of (*R*)-PZQ after the 20 mg/kg dose in adults was low—only 40% of that observed after PZQ administration—complicating the translation of efficacy data from PZQ to (*R*)-PZQ. If (*R*)-PZQ exposure at 20 mg/kg is only 40% of PZQ at 40 mg/kg, a 2.5-fold higher dose, or approximately 50 mg/kg (*R*)-PZQ, might be required to achieve similar efficacy. On the other hand, due to the supra-proportional increase in exposure at 10, 20, and 30 mg/kg, a 50 mg/kg dose might not be necessary. Additionally, it was unknown whether (*R*)-PZQ exposure at the same dose would be comparable between adults and children. To resolve these uncertainties, a dedicated dose-finding phase II study in infected children was needed.

Phase II

Two novel DT formulations, PZQ and (*R*)-PZQ, were initially developed for treating schistosomiasis in preschool-age children and infants. This summary focuses solely on data for the (*R*)-PZQ DT formulation, as development of PZQ DT was discontinued after the phase II study. The open-label, comparative dose-finding phase II trial was conducted in Côte d'Ivoire to assess the efficacy and safety of (*R*)-PZQ DT in children and infants infected with *S. mansoni*, with marketed PZQ included as a comparator. The study consisted of two parts: Part 1 aimed to identify the optimal single dose of (*R*)-PZQ DT that achieved a clinically meaningful cure rate (as measured by the Kato-Katz method), while maintaining an acceptable safety profile in children aged 2–6 years. In Part 2, younger children were included using a dose de-escalation design to determine the appropriate dose for this age group.

In Part 1, 300 children aged 2.1–6.9 years (weighing 10–29.9 kg) were enrolled and randomized into one of five treatment arms (n = 60 per arm): commercially available PZQ at either 3 × 20 or 40 mg/kg (treatment arms 1 and 2) and (*R*)-PZQ DT at 30, 45, or 60 mg/kg (treatment arms 3, 4, and 5). Cure rates of 70% or higher were achieved in all treatment arms, with the highest observed rates in (*R*)-PZQ DT 60 mg/kg (89.7%) and 45 mg/kg (86%), while PZQ 40 mg/kg achieved 74.1%. Treatment-related adverse events (AEs) were reported by 71 participants (16%) and were similar across arms; most were mild and transient, with laboratory abnormalities being the most common treatment-emergent AE. No deaths, treatment discontinuations, or new safety concerns were observed. The assessment from Part 1 identified the optimal dose of (*R*)-PZQ DT as 45 mg/kg, followed by 60 mg/kg. This selection of the optimal dose was based on a predefined, approved utility score to ensure an unbiased assessment of the formulation with the best benefit–risk profile [22].

Since (*R*)-PZQ DT contain 150 mg of active pharmaceutical ingredient (API), a 50 mg/kg dose (closest to 45 mg/kg) was chosen for Part 2. In this part, 24 infants aged 6–24 months (weighing 7.5–14.8 kg) were enrolled and treated in an age-staggered approach: 13–24 months (treatment arm 6, n = 20) and 6–12 months (treatment arm 7, n = 4) each receiving a 50 mg/kg (*R*)-PZQ DT. The primary endpoint was clinical cure rate (CR) measured by the Kato–Katz method 14–21 days posttreatment. Secondary endpoints included CR determined by a point-of-care diagnosis of circulating cathodic antigen (POC-CCA) (Chap. 9), egg reduction rate (ERR), and AEs. CRs exceeded 90% in this population. Treatment-related AEs were consistent in number, severity, and type to those observed in children older than 24 months. However, the small sample size in this age group indicated the need for further inclusion in phase III. ERRs across all treatment arms across Parts 1 and 2 exceeded 95% [22].

Phase III

This open-label, partially randomized phase III study was conducted at two hospitals in Côte d'Ivoire and Kenya, with laboratory staff blinded to treatment groups and baseline values. *S. mansoni* infection was detected using a POC-CCA urine test and confirmed by the Kato-Katz method. Of 2663 prescreened participants, 326 were diagnosed with *S. mansoni* or S. *haematobium* and 288 were enrolled across the study's cohorts (Cohort 1a: n = 100; Cohort 1b: n = 50; Cohort 2: n = 30; Cohort 3: n = 18; Cohort 4a: n = 30; Cohort 4b: n = 60). Eligible participants included children weighing at least 5 kg (ages 3 months to 2 years) or 8 kg (ages 2–6 years). In Cohort 1, children aged 4–6 years with *S. mansoni* infections were randomly assigned in a 2:1 ratio to receive either a single oral dose of 50 mg/kg (*R*)-PZQ DT (Cohort 1a) or 40 mg/kg (Cohort 1b), using a computer-generated randomization list. Cohorts 2 (ages 2–3 years) and 3 (ages 3 months to 2 years) with *S. mansoni* infections, as well as the initial 30 participants in Cohort 4a (ages 3 months to 6 years) with *S. haematobium* infections, were all treated with a single oral dose of 50 mg/kg (*R*)-PZQ DT. After initial assessments, the dose for the *S. haematobium* group was increased to 60 mg/kg in Cohort 4b. The primary efficacy endpoint was CR in Cohorts 1a and 1b at 17–21 days posttreatment, measured in the modified intention-to-treat population using the Clopper–Pearson method. Secondary endpoints included CRs for other age groups and species, ERRs, safety outcomes, and palatability scores. The primary analysis included 280 participants in the modified intention-to-treat population [23] (Fig. 10.2).

The median age of participants was 5.1 years (Interquartile Range [IQR]: 4.1–6.0), with 132 (47%) female and 148 (53%) male participants (Fig. 10.3).

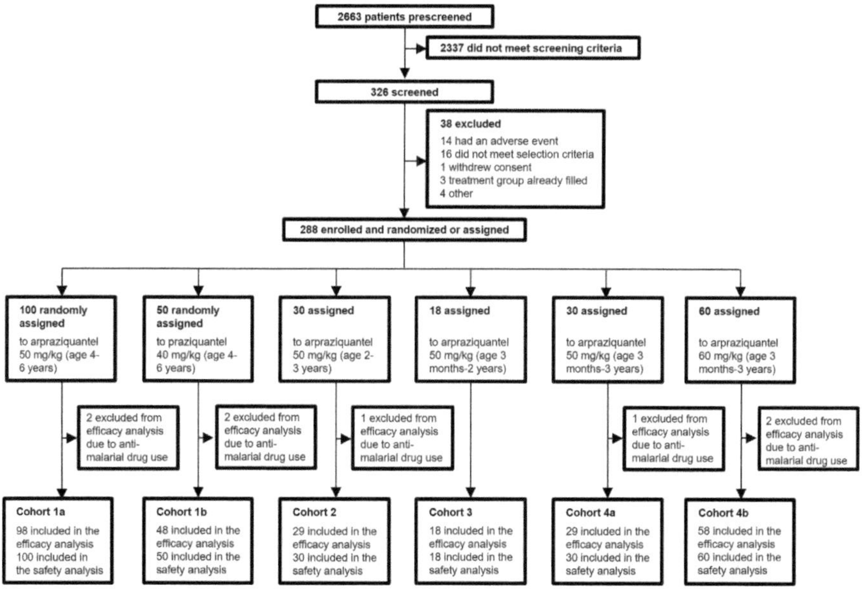

Fig. 10.2 The figure above describes the study design and subject disposition, i.e., the number of subjects prescreened, screened, excluded, enrolled, and randomized. Between September 2, 2019 and August 7, 2021, 2663 participants were prescreened and 326 were diagnosed with *S. mansoni* or *S. haematobium*. Thirty-eight were excluded mostly because of not meeting selection criteria (n = 16) or existing adverse events (n = 14). A total of 288 were enrolled (n = 100 in cohort 1a, n = 50 in cohort 1b, n = 30 in cohort 2, n = 18 in cohort 3, n = 30 in cohort 4a, and n = 60 in cohort 4b). Eight participants (two each in cohorts 1a, 1b, and 4b and one each in cohorts 2 and 4a) received antimalarial drugs during the study and were excluded from the efficacy analyses (n = 280 in the mITT population)

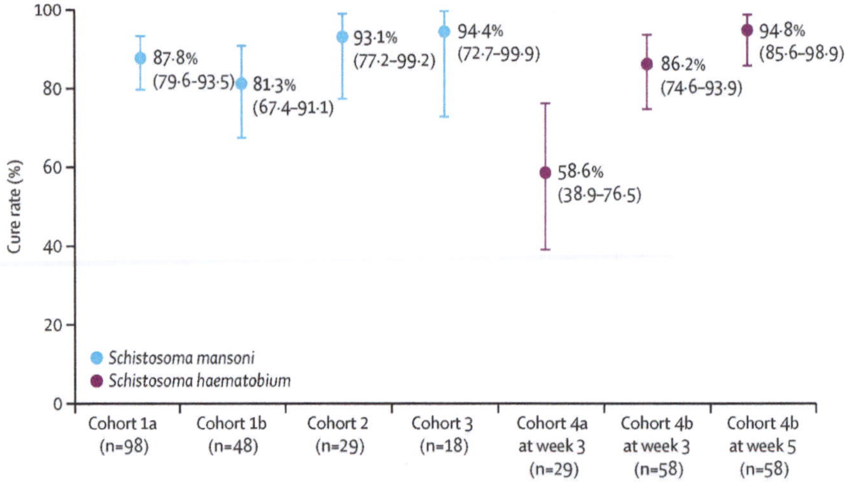

Age	4-6 y	4-6 y	2-3 y	3m-2y	3m-6y	3m-6y	3m-6y
IMP	(*R*)-PZQ	PZQ	(*R*)-PZQ	(*R*)-PZQ	(*R*)-PZQ	(*R*)-PZQ	(*R*)-PZQ
Dose	50 mg/kg	40 mg/kg	50 mg/kg	50 mg/kg	50 mg/kg	60 mg/kg	60 mg/kg

Fig. 10.3 Clinical cure rate by treatment group. Bars show 95% CI

CRs with (*R*)-PZQ DT were comparable to those with PZQ, reaching 87.8% (95% confidence interval [CI]: 79.6–93.5%) in Cohort 1a and 81.3% (95% CI: 67.4–91%) in Cohort 1b. Younger participants also showed high CRs: 94.4% in those aged 3 months to 2 years (Cohort 3) and 93.1% in those aged 2–3 years (cohort 2). For participants infected with *S. haematobium*, CRs at 60 mg/kg were 86.2% at week 3 posttreatment, increasing to 94.8% by week 5, likely due to natural variations in infection dynamics. No new safety concerns were identified during the study. The most common treatment-related AEs were mild and transient, including abdominal pain (14% of participants), diarrhea (9%), vomiting (6%), and drowsiness (7%) [23].

A subgroup of participants consented to provide blood samples to assess plasma concentrations of either (*R*)-PZQ ($n = 74$) or *trans*-4-hydroxypraziquantel ($n = 30$), the main metabolite of PZQ (Chap. 7). Plasma concentration–time profiles for (*R*)-PZQ were observed to be atypical, showing multiple peaks and substantial fluctuations. These patterns likely reflect several factors, including first-pass metabolism via the cytochrome P450 system, the enhancing effect of food on bioavailability, and potential drug–drug interactions. In contrast, *trans*-4-hydroxypraziquantel plasma profiles displayed a more typical pattern, characterized by a single peak followed by a steady decline in concentrations. This major metabolite, given its moderate potency in preclinical assays yet sustained plasma levels, may contribute to clinical efficacy. Overall, the pharmacokinetic parameters for (*R*)-PZQ were similar across age groups, infecting species, and (*R*)-PZQ DT doses of 50 and 60 mg/kg [23].

Summary

The development paths of PZQ and (*R*)-PZQ DT differ significantly, shaped by the time periods in which they were developed and their intended target populations. (*R*)-PZQ DT was specifically developed for children under modern regulatory frameworks designed to address pediatric needs, which were not in place during PZQ's initial development. This summary highlights how methods and priorities in clinical development have evolved. Initially, efforts focused on treating adults, but the emphasis gradually shifted to young children, with a particular focus on chemoprevention.

PZQ and Arpraziquantel are short-lived drugs (half-life below 4 h) and have been shown to be safe at therapeutic doses. For PZQ, an effective dose of 40 mg/kg was identified, with safety demonstrated up to 50 mg/kg or 3 × 25 mg/kg administered every 4 h. PZQ at 40 mg/kg effectively reduced infection intensity across all *Schistosoma* species (*S. haematobium*, *S. mansoni*, *S. japonicum*), with no significant differences observed between preschool- and school-aged children. Arpraziquantel was also effective at single doses of 45, 50, and 60 mg/kg in children aged 3 months to 6 years, demonstrating efficacy against *S. haematobium* and *S. mansoni*.

Competing Interests The phase II and III studies for (*R*)-PZQ were partially funded by external funders (The Global Health Innovative Technology Fund, the European and Developing Countries Clinical Trials Partnership, and the healthcare business of Merck KGaA, Darmstadt, Germany).

The author has no conflicts of interest to declare that are relevant to the content of this chapter.

References

1. US Department of Health and Human Services, National Institutes of Health, National Library of Medicine, and National Center for Biotechnology Information (2024) ClinicalTrials.gov is a place to learn about clinical studies from around the world. https://clinicaltrials.gov/. Accessed 18 Jan 2025
2. European Medicines Agency (2017) Guideline on strategies to identify and mitigate risks for first-in-human and early clinical trials with investigational medicinal products. Committee for Medicinal Products for Human Use https://www.ema.europa.eu/en/documents/scientific-guideline/guideline-strategies-identify-and-mitigate-risks-first-human-and-early-clinical-trials-investigational-medicinal-products-revision-1_en.pdf. Accessed 18 Jan 2025
3. Leopold G, Ungethüm W, Groll E, Diekmann HW, Nowak H, Wegner DH (1978) Clinical pharmacology in normal volunteers of praziquantel, a new drug against schistosomes and cestodes. An example of a complex study covering both tolerance and pharmacokinetics. Eur J Clin Pharmacol 14(4):281–291. https://doi.org/10.1007/bf00560463
4. Castro N, Medina R, Sotelo J, Jung H (2000) Bioavailability of praziquantel increases with concomitant administration of food. Antimicrob Agents Chemother 44(10):2903–2904. https://doi.org/10.1128/AAC.44.10.2903-2904.2000
5. Patzschke K, Pütter J, Wegner LA, Horster FA, Diekmann HW (1979) Serum concentrations and renal excretion in humans after oral administration of praziquantel--results of three determination methods. Eur J Drug Metab Pharmacokinet 4(3):149–156. https://doi.org/10.1007/BF03189418

6. Davis A, Biles JE, Ulrich AM (1979) Initial experiences with praziquantel in the treatment of human infections due to *Schistosoma haematobium*. Bull World Health Organ 57(5):773–779
7. Katz N, Rocha RS, Chaves A (1979) Preliminary trials with praziquantel in human infections due to *Schistosoma mansoni*. Bull World Health Organ 57(5):781–785
8. Katz N, Chaves A, Pellegrino J (1972) A simple device for quantitative stool thick-smear technique in *schistosomiasis mansoni*. Rev Inst Med Trop Sao Paulo 14(6):397–400
9. Zicker F, Katz N, Wolf J (1977) Evaluation of the hatching test for eggs of *Schistosoma mansoni*. Rev Inst Med Trop Sao Paulo 19(3):202–207
10. Ishizaki T, Kamo E, Boehme K (1979) Double-blind studies of tolerance to praziquantel in Japanese patients with Schistosoma japonicum infections. Bull World Health Organ 57(5):787–791
11. Wegner DH (1984) The profile of the trematodicidal compound praziquantel. Arzneimittelforschung 34(9B):1132–1136
12. Rugemalila JB, Asila J, Chimbe A (1984) Randomized comparative trials of single doses of the newer antischistosomal drugs at Mwanza, Tanzania. I. Praziquantel and oxamniquine for the treatment of *Schistosoma mansoni*. J Trop Med Hyg 87(6):231–235
13. Branchini ML, de Pedro JR, Dias LC, Deberaldini ER (1982) Double-blind clinical trial comparing praziquantel with oxamniquine in the treatment of patients with schistosomiasis mansoni. Rev Inst Med Trop Sao Paulo 24(5):315–321
14. McMahon JE (1983) A comparative trial of praziquantel, metrifonate and niridazole against *Schistosoma haematobium*. Ann Trop Med Parasitol 77(2):139–142. https://doi.org/10.1080/00034983.1983.11811688
15. King CH, Mahmoud AA (1989) Drugs five years later: praziquantel. Ann Intern Med 110(4):290–296. https://doi.org/10.7326/0003-4819-110-4-290
16. Danso-Appiah A, Garner P, Olliaro PL, Utzinger J (2009) Treatment of urinary schistosomiasis: methodological issues and research needs identified through a Cochrane systematic review. Parasitology 136(13):1837–1849. https://doi.org/10.1017/S0031182009005939
17. Kramer CV, Zhang F, Sinclair D, Olliaro PL (2014) Drugs for treating urinary schistosomiasis. Cochrane Database Syst Rev 2014:CD000053. https://doi.org/10.1002/14651858.CD000053.pub3
18. Meyer T, Sekljic H, Fuchs S, Bothe H, Schollmeyer D, Miculka C (2009) Taste, a new incentive to switch to (R)-praziquantel in schistosomiasis treatment. PLoS Negl Trop Dis 3:e357. https://doi.org/10.1371/journal.pntd.0000357
19. Meister I, Ingram-Sieber K, Cowan N, Todd M, Robertson MN, Meli C, Patra M, Gasser G, Keiser J (2014) Activity of praziquantel enantiomers and main metabolites against *Schistosoma mansoni*. Antimicrob Agents Chemother 58(9):5466–5477. https://doi.org/10.1128/AAC.02741
20. Zwang J, Olliaro P (2017) Efficacy and safety of praziquantel 40 mg/kg in preschool-aged and school-aged children: a meta-analysis. Parasit Vectors 10:47. https://doi.org/10.1186/s13071-016-1958-7
21. Bagchus WM, Bezuidenhout D, Harrison-Moench E, Kourany-Lefoll E, Wolna P, Yalkinoglu O (2019) Relative bioavailability of orally dispersible tablet formulations of levo- and racemic praziquantel: two phase I studies. Clin Transl Sci 12(1):66–76. https://doi.org/10.1111/cts.12601

22. N'Goran EK, Yalkinoglu O, Kourany-Lefoll E, Tappert A, Hayward B, Yin X, Bezuidenhout D, Huber E, Aka NAD, Ouattara M, Bagchus W (2021) Efficacy and safety of new orodispersible tablet formulations of praziquantel (Racemate and L-praziquantel) in *Schistosoma mansoni*-infected preschool-age children and infants: a randomized dose-finding phase 2 study. Front Trop Dis 2. https://doi.org/10.3389/fitd.2021.679373
23. N'Goran EK, Odiere MR, Aka RA, Ouattara M, Aka NAD, Ogutu B, Rawago F, Bagchus WM, Bödding M, Kourany-Lefoll E, Tappert A, Yin X, Bezuidenhout D, Badenhorst H, Huber E, Dälken B, Saflo OH-A (2023) Efficacy, safety, and palatability of arpraziquantel (L-praziquantel) orodispersible tablets in children aged 3 months to 6 years infected with *Schistosoma* in Côte d'Ivoire and Kenya: an open-label, partly randomised, phase 3 trial. Lancet Infect Dis 23(7):867–876. https://doi.org/10.1016/S1473-3099(23)00048-8

Regulators' Viewpoint 11

Katja Schüz

Abstract

Regulatory requirements in the pharmaceutical and healthcare sectors are essential for safeguarding public health, ensuring product quality, and fostering innovation. These regulations, shaped by historical events affecting patient safety, encompass guidelines from various government agencies and international organizations. The International Council for Harmonisation (ICH) plays a pivotal role in promoting global regulatory harmonization, addressing the complexities arising from divergent national standards. Agencies such as the United States Food and Drug Administration (FDA) and the European Medicines Agency (EMA) enforce stringent safety and efficacy standards for drug approvals, while the World Health Organization (WHO) coordinates global health initiatives and establishes essential medicine lists. The development of praziquantel (PZQ) exemplifies the regulatory process, showcasing the challenges and successes in bringing effective treatments to market. The recent positive scientific opinion of (*R*)-PZQ from the Committee for Medicinal Products for Human Use (CHMP) illustrates the importance of robust regulatory strategies and collaboration among stakeholders to enhance access to essential medicines, particularly in low- and middle-income countries.

Keywords

ICH · EMA · FDA · WHO · Praziquantel · Regulatory dossier · Marketing authorization · List of WHO Prequalified Products

K. Schüz (✉)
Merck Healthcare KGaA, Darmstadt, Germany
e-mail: katja.schuez@merckgroup.com

T. Spangenberg (ed.), *Praziquantel*,
https://doi.org/10.1007/978-3-031-97397-0_11

Introduction

Regulatory requirements are crucial in the pharmaceutical and healthcare sectors, acting as guardians of public health, guarantors of product quality, and catalysts for innovation. These requirements encompass a wide range of regulations, standards, and guidelines established by government agencies, regulatory bodies, and international organizations (Chap. 12). They are shaped by significant events that impact patient safety, as well as addressing potential risks to others, such as the fetus in pregnant women or newborns during breastfeeding. These measures not only aim to prevent disasters but also ensure that medications meet stringent safety and efficacy standards before reaching patients, thereby maintaining the highest levels of patient safety worldwide through continuous evolution and adaptation (Chaps. 1 and 8).

Regulatory requirements are implemented through a combination of laws, regulations, policies, and procedures at national, regional, and international levels. Government agencies such as the Food and Drug Administration (FDA) in the United States of America (USA) and the European Medicines Agency (EMA) in the European Union (EU) are responsible for enforcing these requirements within their respective jurisdictions. The World Health Organization (WHO), while not a regulatory body, plays a critical role globally by setting international standards, providing technical guidance, and supporting regulatory harmonization efforts. Today, the scope of regulatory standards ranges from well-established and clearly communicated guidelines set by stringent health authorities such as the FDA and EMA to more evolving regulations in regions such as Latin America (LATAM) and Asia-Pacific (APAC). The regulatory environment is highly fluid, with numerous frameworks and harmonization initiatives being developed within and across geographical regions. In contrast, some low- and middle-income countries (LMICs) may have limited or underdeveloped regulatory frameworks and often rely on the assessments of Stringent Regulatory Authorities (SRAs) as they strive to adapt to the evolving global standards [1]. Such frameworks cover a broad spectrum of activities and areas within the pharmaceutical and healthcare sectors. These include, for example, preclinical and clinical testing of drugs, good manufacturing practices (GMP) to ensure product quality and consistency, pharmacovigilance and post-market surveillance to monitor product safety, labeling and packaging requirements to provide accurate information to patients and caregivers, and marketing authorization processes to bring and maintain products in the market [2–4]. Additionally, regulatory requirements address ethical considerations, such as patient rights and informed consent, in clinical research and healthcare delivery. Pharmaceutical companies, medical device manufacturers, healthcare providers, and other stakeholders must adhere to specific regulatory frameworks and undergo regulatory review processes to obtain and maintain approval for their products throughout their entire life cycle.

Regulatory Bodies and Health Authorities

A Global Framework: The International Council for Harmonisation (ICH) of Technical Requirements for Pharmaceuticals for Human Use

The importance of for regulatory harmonization became evident in the wake of medicinal product tragedies such as the thalidomide incident in the 1960s. Thalidomide, a drug prescribed to pregnant women for morning sickness, caused severe birth defects in thousands of babies, leading to widespread public outcry and stricter regulations [4, 5]. This tragedy, alongside other incidents, prompted increased regulations in the 1960s and 1970s. However, the globalization of the pharmaceutical industry and the divergence in regulatory requirements across countries led to time-consuming and costly duplications, further highlighting the need for harmonization.

Europe led the charge for regulatory harmonization in the 1980s, demonstrating the feasibility of such efforts through collaborative initiatives. This success prompted further discussions with Japan and the United States, leading to the WHO Conference of Drug Regulatory Authorities in 1989, which catalyzed the development of action plans. These efforts ultimately resulted in the creation of the International Council for Harmonisation of Technical Requirements for Pharmaceuticals for Human Use (ICH). The ICH brings together regulatory authorities and the pharmaceutical industry to address scientific and technical issues in pharmaceuticals, and to develop guidelines that promote global harmonization. These initiatives ensure that safe, effective, and high-quality medicines are developed, registered, and maintained efficiently while adhering to stringent standards. One of the ICH's key objectives are to reduce animal testing without compromising safety and effectiveness, prevent duplication of clinical trials and post-market clinical evaluations, facilitate the development and manufacturing of new medicines, streamline the registration and supervision of new medicines, and enhance post-authorization safety reporting and pharmacovigilance [6].

The ICH is structured with the ICH Assembly as the primary decision-making body, including all ICH Members and Observers. It is supported by the ICH Management Committee (ICH MC), which oversees the ICH Working Groups (ICH WGs) responsible for developing ICH guidelines. The Medical Dictionary for Regulatory Activities (MedDRA) is an international terminology used to classify adverse event data—undesirable effects related to pharmaceutical products. The MedDRA Management Committee (MedDRA MC) manages and maintains MedDRA. The ICH Management Committee currently consists of 13 Regulatory and Industry members and two standing observers [7]. Today, ICH membership extends beyond its founding regulatory members—the EU, the USA, and Japan—to include countries such as Brazil, Egypt, Singapore, and Turkey. Numerous National Health Authorities serve as observers, applying ICH guidelines extensively. Similarly, regional harmonization initiatives, including the Association of Southeast Asian Nations (ASEAN), the East African Community (EAC), and the Global

Harmonization Committee (GHC), along with industry associations and international organizations such as the European Directorate for the Quality of Medicines & HealthCare (EDQM), the Pharmaceutical Inspection Co-operation Scheme (PIC/S), and the Council for International Organizations of Medical Sciences (CIOMS), also hold observer status and adhere to ICH guidelines where applicable [8]. In supporting its members and observers, ICH addresses the entire life cycle of pharmaceutical products through currently 67 technical guidelines and standards, with additional ones under development. These guidelines, along with detailed Q&A documents and training materials, clarify the principles they outline to ensure global applicability. Once adopted by ICH regulators, the guidelines establish a harmonized technical framework that is applied worldwide.

What Is the United States Food and Drug Administration?

The United States Food and Drug Administration (FDA) is a federal agency within the Department of Health and Human Services (HHS) tasked with protecting public health by ensuring the safety, efficacy, and security of human and veterinary drugs, biological products, medical devices, the nation's food supply, cosmetics, and radiation-emitting products. Additionally, the FDA regulates the manufacturing, marketing, and distribution of tobacco products to protect public health and reduce tobacco use among youth. Beyond regulation, the FDA plays a critical role in advancing public health by fostering medical product innovation and ensuring that new drugs and medical devices meet rigorous safety and efficacy standards before entering the market in the USA. It also provides accurate, science-based information to help individuals maintain and improve their health [9]. FDA regulations are published in the Federal Register, the official journal of the federal government, and are codified in Title 21 of the Code of Federal Regulations (CFR). The FDA employs a process known as "notice and comment rulemaking" to introduce new regulations. This process begins with the publication of a proposed rule, which outlines the FDA's rationale and solicits public feedback, typically submitted through the Regulations.gov website [10]. In some cases, the FDA may issue an Advance Notice of Proposed Rulemaking (ANPRM) to gather additional input. After reviewing public comments, the Agency decides whether to finalize the rule, amend the proposal, or terminate the rulemaking process. The final rule, published in the Federal Register, includes the regulatory requirements, addresses public feedback, and becomes part of the CFR [10].

Praziquantel in the United States

Praziquantel (PZQ) was approved by the FDA in 1982 for the treatment of schistosomiasis, and its approval later expanded to include clonorchiasis. PZQ is available for human use in 600 mg tablets, both as a generic formulation and under the brand name Biltricide®. The approved indications, along with essential information for

This label may not be the latest approved by FDA.
For current labeling information, please visit https://www.fda.gov/drugsatfda

HIGHLIGHTS OF PRESCRIBING INFORMATION

These highlights do not include all the information needed to use BILTRICIDE safely and effectively. See full prescribing information for BILTRICIDE.

BILTRICIDE (praziquantel) tablets, for oral use
Initial U.S. Approval: 1982

RECENT MAJOR CHANGES

Warnings and Precautions (5.6) 12/2023

INDICATIONS AND USAGE

Biltricide is an anthelmintic indicated in patients aged one year and older for the treatment of the following infections:

- Schistosomiasis due to all species of schistosoma (for example, *Schistosoma mekongi*, *Schistosoma japonicum*, *Schistosoma mansoni* and *Schistosoma hematobium*), and,
- Clonorchiasis and Opisthorchiasis due to the liver flukes, *Clonorchis sinensis* and *Opisthorchis viverrini*

DOSAGE AND ADMINISTRATION

- Schistosomiasis: 20 mg/kg bodyweight 3 times a day separated by 4 to 6 hours for 1 day only. (2.1)
- Clonorchiasis and Opisthorchiasis: 25 mg/kg 3 times a day separated by 4 to 6 hours for 1 day only. (2.1)
- Take with water during meals. Do not chew or keep segments in the mouth. (2.2)
- For pediatric patients under 6 years of age, the tablets may be crushed or disintegrated and mixed with semi-solid food or liquid. (2.2)
- For additional administration instructions see the full prescribing information.

DOSAGE FORMS AND STRENGTHS

- Tablets: 600 mg (with three scores (notches) on the tablet) (3)

CONTRAINDICATIONS

- Known hypersensitivity to Biltricide or any of its ingredients. (4.1)
- Concomitant administration with strong Cytochrome P450 3A enzyme (CYP 3A) inducers such as rifampin. (4, 5.6 7.1)

WARNINGS AND PRECAUTIONS

- Clinical Deterioration: Potentially life threatening clinical deterioration can occur in patients treated during the acute phase of schistosomiasis. (5.1)
- Central Nervous System (CNS) Effects: Biltricide can exacerbate central nervous system pathology due to schistosomiasis. Consider whether to administer to individuals reporting a history of epilepsy and/or other signs of potential central nervous systems involvement such as subcutaneous nodules suggestive of cysticercosis. (5.2)
- Potential Lack of Efficacy for Acute Schistosomiasis: This has been reported in observational studies (5.3).
- Cardiac Arrhythmias: Bradycardia, ectopic rhythms, ventricular fibrillation, and AV blocks has been observed with Biltricide administration. Monitor patients with cardiac arrhythmias during treatment (5.4).

ADVERSE REACTIONS

The adverse reactions reported were malaise, headache, dizziness, abdominal discomfort (with or without nausea), pyrexia and urticaria. (6)

To report SUSPECTED ADVERSE REACTIONS, contact Bayer HealthCare Pharmaceuticals Inc. at 1-888-842-2937 or FDA at 1-800-FDA-1088 or www.fda.gov/medwatch

DRUG INTERACTIONS

- Moderate CYP 3A Inducers: Avoid concomitant administration of moderate CYP 3A inducers, for example, efavirenz (5.6, 7.1)

USE IN SPECIFIC POPULATIONS

- Pediatrics: Safety has not been established in pediatric patients younger than 1 year of age. (8.4)
- Hepatic Impairment: Monitor patients for adverse reactions when administering the recommended dose of Biltricide to hepatosplenic schistosomiasis patients with moderate to severe liver impairment (Child-Pugh Class B or C). (8.6)

See 17 for PATIENT COUNSELING INFORMATION

Revised: 12/2023

Fig. 11.1 Extract of the Biltricide® labeling in the United States

patients and caregivers, are outlined in the most recent labeling approved by the FDA [11]. This information, including dosage guidelines, safety warnings, and usage instructions, is accessible through the FDA's official database of approved drug products at https://fda.gov/drugsatfda (Fig. 11.1).

What Is the European Medicines Agency?

Established in 1995, the European Medicines Agency (EMA) is a decentralized agency of the European Union (EU) responsible for the scientific evaluation, supervision, and safety monitoring of medicines for human and veterinary use. The EMA ensures that medicinal products on the EU market meet rigorous standards of safety, efficacy, and quality, based on comprehensive scientific assessments conducted in collaboration with EU Member States and the European Economic Area (EEA), which together represent a population of approximately 450 million [12].

As a networking organization, the EMA engages thousands of experts from across Europe who contribute to its seven scientific committees. These committees oversee the life cycle of medicines, covering development, market authorization, and post-market safety monitoring, and are supported by working groups and expert panels drawn from competent authorities within the EU. The EMA also collaborates closely with the European Commission and pharmaceutical industry stakeholders to fulfill its regulatory mandate. The EMA is governed by a Management Board of 36 members acting in the public interest, while its Executive Director, the legal representative of the Agency, oversees operational management, staffing, and the implementation of the annual work program [13]. To maintain robust oversight, the Agency employs regulatory mechanisms that are frequently reviewed and refined. The legislative framework for pharmaceuticals is detailed in The Rules Governing Medicinal Products [14] in the EU, with Volumes 1 and 5 focusing on human and veterinary medicinal products, respectively, and supplementary guidelines provided in additional volumes. To promote transparency, the EMA provides clear and impartial information about medicines, including public scientific assessment reports and summaries in lay language. Its collaborative structure ensures comprehensive scientific consultation within the European medicines regulatory network, enhancing public health across the EU [13].

Prior to the submission of a Marketing Authorisation Application (MAA), EMA offers various mechanisms for scientific input. These include scientific advice for full development programs, special protocol assistance for specific clinical study proposals, and pre-submission interactions to prepare for successful regulatory filings. For medicines targeting rare diseases or unmet medical needs, EMA grants designations such as Committee for Orphan Medicinal Products (COMP) or Priority Medicines (PRIME) [15, 16]. PRIME is a regulatory support program designed to enhance the development of medicines addressing unmet medical needs. It offers early and proactive regulatory support, priority access to scientific advice, and accelerated assessments to optimize development plans and improve access to innovative treatments [16].

After a Marketing Authorisation (MA) is granted, EMA implements procedures to monitor post-market safety and assess proposed changes to the MA throughout the product's life cycle [17]. Additionally, EMA offers a unique regulatory pathway, the EU-Medicines for All (EU-M4all) procedure—formerly known as the Article 58 procedure under Regulation (EC) No. 726/2004. This pathway facilitates scientific review for medicines intended for use outside the EU, often in LMICs, in collaboration with the WHO and local health authorities. The procedure culminates in a scientific opinion issued by the Committee for Medicinal Products for Human Use (CHMP) but does not result in EU marketing authorization [18].

As of July 2024, 16 products have received a positive CHMP opinion under the EU-M4all pathway. Notably, in December 2023, (*R*)-praziquantel, also known by its international nonproprietary name Arpraziquantel, a pediatric formulation developed by the Pediatric Praziquantel Consortium, obtained a positive scientific opinion under this procedure. This formulation, consisting solely of the (*R*)-enantiomer, exemplifies the pathway's capacity to facilitate access to essential medicines for LMICs [19].

Why Did (*R*)-Praziquantel Receive a Positive CHMP Opinion?

The EMA concluded that the benefits of (*R*)-praziquantel, (*R*)-PZQ, outweigh its risks. A pivotal clinical trial confirmed its effectiveness in treating schistosomiasis caused by *Schistosoma mansoni* or *Schistosoma haematobium*, matching the efficacy of PZQ. The formulation is specifically designed for young children aged 3 months to 6 years, offering easier dosing and improved palatability, with side effects considered mild to moderate. The approved label endorses the use of (*R*)-PZQ 150 mg dispersible tablets for this age group. This positive opinion enabled the inclusion of (*R*)-PZQ 150 mg on the WHO List of Prequalified Medicines. On May 3, 2024, the WHO officially added the product to the list, following the EMA's scientific opinion issued in December 2023 [20].

What Is the Role of the World Health Organization?

Established in 1948, the World Health Organization (WHO) is a specialized agency of the United Nations. It serves as a global coordinator, connecting nations, partners, and communities to address health emergencies like COVID-19 and Zika. WHO also combats major communicable diseases such as HIV, malaria, and tuberculosis, while addressing chronic conditions like diabetes, heart disease, and cancer. Additionally, WHO drives global efforts to expand universal health coverage and promote healthier lives, from pregnancy care through old age, guided by science-based policies and programs [21, 22]. WHO is governed by the World Health Assembly, which consists of delegations from all 194 Member States. The Assembly determines policies, appoints the Director-General, supervises financial policies, and approves the program budget. The Executive Board, with 34 technically qualified members serving three-year terms, implements the Assembly's decisions, advises on policies, and facilitates its functions. The Director-General, nominated by the Executive Board and appointed by the Assembly, oversees WHO's global health initiatives. WHO operates through six regional offices, each responsible for coordinating activities within its designated area:

- African Region (AFRO)—Brazzaville, Republic of Congo
- Region of the Americas (AMRO/PAHO)—Washington, D.C., USA
- South-East Asia Region (SEARO)—New Delhi, India
- European Region (EURO)—Copenhagen, Denmark
- Eastern Mediterranean Region (EMRO)—Cairo, Egypt
- Western Pacific Region (WPRO)—Manila, Philippines [23]

While WHO does not regulate drugs directly like national authorities (e.g., the FDA or EMA), it plays a vital role in global medicine oversight through key activities such as establishing international standards, prequalifying medicines, providing technical assistance, monitoring pharmacovigilance, facilitating collaborative procedures, promoting global regulatory harmonization, and developing model lists of essential medicines [24].

Building on its critical role in global medicine oversight, WHO publishes the Model List of Essential Medicines, which identifies medicines considered essential for addressing the most important needs within health systems. This list serves as a guide for countries to select and prioritize medicines for their populations. Updated

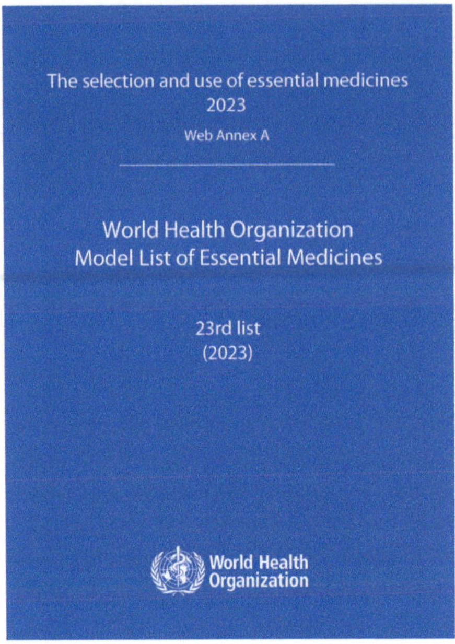

WHO Model List of Essential Medicines – 23rd List (2023)

6. ANTI-INFECTIVE MEDICINES	
6.1 Anthelminthics	
6.1.1 Intestinal anthelminthics	
albendazole	**Tablet (chewable, scored):** 400 mg.
ivermectin	**Tablet:** 3 mg.
levamisole	**Tablet:** 50 mg; 150 mg (as hydrochloride).
mebendazole	**Tablet (chewable):** 100 mg; 500 mg.
niclosamide	**Tablet (chewable):** 500 mg.
praziquantel	**Tablet:** 150 mg, 500 mg **Tablet (scored):** 600 mg.
pyrantel	**Tablet (chewable):** 250 mg (as embonate or pamoate).
6.1.2 Antifilarials	
albendazole	**Tablet (chewable, scored):** 400 mg.
diethylcarbamazine	**Tablet:** 50 mg; 100 mg (dihydrogen citrate).
ivermectin	**Tablet:** 3 mg.
6.1.3 Antischistosomals and other antitrematode medicines	
praziquantel	**Tablet:** 150 mg, 500 mg. **Tablet (scored):** 600 mg.
triclabendazole	**Tablet (scored):** 250 mg.
Complementary List	
*oxamniquine**	***Capsule:*** *250 mg.* ***Oral liquid:*** *250 mg/5 mL.* **For use when praziquantel treatment fails.*
6.1.4 Cysticidal medicines	
Complementary List	
albendazole	***Tablet (chewable):*** *200 mg [c].* ***Tablet (chewable, scored):*** *400 mg.*
mebendazole	***Tablet (chewable):*** *100 mg [c], 500 mg.*
praziquantel	***Tablet:*** *150 mg, 500 mg.* ***Tablet (scored):*** *600 mg.*

WHO Model List of Essential Medicines – 23rd List (2023) page 7

Fig. 11.2 WHO's 23rd Essential Medicines List, page 7

biennially by the Expert Committee on Selection and Use of Essential Medicines, the list ensures alignment with the latest evidence and evolving health needs. The most recent updates of the list, released in July 2023, include the 23rd Essential Medicines List (EML) and the ninth Essential Medicines List for Children (EMLc) [25] (Fig. 11.2).

Praziquantel 600 mg, World Health Organization Pre-qualified

On February 22nd, 2024, the WHO Prequalification Unit—Medicines Assessment Team (PQT/MED) approved PZQ 600 mg tablets (Cesol® 600), officially adding them to its List of Prequalified Products. PZQ is included on the list in three strengths: 150 mg, 500 mg, and 600 mg [26]. As of September 5th 2025, (R)-PZQ 150 mg dispersible tablets is on the 10th WHO Model List of Essential Medicines for Children [https://www.who.int/publications/i/item/B09475] [51] (Fig. 11.3).

Regulatory Know-How

Navigating the diverse regulatory landscapes across different regions is a complex and often challenging task. Although extensive information is publicly available, interpreting guidelines and requirements remains a significant challenge. Many

WHO Reference Number	INN, Dosage Form and Strength	Product Type	Therapeutic Area	Applicant	Dosage Form	Basis of Listing	Date of Prequalification
EMA Art 58 H-W-4252-1	Arpraziquantel Tablet, Dispersible 150mg	Finished Pharmaceutical Product	Neglected tropical diseases	Merck Europe B.V.	Tablet, Dispersible	Alternative Listing	
NT013	Praziquantel Tablet, Film-coated 600mg	Finished Pharmaceutical Product	Neglected tropical diseases	Merck Healthcare KGaA	Tablet, Film-coated	Prequalification - Full	22 Feb, 2024

Fig. 11.3 Extract from the WHO list of prequalified products [27, 28]

organizations turn to external vendors and specialized tools to gather, organize, and distill essential regulatory insights. Examples include:

- *Cortellis*: A comprehensive platform offered by Clarivate Analytics, providing integrated regulatory intelligence with updates and insights across various regulatory domains [29].
- *Regulatory Affairs Professionals Society (RAPS)*: A key resource offering educational materials, publications, and regulatory news to support professionals navigating global regulatory challenges [30].
- *Vivpro AI*: An AI-powered platform designed to extract and synthesize regulatory information from validated sources, efficiency in regulatory compliance [31].

These tools play a critical role in helping organizations remain informed and compliant in evolving regulatory environments. Nevertheless, many companies continue to allocate significant resources to develop and maintain regulatory intelligence as well as inhouse expertise. They also engage with international associations and health authorities to influence global policies through collaboration and partnerships. Regulatory affairs involve diverse responsibilities, ranging from deep scientific discussions and navigating political considerations to addressing administrative requirements. These administrative hurdles, though often overlooked, can contribute to delays in regulatory processes, particularly in international markets. Addressing such challenges requires a balance between leveraging advanced tools and fostering strategic collaborations.

Schistosomiasis, such as other neglected tropical diseases (NTDs), is a significant public health concern, particularly in less developed countries [32]. Addressing such diseases effectively requires early consideration of Target Product Profiles (TPPs), which outline the desired attributes of a pharmaceutical product, including its intended use, target population, route of administration, dosage, efficacy, and safety (Chap. 3). Regulatory strategies should also be developed early to guide submission planning, ensure compliance with international standards, and identify potential challenges. Products intended for use in resource-limited settings must balance suitability for these environments with adherence to the stringent standards required by global regulatory authorities. While local requirements must be addressed, international standards remain the benchmark for ensuring the development and approval of safe, effective, and high-quality drugs.

Regulatory Procedure

In the United States, the Center for Drug Evaluation and Research (CDER), part of the FDA, oversees the approval and regulation of drugs, ensuring safety, efficacy, and quality through offices such as the Office of New Drugs (OND) and the Office of Generic Drugs (OGD). CDER evaluates Investigational New Drug (IND) applications—which include preclinical, manufacturing, and clinical study protocols—before allowing clinical trials to proceed. For market approval, it reviews New Drug Applications (NDAs) for new medicines, while generic drugs are approved through Abbreviated New Drug Applications (ANDAs), which must demonstrate bioequivalence to their branded counterparts, providing cost-effective alternatives [33].

In the EU, most new innovative medicines are evaluated by the EMA and authorized by the European Commission before they can be marketed and made available to patients. However, many older medicines that are currently available were authorized at the national level, as they were marketed before the establishment of the EMA. For PZQ, which was first evaluated in Germany, this context is particularly important as its early European regulatory history sets the stage for understanding the more recent EMA procedures. There are two primary routes for authorizing medicines in the EU. The centralized procedure is mandatory for human medicines containing a new API (Active Pharmaceutical Ingredient) – the biologically active component in a drug product responsible for its intended therapeutic effect. Today, this centralized procedure is the primary route for authorizing most new medicines in the EU. Once authorized by the EMA, these medicines can be marketed across all EU member states. Most generic medicines and over-the-counter drugs are still assessed and authorized at the national level in the EU, with information about these procedures typically available on the websites of the national competent authorities. For medicines that fall outside the scope of the centralized procedure but require marketing authorization in multiple EU member states, companies can use alternative pathways. The Mutual Recognition Procedure (MRP) allows one or more EU countries to recognize an existing national marketing authorization granted by another Member State. Alternatively, the decentralized procedure enables a company to apply for marketing authorization simultaneously in multiple Member States for a medicine that has not yet been authorized in any EU country. Both pathways provide streamlined processes to facilitate market access across the EU while ensuring compliance with regulatory standards [34].

International Markets

It is advisable to align the data package with the specific regulatory requirements of each target country early in the development process. While some countries provide comprehensive guidance and guidelines in English, others lack clear or accessible documentation, making early preparation essential. In low- and middle-income countries (LMICs), regulatory processes can be prolonged due to limited resources, technical capacity, and underdeveloped regulatory systems [35]. These delays can hinder timely access to quality-assured, safe, and effective medicines, impacting public health outcomes and life expectancy. To address these challenges, many LMICs adopt the principle of reliance to enhance the efficiency of their regulatory systems. Reliance allows regulatory authorities to leverage the scientific evaluations and decisions of trusted bodies, such as Stringent Regulatory Authorities (SRAs), while retaining their independent decision-making authority [36]. This approach can be applied across all regulatory activities, including marketing authorization, enabling faster approvals based on prior assessments by other competent institutions. In support of these efforts, the WHO introduced the SRA Collaborative Registration Procedure (CRP) in 2015 [37]. This initiative accelerates national regulatory assessments and approvals for medicines, including small molecules, biologicals, generics, biosimilars, and vaccines, in LMICs. The CRP applies to products previously reviewed or approved by an SRA and deemed essential for public health, even if they fall outside the scope of WHO prequalification. By leveraging SRA evaluations, the CRP streamlines regulatory processes, ensuring faster access to critical medicines in resource-limited settings.

Regulatory Approval—Regulatory Dossier

Data collected during the drug development process must be accurately described in the registration dossier, with specific requirements varying by region [38]. For instance, in the EU, expert reports and tabulated summaries are mandatory, while written summaries are strongly encouraged to provide additional clarity [39]. In the USA, the FDA provides detailed guidance on the format and content of the New Drug Application (NDA), outlining the information needed to demonstrate a product's safety and efficacy [40]. Similarly, in Japan, applicants are required to prepare the GAIYO, a document summarizing technical information about the product in a structured format that aligns with national standards [41, 42]. To streamline these diverse requirements and minimize the need for multiple dossier formats, the International Council for Harmonisation (ICH) developed the ICH M4 (R4) guideline [43]. This guideline standardizes the structure of registration dossiers through the Common Technical Document (CTD) format, which is accepted in the USA, EU, and Japan. The CTD provides a unified format for presenting data, ensuring consistency, clarity, and efficiency in the evaluation process. It also reduces the time and resources needed to prepare submissions and simplifies the exchange of regulatory information between authorities. Additionally, the CTD format supports the preparation of electronic submissions, further enhancing efficiency. The CTD is divided into five modules, with Module 1 being region-specific and Modules 2–5 harmonized across regions (Fig. 11.4):

- *Module 1: Administrative and Prescribing Information*
  Contains region-specific documents, such as application forms and proposed labeling. The format and content of this module are determined by the relevant regulatory authority.
- *Module 2: Common Technical Document Summaries*
  Begins with an introduction summarizing the pharmaceutical, including its pharmacologic class, mode of action, and clinical use. This module also includes: CTD Table of Contents, CTD Introduction, Quality Overall Summary, Nonclinical Overview, Clinical Overview, Nonclinical Written and Tabulated Summaries, and Clinical Summary.
- *Module 3: Quality*
  Presents detailed information about the manufacturing, characterization, and control of the pharmaceutical substance and product.
- *Module 4: Nonclinical Study Reports*
  Includes data on pharmacology, toxicology, and other nonclinical studies to support safety and efficacy.
- *Module 5: Clinical Study Reports*
  Contains clinical trial data, including results from human studies to demonstrate safety, efficacy, and benefit–risk assessment.

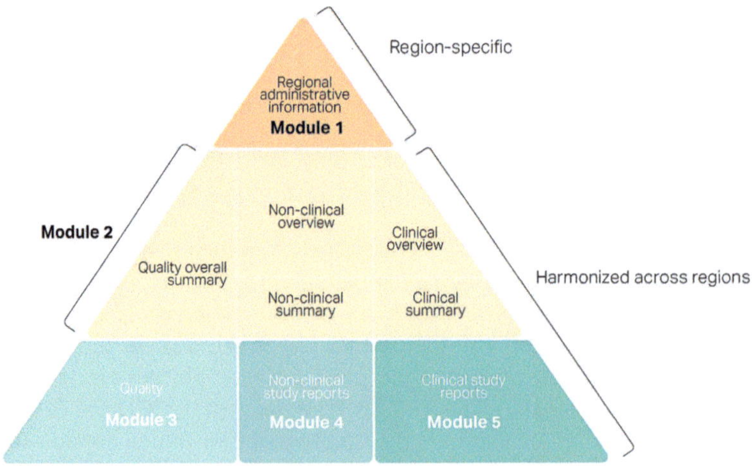

Fig. 11.4 Registration dossiers structured using the Common Technical Document (CTD) format

By adhering to the CTD format, pharmaceutical companies can meet regulatory requirements across multiple regions, making it a critical tool for global drug development and approval. Applying these standards from the outset of nonclinical experiments helps avoid later rewrites and ensures the dossier is "submission-ready." Using document templates, adhering to style guides for consistency, and meeting technical requirements such as electronic platform compatibility, citation accuracy, and proper hyperlinking streamlines the process, leading to a more efficient and compliant CTD dossier [44].

Regulatory Strategies and Health Authority Interactions

A regulatory strategy is a foundational assessment that defines critical pathways and objectives, ensuring effective planning for a product's market entry, life cycle management, and eventual discontinuation. These strategies are vital for ensuring compliance with regulations set by regulatory authorities and guide the entire development, testing, manufacturing, and marketing process. From early research to commercialization, a well-structured regulatory strategy helps identify and address risks, ensuring regulatory approval and successful product launch while maintaining patient safety and adherence to guidelines. Early regulatory engagement is crucial, as interacting with regulatory authorities in the development process allows companies to gain valuable insights into regulatory expectations. This early dialogue can inform development plans, study designs, and regulatory submission strategies. In addition, regulatory strategies should assess potential risks and develop mitigation plans. This may involve addressing safety concerns, demonstrating product efficacy, and ensuring compliance with standards throughout development. In clinical development, regulatory strategies are vital for planning robust studies that support regulatory submissions. These plans must include the selection of appropriate endpoints and adherence to Good Clinical Practice (GCP) guidelines [45]. For global rollouts, strategies should also consider ethnic sensitivity, patient diversity requirements, and the need for clinical evidence in local populations. Regulatory submissions, such as Investigational New Drug (IND) applications [46], New Drug Applications (NDAs) [47], and Marketing Authorization Applications (MAAs) [17], are key components of a regulatory strategy. These submissions guide the preparation of necessary documentation for obtaining regulatory approval. Beyond approval, life cycle management is also essential, covering post-marketing activities such as conducting post-approval studies, managing labeling changes, and addressing safety concerns or seeking additional indications. In today's global pharmaceutical market, regulatory strategies must consider differences in requirements across regions and countries. Developing global plans and harmonizing data requirements helps navigate international pathways efficiently. Early definition of target territories and a comprehensive project and regulatory strategy are vital.

Ultimately, a well-developed regulatory strategy serves as a comprehensive framework for guiding a product from early development through commercialization. This document, agreed upon by cross-functional teams, ensures alignment on key elements such as development milestones, risk management, accelerated approval options, and the creation of a Target Product Profile (TPP). Additionally, it defines the strategy for global rollout, taking into account region-specific requirements and establishing a clear timeline for launch and life cycle management.

Critical Elements for a Successful Strategy

It is recommended to clarify whether the drug's path to registration for the target disease and patient population (indication) is straightforward, characterized by established guidelines, precedents, or program-specific agreements with health authorities, or if it will require more negotiation due to the absence of guidelines, recent drug approvals, or validated clinical endpoints. Key considerations include:

- Ensuring compliance with the standards set by the International Council for Harmonisation (ICH), as well as USA and EU guidelines on indications, non-clinical studies, and chemistry, manufacturing, and controls (CMC). Any significant differences in these requirements should be noted and updated regularly to reflect changes in the regulatory landscape.
- Evaluating the availability of literature to support clinical endpoints, particularly in the absence of specific guidance documents or previous approvals.
- Reviewing program-specific advice received from health authorities, including prior consultations on similar product types or issues, and seeking clarification on clinical endpoint requirements.
- Understanding the approval process for competitive drugs, including challenges faced by other drugs in similar therapeutic areas.
- Identifying any relevant global regulatory issues, such as transmissible spongiform encephalitis, that may impact the approval process.

For any chosen regulatory pathway, it is highly advisable to confirm the development program with the relevant health authorities through direct consultations. Many regulatory agencies worldwide offer opportunities for scientific advice consultations, which can help refine risk mitigation strategies and ensure the available data aligns with regulatory expectations. This proactive engagement is particularly useful when data gaps or inconsistencies are identified, allowing for early resolution. Integrating risk assessments and mitigation plans into regulatory strategies is crucial, especially when navigating unfamiliar disease areas, new drug classes, or entering new countries. Geopolitical factors, including cultural context, should also be considered when developing effective regulatory plans.

Praziquantel and the Development of (*R*)-Praziquantel

Praziquantel (PZQ) oral tablets are the current standard treatment for schistosomiasis. Developed jointly by Bayer and Merck KGaA in the 1970s, PZQ was first registered for medical use in the early 1980s. Its initial approval came in Germany in 1982, meeting the regulatory requirements defined at the time. Following this, PZQ received approval in other countries for the treatment of various parasitic infections, including schistosomiasis and cestode infections. The medication is marketed as Biltricide® 600 mg by Bayer and as Cisticid® 500 mg, Cisticid® 600 mg, and Cysticide® 500 mg by Merck Healthcare KGaA for human use. Additionally, various generic PZQ products are now available worldwide [48].

Since 2007, Merck has supported the WHO in combatting schistosomiasis in Africa by donating over 1 billion tablets of PZQ 600 mg. Approximately 400 million patients, primarily school-age children, have been treated with these donations. PZQ 600 mg tablets are listed in the WHO Model List of Essential Medicines (21st list, 2019) and the WHO Model List of Essential Medicines for Children (seventh list, 2019).

In 2012, the Pediatric Praziquantel Consortium began developing an innovative PZQ formulation for pediatric patients. While the existing PZQ tablets are approved for children aged 6 years and older, the new formulation, containing only the active enantiomer (*R*)-PZQ, is designed for children aged 3 months to 6 years. The smaller tablets, which can be dissolved in water, also offer improved palatability due to a taste-masking excipient. A clinical development plan was created to establish proof of efficacy, safety, and clinical benefit for (*R*)-PZQ, with the goal of obtaining marketing authorization. A substantial preclinical data package was followed by clinical studies in healthy volunteers and children during phase I, II, and III trials, conducted in endemic countries such as Kenya and Ivory Coast [20]. Given the challenges of conducting clinical development in countries with unstable political systems and evolving regulatory frameworks, the program was carefully prepared to meet unexpected requirements. Also the program underwent discussions with multiple stringent regulatory authorities to ensure high standards of quality, safety, and efficacy. Scientific advice meetings with German Health Authorities (BfArM), the WHO Prequalification Team, and the EMA during the EU-M4all procedure provided essential guidance, culminating in a positive scientific opinion from the CHMP issued in December 2023. The newly obtained and availaible legacy data for PZQ were combined into a single regulatory dossier because the existing documentation, based solely on its established use in the field and supported by bibliographic references, no longer met current regulatory standards. To address this, the EMA allows a "bridging approach" under its hybrid application process, as outlined in Article 8(3) of Directive 2001/83/EC [49]. This application type is designed specifically for known active substances.

After almost 10 years of efforts invested in the development of a new tablet formulation, conducting clinical studies in endemic countries, consolidating data into an updated regulatory dossier, and undergoing review under the EU-M4all procedure, the positive scientific opinion from the EMA was issued in December 2023

[20]. This led to the inclusion of (*R*)-PZQ in the WHO List of Prequalified Medicines [50], representing a major step toward the ultimate goal of drug development and registration: providing safe and effective treatment options.

Summary

During the initial development of praziquantel (PZQ) in the late 1970s, the regulatory landscape was less advanced compared to today. Approvals were obtained through national health authorities using regulatory procedures, and dossiers and documentation were not standardized according to frameworks such as the Common Technical Document (CTD). Over the years, joint efforts by the International Council for Harmonisation (ICH) and various national health authorities have aimed to harmonize standards, establish common requirements, and implement stringent procedures to ensure the availability of safe and effective drugs for patients worldwide. In addition to harmonized regulatory standards, there are extensive requirements for clinical data, encompassing preclinical studies, clinical trials, and post-marketing surveillance studies to monitor the safety and efficacy of drugs beyond the development phase. Furthermore, the evaluation of manufacturing processes and quality control is a critical pillar of regulatory oversight, with comprehensive guidance issued by health authorities, such as the European Medicines Agency (EMA).

Regulatory bodies, such as the United States Food and Drug Administration (FDA), the EMA, and national competent authorities, are responsible for the scientific evaluation, supervision, and safety monitoring of medicines to ensure their safety, efficacy, and quality within their respective jurisdictions. Defining the target markets early in the regulatory strategy is crucial, as requirements can vary significantly across regions. Understanding the specific guidelines and expectations of regulatory bodies in the intended approval regions is essential. Failure to comply with these requirements may lead to delays, requests for additional data, or even rejection of the drug application in the target market.

Throughout the drug development process, all clinical trials must adhere to Good Clinical Practice (GCP) standards, including proper study design, ethical considerations, and accurate data collection. Comprehensive data on efficacy and safety are critical to demonstrating a drug's effectiveness against target parasites. For example, during the development of Arpraziquantel, which contains the more active (*R*)-enantiomer, a tailored regulatory strategy was essential. This strategy combined historical data for PZQ with newly generated, state-of-the-art clinical data to meet modern regulatory standards. Extensive engagement with competent authorities helped refine the approach, culminating in the submission via the EMA's EU-4all procedure. The combined data were presented in a single regulatory dossier, as existing documentation relying solely on established use and bibliographic references was no longer considered sufficient. The dossier was submitted as a hybrid application under Article 8(3) of Directive 2001/83/EC, resulting in a positive scientific opinion issued by the EMA in December 2023.

Competing Interests K.S. is an employee of Merck Healthcare KGaA.

References

1. World Health Organization (2024) https://www.who.int/news-room/questions-and-answers/item/who-listed-authorities. Accessed 09 Aug 2024
2. European Medicines Agency (2022) https://www.ema.europa.eu/en/human-regulatory-overview/research-development/compliance-research-development/good-manufacturing-practice. Accessed 09 Aug 2024
3. World Health Organization (2018) https://www.who.int/teams/health-product-policy-and-standards/standards-and-specifications/norms-and-standards/gmp. Accessed 09 Aug 2024
4. The United State Food and Drug Administration (2023) https://www.fda.gov/drugs/pharmaceutical-quality-resources/current-good-manufacturing-practice-cgmp-regulations. Accessed 09 Aug 2024
5. Vargesson N, Stephens T (2021) Thalidomide: history, withdrawal, renaissance, and safety concerns. Expert Opin Drug Saf 20(12):1455–1457. https://doi.org/10.1080/14740338.2021.1991307
6. International Council for Harmonisation (2024) https://www.ich.org/page/history. Accessed 19 July 2024
7. International Council for Harmonisation (2024). https://www.ich.org/page/management-committee. Accessed 19 July 2024
8. International Council for Harmonisation (2024) https://www.ich.org/page/members-observers. Accessed 19 July 2024
9. The United State Food and Drug Administration (2023) https://www.fda.gov/about-fda/what-we-do. Accessed 23 July 2024
10. The United State Food and Drug Administration (2023) https://www.fda.gov/regulatory-information/fda-rules-and-regulations. Accessed 23 July 2024
11. The United State Food and Drug Administration (2023) https://www.accessdata.fda.gov/scripts/cder/daf/index.cfm?event=overview.process&ApplNo=208820. Accessed 23 July 2024
12. European Medicines Agency (2023) https://www.ema.europa.eu/en/about-us/history-ema. Accessed 19 July 2024
13. European Medicines Agency (2023) https://www.ema.europa.eu/en/about-us/who-we-are. Accessed 19 July 2024
14. https://health.ec.europa.eu/medicinal-products/eudralex_en
15. European Medicines Agency (2023) https://www.ema.europa.eu/en/committees/committee-orphan-medicinal-products-comp. Accessed 19 July 2024
16. European Medicines Agency (2023) https://www.ema.europa.eu/en/human-regulatory-overview/research-development/prime-priority-medicines. Accessed 19 July 2024
17. European Medicines Agency (2023) https://www.ema.europa.eu/en/human-regulatory-overview/marketing-authorisation. Accessed 19 July 2024
18. European Medicines Agency (2023) https://www.ema.europa.eu/en/committees/committee-medicinal-products-human-use-chmp. Accessed 19 July 2024
19. European Medicines Agency (2023) https://www.ema.europa.eu/en/news/new-treatment-young-children-parasitic-disease-schistosomiasis. Accessed 19 July 2024
20. European Medicines Agency (2024) https://www.ema.europa.eu/en/opinion-medicine-use-outside-EU/human/arpraziquantel. Accessed 19 July 2024
21. World Health Organization (2024) https://www.who.int/about. Accessed 20 July 2024
22. World Health Organization (2024) https://www.who.int/about/history. Accessed 20 July 2024
23. World Health Organization (2024) https://www.who.int/about/structure. Accessed 20 July 2024
24. World Health Organization (2024) https://www.who.int/our-work. Accessed 20 July 2024
25. World Health Organization (2024) https://www.who.int/groups/expert-committee-on-selection-and-use-of-essential-medicines/essential-medicines-lists. Accessed 20 July 2024

26. World Health Organization (2024) https://www.who.int/publications/i/item/WHO-MHP-HPS-EML-2023.02. Accessed 20 July 2024
27. World Health Organization (2024) https://extranet.who.int/prequal/news/mercks-praziquantel-prequalified. Accessed 20 July 2024
28. World Health Organization (2024). https://extranet.who.int/prequal/medicines/prequalified/finished-pharmaceutical-products. Accessed 20 July 2024
29. Clarivate (2024) https://clarivate.com/life-sciences-healthcare/cortellis/. Accessed 14 Aug 2024
30. Regulatory Affairs Professionals Society (2024) https://www.raps.org/join-raps?GA_network=g&GA_device=c&GA_campaign=2071988053&GA_adgroup=78095827164&GA_target=&GA_placement=&GA_creative=595247874160&GA_extension=&GA_keyword=raps%20regulatory&GA_loc_physical_ms=9189258&GA_landingpage=https://www.raps. Accessed 14 Aug 2024
31. Vivpro (2024) https://vivpro.ai/. Accessed 14 Aug 2024
32. World Health Organization (2024) https://www.who.int/news-room/fact-sheets/detail/schistosomiasis. Accessed 20 July 2024
33. The United State Food and Drug Administration (2023) https://www.fda.gov/drugs/development-approval-process-drugs/how-drugs-are-developed-and-approved. Accessed 23 July 2024
34. European Medicines Agency (2023) https://www.ema.europa.eu/en/human-regulatory-overview. Accessed 19 July 2024
35. World Health Organization (2020) https://apps.who.int/gho/data/view.main.LIFEWBLMI?lang=en. Accessed 20 July 2024
36. World Health Organization (2016) https://extranet.who.int/prequal/sites/default/files/document_files/75%20SRA%20clarification_Feb2017_newtempl.pdf. Accessed 20 July 2024
37. World Health Organization (2022) https://www.who.int/news/item/15-02-2022-who-collaborative-registration-procedure-using-stringent-regulatory-authorities-medicine-evaluation-reliance-in-action. Accessed 20 July 2024
38. International Council for Harmonisation (2024) https://www.ich.org/page/ctd. Accessed 23 July 2024
39. Public Health European Commission (2003) https://health.ec.europa.eu/document/download/672171e7-8f79-41a9-a523-c82d9c07a0eb_en. Accessed 22 July 2024
40. The United State Food and Drug Administration (2001) https://www.fda.gov/regulatory-information/search-fda-guidance-documents/submitting-marketing-applications-according-ichctd-format-general-considerations-2001. Accessed 23 July 2024
41. Pharmaceutical and Medical Devices Agency (2024) https://www.pmda.go.jp/english/. Accessed 23 July 2024
42. Japan Approvals Database a Jouhou Koukai Business (2008) http://www.jouhoukoukai.com/jad_subscribers/misc/gaiyo/gaiyo.htm. Accessed 23 July 2024
43. European Medicines Agency (2023) https://www.ema.europa.eu/en/documents/scientific-guideline/ich-guideline-m4-r4-common-technical-document-ctd-registration-pharmaceuticals-human-use-organisation-ctd-step-5_en.pdf. Accessed 19 July 2024
44. International Council for Harmonisation (2024) https://www.ich.org/page/ctd. Accessed 19 July 2024
45. International Council for Harmonisation (2024) https://www.ich.org/page/efficacy-guidelines. Accessed 19 July 2024
46. National Institutes of Health (2024) https://www.cc.nih.gov/orcs/ind/what-is-an-ind. Accessed 21 July 2024
47. The United State Food and Drug Administration (2022) https://www.fda.gov/drugs/types-applications/new-drug-application-nda. Accessed 23 July 2024
48. Waechtler A, Cezanne B, Maillard D, Sun R, Wang S, Wang J, Harder A (2023) Praziquantel - 50 years of research. Chem Med Chem 18(12). https://doi.org/10.1002/cmdc.202300154

49. European Medicines Agency (2024) https://www.ema.europa.eu/en/human-regulatory-overview/marketing-authorisation/generic-hybrid-medicines/generic-hybrid-applications. Accessed 19 July 2024
50. World Health Organization (2024) https://extranet.who.int/prequal/medicines/ema-art-58-h-w-4252-1. Accessed 20 July 2024
51. The selection and use of essential medicines (2025) WHO Model List of Essential Medicines for Children, 10th list

12 Proprietary Aspects of Praziquantel

Arno Hartmann

Abstract

Praziquantel was jointly developed by Bayer AG (Elberfeld) and E. Merck (now Merck KGaA, Darmstadt) in the early 1970s. The two companies established a contractual agreement focused on human tropical medicine to formalize this partnership. Under the terms of the agreement, they cooperated on securing Intellectual Property Rights and formulating an effective filing strategy.

Keywords

Praziquantel · Patent · Intellectual property · Trade secret · Trademark · International nonproprietary name · Invention

Introduction

This chapter discusses the necessary and useful measures to protect experimental achievements, results, know-how, trade secrets, and intellectual property (IP) rights, including patents. It provides a general overview of IP law and discusses specific considerations related to praziquantel (PZQ) and its enantiomers (optical antipodes), as illustrated in Fig. 12.1 and previously introduced in Chap. 4.

A. Hartmann (✉)
Merck Healthcare KGaA, Darmstadt, Germany
e-mail: Arno.Hartmann@merckgroup.com

T. Spangenberg (ed.), *Praziquantel*,
https://doi.org/10.1007/978-3-031-97397-0_12

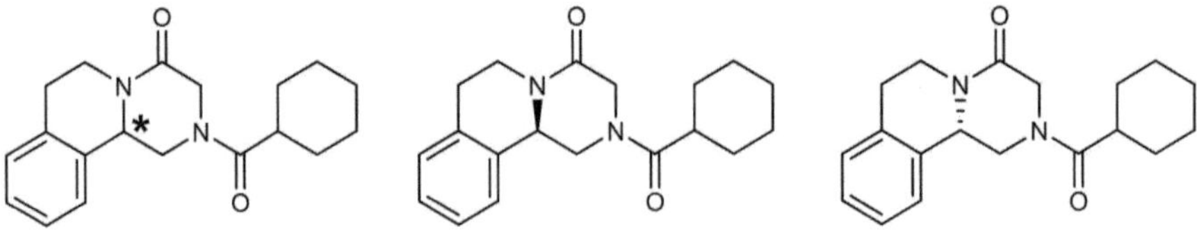

praziquantel (PZQ) **(*R*)-PZQ** **(*S*)-PZQ**

Fig. 12.1 Stereochemistry of PZQ. Structure of PZQ, highlighting the chiral center (asterisked, *). The individual enantiomers—(*R*)-PZQ (middle) and (*S*)-PZQ (right)—are also shown

What are Intellectual Property Rights and Why Are They Important?

Let us start by exploring the basic principles of IP before examining the relevance of IP rights, particularly patents. The four major types of IP rights are patents, copyrights, trademarks, and trade secrets. Although these rights differ in purpose and application, they all serve to protect creations of the human mind. Specialized categories of IP also include plant varieties, industrial designs, geographical indications, and semiconductors. Among these, patent rights are the most significant. A patent grants an inventor or patent holder an exclusive right, conferred by the state where the patent is filed. This right is typically valid for 20 years, provided that maintenance fees are paid. It allows the owner to prevent others from making, using, selling, offering for sale, or importing the patented invention within the territory where the patent is granted. Any unauthorized use constitutes patent infringement, though certain exceptions, such as the research exemption or the Bolar provision, permit limited use without infringement.

The importance of patent rights is evident, as generating inventions and developing related products is often costly and requires significant investment. Patent protection grants exclusivity for a limited time to the patent holder, safeguarding the invention and establishing its economic value. By providing market exclusivity, patents increase the potential profitability of innovations, enabling a return on investment (ROI). During the patent's lifetime, it keeps competitors off the market and ensures a ROI and profit. Patent rights are competitive tools that can be exploited in various ways. They may serve as strategic assets for internal developments or products, or generate income through licensing agreements or sales.

Criteria for Patentability and Types of Patents in the Healthcare Sector

A patentable invention must fulfill certain criteria to be accepted by patent offices and to have a realistic chance of being granted. Therefore, the invention and its corresponding patent application must be technical in nature, novel, inventive (i.e., involve an inventive step), useful, and nonobvious. This generally means that the invention must have industrial applicability. For example, scientific theories, equations, mathematical methods, and mental processes are not patentable. Inventive step or inventiveness means that the solution to an unsolved technical problem is not obvious to a person skilled in the relevant field (e.g., a scientist). Pharmaceutical patent rights can cover various categories, including compounds, enantiomers, diastereomers, combinations, specific medical uses or indications, formulations (galenic specialties), dosing regimens, modes of administration, manufacturing methods, crystalline forms (polymorphic forms), specific salts or esters, biomarkers, prodrugs, and other related innovations.

What Is the Difference Between Patent Rights, Copyright, Trademarks and Trade Secrets?

Trade secrets are considered intellectual property rights, but they differ significantly from patent rights. Trade secrets are kept highly confidential and are not disclosed to the public. In contrast, patent rights, and the inventions they protect, are initially treated as confidential to preserve their novelty, a key requirement for patentability. Once a patent application is filed, it is typically published 18 months later, as long as the application is still pending. As a result, patent rights become public, while trade secrets remain confidential, with the owner taking specific measures to maintain their secrecy. Copyrights and trademarks are also IP rights, but they differ significantly from patent rights, which protect technical inventions. Copyrights are automatically granted upon the creation of original works, such as books, music, or other artistic creations, and do not require formal registration (though registration can offer additional protection). Trademarks protect brand names, logos, slogans, and other identifiers, which can include two- or three-dimensional words, phrases, symbols, or designs [1].

From an Invention to a Patent: The Patent Life Cycle

A patent application can be filed when the related invention meets the criteria for patentability. Developing the ability to recognize an invention is crucial once sufficient data have been generated to support it. Determining the right moment to file can be challenging, as it requires ensuring that the data are comprehensive enough to justify a meaningful invention report without encompassing multiple inventions.

Inventions, patent applications, and valid patent rights are important features to protect all kinds of technologies, including pharmaceuticals, which support technological advancement, the global economy, and, most importantly, patient care. This section will focus on pharmaceutical inventions, specifically the process of obtaining different types of patent rights, with an initial emphasis on compound protection and the drug praziquantel (PZQ).

A pharmaceutical invention typically occurs several years before the first patent application is filed. First, an experimental concept is needed along with a clear synthetic plan for compounds to be made. A biological target or a disease to be targeted by a certain active pharmaceutical ingredient needs to be identified. This normally requires a clear understanding of the biochemical pathways involved and the causality of a disease that leads to a disorder in an organism. During this phase, researchers must consider suitable chemical scaffolds, screening processes, available assays, and sources for initial testing compounds, such as compound libraries. Before early preclinical investigations can begin, the concept must be translated into practical experimental work. This early phase corresponds to the time frame from year −5 to year 0 in Fig. 12.2. Potential inventions often emerge during this early phase when series of small organic molecules are tested, derivatives of the most potent compounds are synthesized, and initial test results demonstrate the activity of specific compound classes. Determining when an invention is complete and fulfills the criteria for patentability requires experience and skills. The initial patent application must also meet formal requirements, including the proper format for submission to the chosen patent office [2]. In most jurisdictions, a patent application can be filed either on paper or online in electronic format. The first application may depend on the inventors' nationality and the jurisdiction where the invention was made. All patent applications must include a description of the invention, a set of claims, experimental data or examples, an abstract, and, if necessary, illustrations such as tables, drawings, or diagrams. A list of inventors is also required but can often be submitted later, as defined in the legal framework. The description should include the known state of the art and a summary of the invention's setup and its improvements over existing technologies. In the patent claims, the core of the invention is described in clear terms, defining the scope of the patent protection sought.

The first filing of a patent application (Year 0, Fig. 12.2) establishes the priority filing and enables the applicant to claim the filing date as the priority date for subsequent filings in other jurisdictions to extend patent protection. These are often

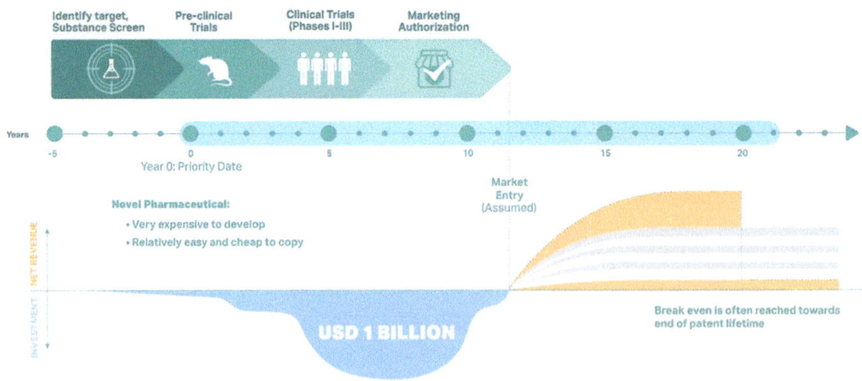

Fig. 12.2 Representative life cycle of a patent for a pharmaceutical product

referred to as foreign filings and must be submitted within 1 year of the first filing date to maintain priority. The priority date, which is the date the invention was made, applies to all subsequent foreign filings. This date is crucial because it determines whether prior art was made public before or after your invention, potentially challenging its patentability. The final opportunity to claim priority from the first filing is the first anniversary of the priority application, exactly 12 months after the filing date. If this deadline is missed, the priority can no longer be claimed, and it is lost. Most applicants, especially in the healthcare sector, choose the Patent Cooperation Treaty (PCT) process for their foreign filings. The PCT provides a unified procedure for filing patent applications worldwide. Established in 1970, the PCT is an international treaty with more than 150 Member States. All major countries or regions, such as the United States (US), Europe, Japan, Canada, South Korea, Australia, China, and India, which contribute significantly to the global pharmaceutical market, are PCT members. Exceptions include Argentina and Taiwan. Using the PCT route, it is possible to seek patent protection simultaneously in numerous countries through a single international filing process, rather than filing separately in multiple jurisdictions. The final examination and granting of patents are under the control of national or regional patent offices once the international application enters the "national phases." National or regional phase entry must be initiated within 30 months from the priority date (or 18 months from the international filing). Regardless of the chosen filing strategy—whether national, regional, or via the PCT (international)—the patent application becomes public 18 months after the priority filing. This is when the application becomes searchable in databases and is no longer confidential. Depending on the filing strategy, the application may be published with or without a search report that includes prior art references. These references must be addressed by the applicant when the patent office renders its official decisions, known as office actions. Office actions are formal communications from the patent office that outline the examiner's evaluation of the application, including any objections or rejections based on issues such as novelty, inventiveness, or clarity. The applicant must respond to these office actions, often by amending the application, to address the concerns raised by the examiner. All PCT applications include an International Search Report, which identifies the prior art found by the PCT authority during the search. This report is required for country-specific examinations once the application enters the national or regional phases.

Figure 12.3 is an excerpt from a PCT application, showing the first page of WO 2016/078765 A1, including the bibliographic data and the international search report generated by the World Intellectual Property Organization (WIPO). The application covers a "Method for the Production of Praziquantel and Precursors Thereof" and is discussed later in the section "The Praziquantel Patent Portfolio and Its History." It was published on May 26, 2016, about 18 months after its priority date, along with the international search report. The report, in this case, is relatively simple, citing only one literature reference and no patent references. Moreover, the only category of ascertained documents found belongs to category "A," which indicates that these documents define the general state of the art and are not considered

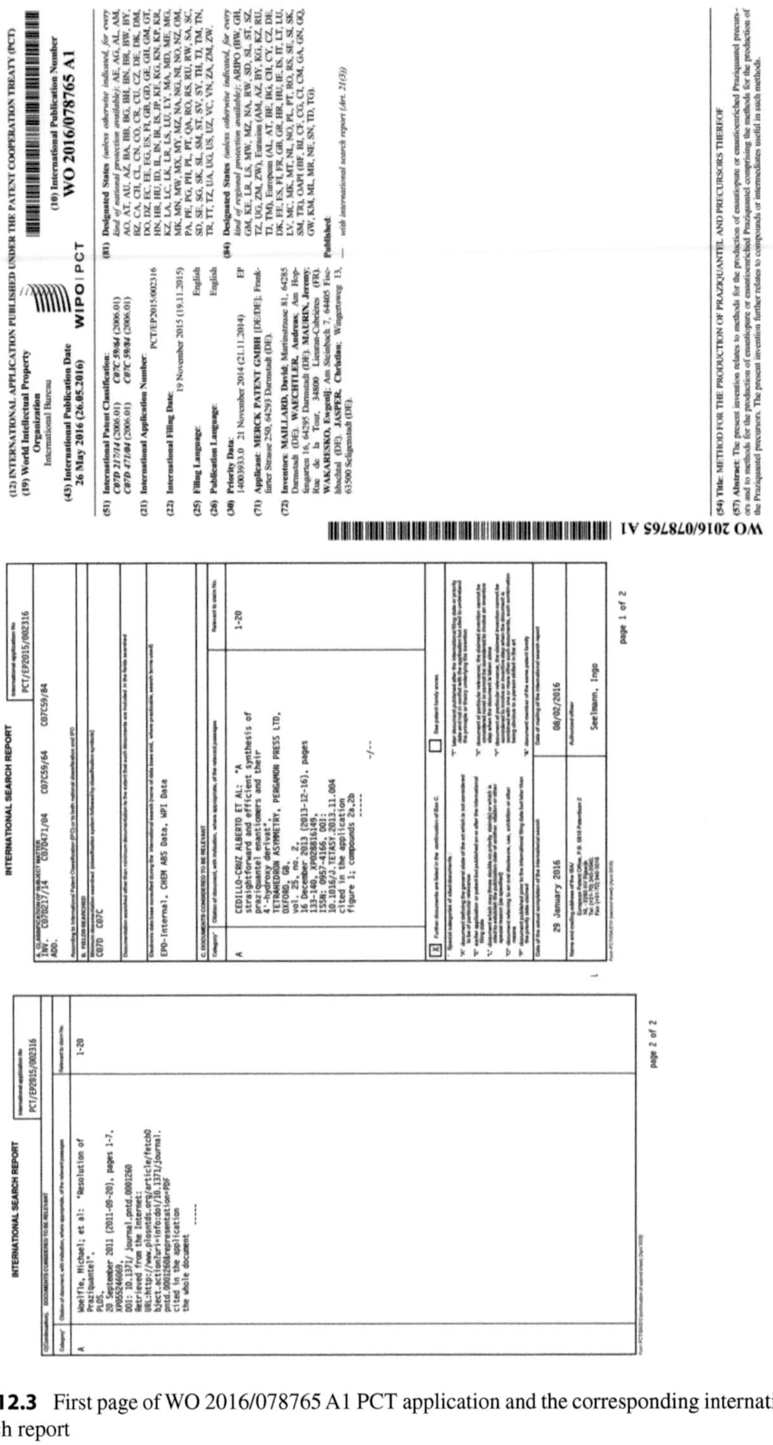

Fig. 12.3 First page of WO 2016/078765 A1 PCT application and the corresponding international search report

particularly relevant to the examination or patentability of the invention covered by WO 2016/078765 [3]. The more critical categories, which do not appear here, are categories "X" and "Y." Category "X" includes documents that may jeopardize patentability because, in light of the reference, the claimed invention may lack novelty or an inventive step. Category "Y" includes documents that may also jeopardize patentability, but only in combination with other references, which may render the invention obvious to a person skilled in the art.

The PCT route became available as a filing strategy in 1970. However, most of the early member states adopted the PCT system in the late 1970s. For example, the US joined the PCT in January 1978, along with Germany, Switzerland, and the United Kingdom. France, Sweden, Denmark, Luxembourg, Japan, Brazil, and the Russian Federation, became members later that same year. Currently, the PCT has 157 member states [4]. This is why the first and probably most relevant patent application for PZQ was not filed through the PCT route, but rather through national and regional filings. The priority application, DE2362539.0, which claims the compound along with other aspects, was filed with the German Patent Office at that time, as the PCT system had not yet been introduced. This application will be discussed in more detail in the section titled "The Praziquantel Patent Portfolio and Its History."

The development timeline for a pharmaceutical product, such as PZQ, often exceeds 10 years before marketing authorization is granted. In brief, the process begins with the drug discovery phase, followed by early preclinical investigations that involve target identification, compound screening, and the identification and optimization of lead candidates for further development. This is followed by preclinical and clinical trials (Chaps. 6, 7, 8, and 10). It is important to note that the average cost for developing a drug is substantial, often approaching or exceeding 1 billion US dollars. As a result, the break-even point—when income from product sales equals the investment—is typically reached just before patent protection expires (Fig. 12.2). To maximize ROIs, drugmakers rely on patent term extensions (PTEs) or supplementary protection certificates (SPCs), which offer up to 5 additional years of exclusivity. PTEs or SPCs require a separate application, which can be filed as long as a valid patent exists, and a market authorization (MA) has been granted for the tested novel pharmaceutical drug (Chap. 11). PTEs or SPCs must be filed within a specific timeframe, provided two conditions are met: the patent is granted and still in force, and a valid MA is available. The application is reviewed by national or regional patent offices and granted if all criteria outlined in the relevant regulations or patent law are satisfied. PTEs are available in countries such as the US, Canada, Japan, and Australia, while SPCs are available in Europe. However, not all jurisdictions offer patent term extensions, as it is the case in China and many countries in South America and Asia, where such extensions are not included in their national patent laws.

International Nonproprietary Name (INN)

Drugs and active ingredients, which are intended to be marketed as pharmaceutical drugs or that serve as the active pharmaceutical ingredient (API, Chap. 4) of such drugs, require a unique name that is recognized globally. This name must be precise to allow for harmonized identification, helping to avoid confusion or prescribing errors. The INN process, which establishes the rules and standards for drug nomenclature has been coordinated by the World Health Organization (WHO) since 1953. The INN system applies to a wide range of substances, including biologicals, small molecules, and similar products. The use of INNs is required by many authorities, such as the European Medicines Agency (EMA), the United States Food and Drug Administration (FDA) and is considered in both national and international legislation [5].

An INN application must be filed by the innovator or originator (product owner) and submitted to the WHO for further examination. The applicant must provide a completed application form with the required information. The requirements for an INN submission are very strict, and applicants must follow certain rules and use designated word stems that identify groups of substances with similar pharmacological activity when selecting name proposals [6]. For every class of compounds, different word stems must be used. After successful submission and the payment of a fee, a WHO expert advisory panel examines the application and proposes, if all formalities and requirements are met, an INN. The INN is then published in an official WHO gazette, allowing the public to provide comments or objections within a defined 4-month period. In case there are no objections filed, the proposed INN becomes a recommended INN and is published in the WHO Drug Information. Publications of newly selected recommended INNs occur twice a year. The recommended INN for the chemical substance (*RS*)-2-(cyclohexylcarbonyl)-2,3,4,6,7,11b-hexahydro-1*H*-pyrazino[2,2-a]isochinolin-4-on ($C_{19}H_{24}N_2O_2$) according to the IUPAC nomenclature is praziquantel (PZQ) and may be derived from the component of the chemical name, as follows: P(y)razi(ne iso)qu(inoline)ant(h)el(minitic) [7].

Importantly, INN is not synonymous with trade name. Trade names are used to market a certain product, such as a pharmaceutical drug. For example, PZQ was first launched in 1975 by Bayer for veterinary use under the trade names Droncit® and Cestocur® by Bayer AG. Other trade names for veterinary products include Anipracit, Aniprazol KH, Band-ex, Bihelminth, Broadline, Caniquantel, Docatel, Dolpac, Drontal, Equest Pramox, Equimax, Eqvalan duo, Fenprasel, Fenquantel, Furexel Combi, Milbactor, Milbemax, Nexgard Combo, Plerion, Prazifen-Kombi, Prazinex, Praziquasel, Profender, Professional Tremazol, Strantel, Tremazol, Vermis-Ex, and VetBancid. PZQ as a medication for humans was introduced in 1980 under the trade names Biltricide®, Cesol®, and Cycticide® (Fig. 12.4).

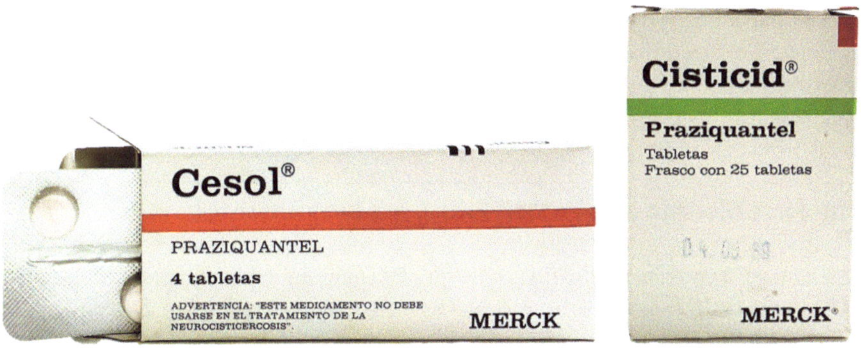

Fig. 12.4 Examples of trade names for praziquantel

The Praziquantel Patent Portfolio and the History Thereof

This section reflects on the PZQ patent portfolio, which is controlled or jointly owned by third parties, i.e., Bayer AG, by Merck KGaA or Merck Healthcare KGaA as a beneficial owner. The legal representation of all patent rights related to PZQ is handled by Merck Patent GmbH, a fully owned subsidiary of Merck KGaA. PZQ resulted from a joint research collaboration and biological screening agreement between Bayer AG and E. Merck (now Merck KGaA; Darmstadt) in the early 1970s. The first collaboration agreement was signed between the parties in March 1973 and focused on the field of tropical medicine (the Field) to find, develop, and market small organic molecules, the so-called new chemical entities (NCEs), for the treatment of mycotic diseases or infections caused by tropical parasites in humans. The initial idea of the parties was that Merck would synthesize these organic small molecules, which might be suitable candidates for screening in the field, while Bayer would provide the respective assays and screening tools.

Agreements of this type are already relatively complex and must consider and regulate various aspects. They need clear definitions for the important features of the contract, a clear working plan and distribution of responsibilities, term and termination rules, confidentiality obligations, cost sharing, and an agreement on how potential inventions are handled, i.e., who is the applicant, what the filing strategy and procedure are, who prepares the patent application, and in which countries are the patent applications filed. Other aspects include the exploitation of the results.

As a side note, it is quite interesting to remember how different communication and information exchange was in the 1970s in comparison to today, where one can easily set up virtual meetings, exchange e-mails frequently, communicate rapidly, and share large amounts of data electronically. For the younger readers among us it may be worthwhile digging a little bit deeper into the working climate of about 50 years ago.

The parties of the initial joint research collaboration and biological screening agreement realized that it might be worthwhile to expand the agreement to include veterinary uses and exploitation of the results. So, they concluded two additional contracts for this purpose in March 1976.

The First Meaningful Patent Filing for Praziquantel

One of the first patent applications covering PZQ among other compounds was filed in December 1973 and registered as German Patent Application *DE2362539.0*, which was the priority application with a priority of December 17, 1973. It was made public approximately 20 months later, on August 28, 1975. The German patent was granted on May 7, 1986. According to the abstract, the patent covers various 2-Acyl-4-oxo-hexahydro-4*H*-pyrazino [2, I-a]isoquinoline derivatives of the general formula depicted in Fig. 12.5.

In Fig. 12.5, –C(O)–R represents an acyl moiety of up to 26 carbon atoms and its physiologically acceptable acid addition and quaternary ammonium salts are anthelmintics and can be produced by reacting 4-oxo-1,2,3,6,7,11b-hexahydro-4*H*-pyrazino [2, 1-a]isoquinoline with an acid or a reactive functional derivative thereof [8].

As described previously, an applicant normally seeks protection in several jurisdictions through the so-called foreign applications in which the priority date of the first application, here DE 2362539.0, is claimed (Fig. 12.6). In this case a certain importance was anticipated, and the application for PZQ was filed in many countries listed using the standard two-letter country codes:

AR; AT; AU; BE; CA; CH; CS; DD; DE; DK; ES; FI; FR; GB; GR; GT; HK; HU; IE; IL IN; JP; KE; KR; LU; NL; NO; OA; PH; PL; RO; SE; SU; US; YU; and ZA [9].

It is also important to realize that the application was filed approximately 50 years ago and, therefore, the geographical scope and the two-letter country codes have changed. For example, DD stood for German Democratic Republic, which no longer exists.

The corresponding US equivalent of *DE2362539.0* was filed on December 16, 1974, and was granted as US Patent No. 4,001,411 on January 4, 1977 (Fig. 12.7), more than 9 years earlier than the German patent [10].

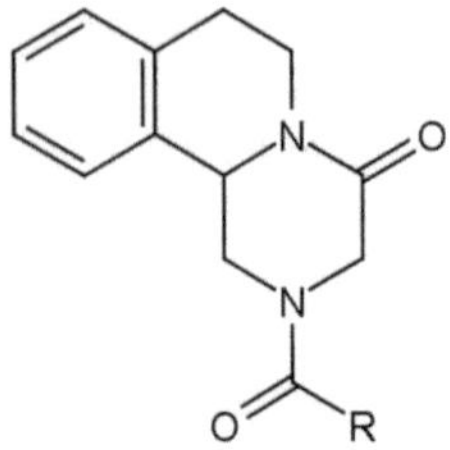

Fig. 12.5 Markush formula of praziquantel derivatives

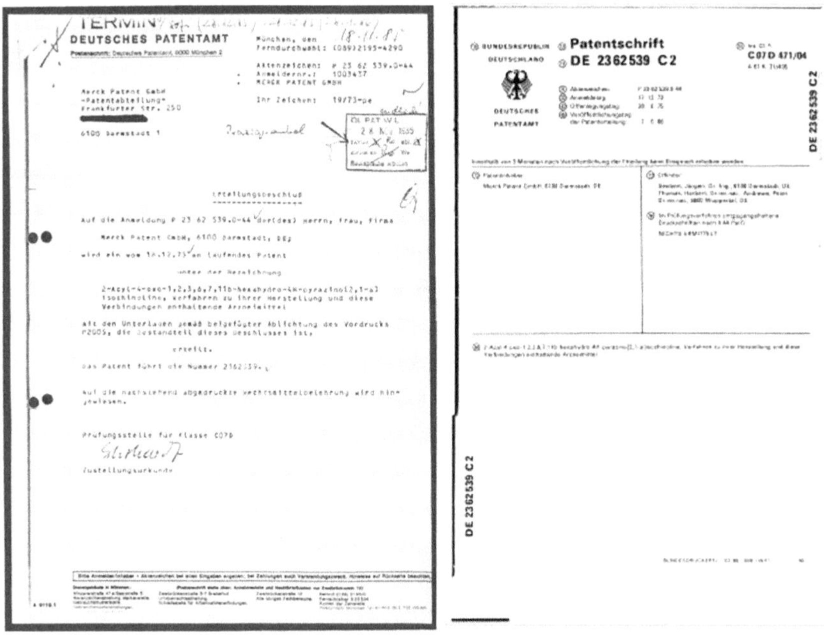
DEUTSCHES PATENTAMT

Patentschrift
DE 2362539 C2

DE 2362539 C2

Fig. 12.6 The bibliographic data and the decision of granting the German Patent Application DE2362539.0

The claim set consists of a total of 16 claims, 14 of which are compound claims, one covering a pharmaceutical composition, and one a method of treatment claim for managing cestode and trematode infestations. The method involves administering to an animal infested with cestodes or trematodes an anthelmintically effective amount of a compound of claim 1 (Fig. 12.7b).

However, several additional layers of patent protection surround the chemical substance PZQ, collectively forming the PZQ patent portfolio.

EP 3221296 B1 covers a "Method for the Production of Praziquantel and Precursors Thereof." This patent right is a granted patent as indicated by the classification B1 following the patent number on the title page below. In contrast, pending applications are classified as A1. The patent was derived from a PCT application *WO 2016/078765 A1* based on a European priority filing from November 21, 2014 (Fig. 12.3).

This invention relates to methods for the production of the enantiopure or enantioenriched PZQ precursors and to methods for the production of enantiopure or enantioenriched PZQ comprising the methods for the production of the PZQ precursors. The invention further relates to compounds or intermediates useful in such methods. The justification for this invention was that the anthelmintic PZQ has been registered, approved, and commercialized as a racemate. However, the more active molecule (eutomer) is the (*R*)-enantiomer [11]. PZQ has a bitter taste. This led to acceptance issues—in particular, in the treatment of young children. The

A

United States Patent [19] [11] **4,001,411**
Seubert et al. [45] **Jan. 4, 1977**

[54] **2-ACYL-4-OXO-PYRAZINO-ISOQUINOLINE DERIVATIVES AND PROCESS FOR THE PREPARATION THEREOF**

[75] Inventors: **Jürgen Seubert**, Darmstadt; **Herbert Thomas**; **Peter Andrews**, both of Wuppertal, all of Germany

[73] Assignee: **Merck Patent Gesellschaft mit beschraenkter Haftung**, Darmstadt, Germany

[22] Filed: **Dec. 16, 1974**

[21] Appl. No.: **533,467**

[30] **Foreign Application Priority Data**

Dec. 17, 1973 Germany 2362539

[52] U.S. Cl. .... **424/250**; 260/250 BC; 260/268 TR; 424/200; 424/232

[51] **Int. Cl.**[2] **C07D 241/36**

[58] **Field of Search** .... 260/268 TR; 424/250

[56] **References Cited**

OTHER PUBLICATIONS

Burger, Medicinal Chemistry 3rd Ed., Part I, p. 638.
Burger, Medicinal Chemistry 3rd Ed., Part II, p. 1588.

Primary Examiner—R. Gallagher
Assistant Examiner—Jose Tovar
Attorney, Agent, or Firm—Millen, Raptes & White

[57] **ABSTRACT**

2-Acyl-4-oxo-hexahydro-4H-pyrazino[2,1-a]isoquinoline derivatives of the formula

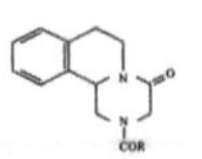

wherein COR is the acyl radical of an up to 26 carbon atom acid and their physiologically acceptable acid addition and quaternary ammonium salts, are anthelmintics and can be produced by reacting 4-oxo-1,2,3,6,7,11b-hexahydro-4H-pyrazino[2,1-a]isoquinoline with an acid or a reactive functional derivative thereof.

16 Claims, No Drawings

B

4,001,411

45

EXAMPLE H

Premix for Purposes of Veterinary Medicine, Suitable for Mixing with a Feed Material as the Vehicle to Obtain a Medical Animal Food

a. 25% Premix (preferably for larger animals):

25 kg. of 2-cyclohexylcarbonyl-HPI is mixed with 75 kg. of fine bran (wheat middlings) and/or lactose.

b. 5% Premix (preferably for smaller animals):

5 kg. of 2-cyclohexylcarbonyl-HPI are processed analogously to (a).

c. Example for the use of a premix produced according to (a) for combating Moniezia genera in the cattle intestine.

To obtain a suitable medical feed, 1 kg. of the premix produced according to (a) is combined with 9 kg. of a conventional feed concentrate. To combat Moniezia infestation, 400 g. of this medical feed, containing 10,000 mg. of 2-cyclohexylcarbonyl)-HPI, is administered to adult cattle, per head.

Analogously to Examples A-H, it is also possible to process, in place of 2-cyclohexylcarbonyl-HPI, the other active agents of Formula 1 or the physidogically acceptable salts thereof to obtain pharmaceutical preparations.

The preceding examples can be repeated with similar success by substituting the generically and specifically described reactants and/or operating conditions of this invention for those used in the preceding examples.

From the foregoing description, one skilled in the art can easily ascertain the essential characteristics of this invention, and without departing from the spirit and scope thereof, can make various changes and modifications of the invention to adapt it to various usages and conditions.

What is claimed is:

1. A compound of the formula

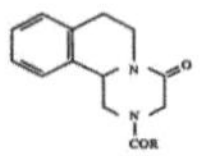

wherein R is (a) cycloalkyl of 3–7 carbon atoms, or a corresponding group mono-substituted by fluorine, chlorine, nitro, hydroxy, amino, formylamino, acetylamino, pentanoylamino, hexanoylamino, octanoylamino, oleoylamino, methoxyacetylamino, methylamino, dimethylamino or allylamino;

b. alkyl of up to 7 carbon atoms or a corresponding group substituted by methoxy or ethoxy; (c) oxocyclobutyl, oxocyclopentyl or oxocyclohexyl; or (d) cyclohexenyl; and the physiologically acceptable acid addition salts thereof.

46

2. A compound of claim 1 wherein R is alkyl of 1–6 carbon atoms.

3. A compound of claim 1 wherein R is cycloalkyl of 3–7 carbon atoms.

4. a compound of claim 1, 2-cyclohexylcarbonyl-4-oxo-1,2,3,6,7,11b-hexahydro-4H-pyrazino[2,1-a]isoquinoline.

5. A compound of claim 1, 2-propionyl-4-oxo-1,2,3,6,7,11b-hexahydro-4H-pyrazino[2,1-a]isoquinoline.

6. A compound of claim 1, 2-isobutyryl-4-1,2,3,6,7,11b-hexahydro-4H-pyrazino[2,1-a]isoquinoline.

7. A compound of claim 1, 2-cyclopropylcarbonyl-4-oxo-1,2,3,6,7,11b-hexahydro-4H-pyrazino[2,1-a]isoquinoline.

8. A compound of claim 1, 2-cyclobutylcarbonyl-4-oxo-1,2,3,6,7,11b-hexahydro-4H-pyrazino[2,1-a]isoquinoline.

9. A compound of claim 1, 2-cyclopentylcarbonyl-4-oxo-1,2,3,6,7,11b-hexahydro-4H-pyrazino[2,1-a]isoquinoline.

10. A compound of claim 1, 2-(3-cyclohexenylcarbonyl)-4-oxo-1,2,3,6,7,11b-hexahydro-4H-pyrazino[2,1-a]isoquinoline.

11. A compound of claim 1, 2-(4-oxocyclohexylcarbonyl)-4-oxo-1,2,3,6,7,11b-hexahydro-4H-pyrazino[2,1-a]isoquinoline.

12. A compound of claim 1, 2-(4-hydroxycyclohexylcarbonyl)-4-oxo-1,2,3,6,7,11b-hexahydro-4H-pyrazinl[2,1-a]isoquinoline.

13. A compound of claim 1, 2-cycloheptylcarbonyl-4-oxo-1,2,3,6,7,11b-hexahydro-4H-pyrazino[2,1-a]isoquinoline.

14. A compound of claim 1, 2-acetyl-4-oxo-1,2,3,6,7,11b-hexahydro-4H-pyrazino[2,1-a]isoquinoline.

15. A pharmaceutical composition consisting essentially of an anthelmintically effective amount per unit dosage of a compound of claim 1 in admixture with a pharmaceutically acceptable carrier.

16. A method of treating cestode and trematode infestations which comprises administering to an animal infested with cestodes or trematodes an anthelmintically effective amount of a compound of claim 1.

* * * * *

Fig. 12.7 (**a**) US Patent No. 4,001,411. The abstract provides a short description of the invention, refers to the number of claims and the foreign application priority data among other information, such as inventors, assignee, and cited references. Moreover, it shows the Markush structure of the claimed compounds. (**b**) Set of claims granted in the US

(*R*)-praziquantel, (*R*)-PZQ, eutomer is considered to have a less bitter taste than the (*S*)-praziquantel, (*S*)-PZQ, distomer [12]. This caused a strong demand for a cost-efficient manufacturing process for enantiomerically pure (*R*)-PZQ, which is briefly described by the quotation from *EP 3221296 B1* [13].

Significant efforts have been made over the last decades to develop a manufacturing process for (*R*)-PZQ (Chap. 4). This process can be divided into two groups: enantioselective synthetic routes and racemic processes combined with chiral resolution. While a few enantioselective processes have been reported, none of them are cost efficient. Woelfle et al. described a chiral resolution of PZQ precursor, praziquanamine (1,2,3,6,7,11b-hexahydro-pyrazino[2,1-a]isoquinolin-4-one), using (−)-dibenzoyl-L-tartaric acid to obtain the desired enantiomer [14]. This resolution achieves rather low yields because two crystallization steps are necessary to reach sufficiently high optical purity. Alberto Cedillo Cruz et al. described a chromatographic separation of the diastereomers naproxen-(*R*)/(*S*)-praziquanamide, ((11bS)- and (11bR)-[(2S)-2-(6-methoxy-2-naphthalenyl)-1-oxopropyl]-1,2,3,6,7,11b-hexahydro-2-4*H*-pyrazino[2,1-a]isoquinolin-4-one), synthesized from (*S*)-naproxen-acid chloride and praziquanamine, using an achiral phase. In order to obtain (*R*)-praziquanamine, the covalent bond in (11b*R*)-[(2*S*)-2-(6-methoxy-2-naphthalenyl)-1-oxopropyl]-1,2,3,6,7,11b-hexahydro-2-4*H*-pyrazino[2,1-a]isoquinolin-4-one must be cleaved under harsh conditions (85%

phosphoric acid, 150 °C) [15]. This process was laborious and proved economically unviable.

The present invention covered by *EP 3221296 B1* and corresponding equivalents provides an efficient method for the chiral resolution of a PZQ precursor, in particular, praziquanamine and *N*-{[(1*R*)-1,2,3,4-tetrahydro-1-isoquinolin-1-yl]methyl} acetamide, commonly referred to as praziquanacetamide. The new process works reliably even on a large scale and can be utilized to produce enantiopure or enantioenriched (*R*)-PZQ. This challenge has been surprisingly overcome by a method that involves the precipitation of a diastereomeric salt formed by one of the enantiomers (*S*)-(I) or (*R*)-(I) with an enantiopure or enantioenriched chiral 2-aryl-propionic acid containing at least 14 carbon atoms (Chap. 4) [13].

The chiral resolution of a mixture of the enantiomers (*S*)-(I) and (*R*)-(I) of the well-known PZQ precursor praziquanamine (I), as described in the invention, provides a simple and efficient procedure that can be integrated into existing processes for PZQ (Fig. 12.8). This allows the preparation of (*R*)-PZQ by an appropriate adaptation of established Shin-Poong or Merck processes, which are highly optimized on large scale for PZQ. The enantiopure or enantiomerically enriched (*R*)-praziquanamine obtained by this method can be used to produce enantiopure or enantiomerically enriched (*R*)-PZQ in large quantities. Additionally, the invention addresses the production of enantiopure or enantiomerically enriched *N*-substituted [(1*R*)-1,2,3,4-tetrahydro-1-isoquinolin-1-yl]methanamine (*N*-substituted (*R*)-praziquandiamine), comprising the following step: chiral resolution of the enantiomers (*S*)-(II) and (*R*)-(II) of *N*-substituted praziquandiamine via precipitation of a

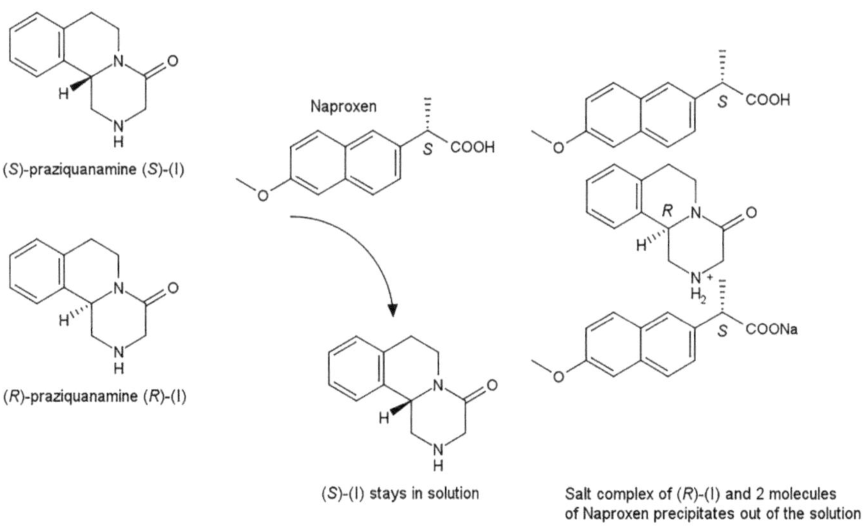

Fig. 12.8 (*S*)-Naproxen is added to the racemic praziquanamine allowing the (*R*)-(I) enantiomer to precipitate out of the solution. The salt is recovered by filtration

diastereomeric salt formed by one of the enantiomers (*S*)-(II) or (*R*)-(II) with an enantiopure or enantiomerically enriched 2-aryl-propionic acid containing at least 14 carbon atoms, wherein R2 denotes *tert*-butyloxycarbonyl, benzoyl, cyclohexanoyl, or acetyl, with acetyl (–(C=O)–CH_3) being the preferred group.

In this case the applicant, Merck Patent GmbH, filed the patent application as a PCT application before it was nationalized in multiple jurisdictions. *EP 3221296 B1* was the first filing, the so-called priority application, with a priority date of November 21, 2014. For this case a certain importance was anticipated, and the application was filed in many countries listed by their respective two-letter codes:

AR; AT; AU; BE; CA; CH; CS; DD; DE; DK; ES; FI; FR; GB; GR; GT; HK; HU; IE; IL IN; JP; KE; KR; LU; NL; NO; OA; PH; PL; RO; SE; SU; US; YU; and ZA [9]

EP 3221311 B1 (Fig. 12.9) covers another production method entitled "Method for the Production of Praziquantel." In more detail the present invention relates to a method for the racemization of enantiomerically pure or enantiomerically enriched PZQ under basic conditions and to a method for the production of PZQ in enantiopure or enantiomerically enriched form, using this racemization method. The patent application was initially filed by Merck Patent GmbH as a European priority filing with the European Patent Office on November 21, 2014, followed by a PCT application WO 2016/078758. For this case the application was nationalized in the following countries listed by their two-letter codes:

AU; BR; CA; CH; CN; DE; ES; FR; GB; IL; IN; IT; JP; US; and ZA [9]

This invention describes an efficient, reliable, and cost-effective method for the recycling of (*S*)-PZQ, which can be utilized in a method for the production of PZQ in enantiopure or enantiomerically enriched form. This has surprisingly been solved by a method for the racemization of enantiomerically pure or enantiomerically enriched PZQ. The racemization method according to the invention provides a simple and efficient recycling process and is an economical prerequisite for preparing enantiomerically pure or enantiomerically enriched PZQ (in particular, (*R*)-PZQ) by chiral separation of the racemate, which can be produced by known highly efficient processes). These existing processes, e.g., the original Merck processes, are well established and highly optimized even on a large scale [16].

EP 3344605 B1 covers another production method entitled "Method for the Production of Praziquantel and Precursors Thereof" in more detail because the title can be somewhat misleading as the patent focuses primarily on the "Enantioselective Synthesis of (*R*)-(−)-*N*-(1,2,3,4-tetrahydroisoquinolin-1-yl-methyl)-acetamide and Its Derivatives." This patent right is a granted patent, identified by the classification B1 following the patent number, as shown on the title page below. The patent was derived from PCT application, *WO 2017/036577*, which was based on a European priority filing from September 1, 2015.

(19) Europäisches Patentamt / European Patent Office / Office européen des brevets

(11) **EP 3 221 311 B1**

(12) **EUROPEAN PATENT SPECIFICATION**

(45) Date of publication and mention of the grant of the patent:
30.10.2019 Bulletin 2019/44

(21) Application number: **15807798.2**

(22) Date of filing: **17.11.2015**

(51) Int Cl.:
C07D 471/04 (2006.01)

(86) International application number:
PCT/EP2015/002296

(87) International publication number:
WO 2016/078758 (26.05.2016 Gazette 2016/21)

(54) **METHOD FOR THE PRODUCTION OF PRAZIQUANTEL**
VERFAHREN ZUR HERSTELLUNG VON PRAZIQUANTEL
PROCÉDÉ DE PRODUCTION DE PRAZIQUANTEL

(84) Designated Contracting States:
AL AT BE BG CH CY CZ DE DK EE ES FI FR GB GR HR HU IE IS IT LI LT LU LV MC MK MT NL NO PL PT RO RS SE SI SK SM TR

(30) Priority: **21.11.2014 EP 14003934**

(43) Date of publication of application:
27.09.2017 Bulletin 2017/39

(73) Proprietor: **Merck Patent GmbH**
64293 Darmstadt (DE)

(72) Inventors:
- **WAECHTLER, Andreas**
64295 Darmstadt (DE)
- **SALEH-KASSIM, Hadia**
57540 Petite Rosselle (FR)
- **JASPER, Christian**
63500 Seligenstadt (DE)
- **KOLB, Joern**
64291 Arheiligen (DE)
- **MAILLARD, David**
64285 Darmstadt (DE)

(56) References cited:
- **EBBERS E J ET AL: "Controlled Racemization of Optically Active Organic Compounds: Prospects for Asymmetric Transformation", TETRAHEDRON, ELSEVIER SCIENCE PUBLISHERS, AMSTERDAM, NL, vol. 53, no. 28, 14 July 1997 (1997-07-14) , pages 9417-9476, XP004105880, ISSN: 0040-4020, DOI: 10.1016/S0040-4020(97)00324-4**
- **MUNEER AHAMED ET AL: "The Outcome of the Oxidations of Unusual Enediamide Motifs Is Governed by the Stabilities of the Intermediate Iminium Ions", PLOS ONE, vol. 7, no. 10, 19 October 2012 (2012-10-19), page e47224, XP055246218, DOI: 10.1371/journal.pone.0047224 cited in the application**

EP 3 221 311 B1

Note: Within nine months of the publication of the mention of the grant of the European patent in the European Patent Bulletin, any person may give notice to the European Patent Office of opposition to that patent, in accordance with the Implementing Regulations. Notice of opposition shall not be deemed to have been filed until the opposition fee has been paid. (Art. 99(1) European Patent Convention).

Printed by Jouve, 75001 PARIS (FR)

Fig. 12.9 First page of *EP 3221311 B1* from an available patent database [1]

This invention also relates to an alternative method for producing precursors for the synthesis of enantiopure or enantioenriched PZQ. It differs from *EP 3221296 B1* and is considered as an alternative route of synthesis as well as an improvement. The object of the aforementioned invention was to provide an efficient method for the preparation of PZQ and analogues thereof that is suitable to provide enantiomerically enriched or enantiomerically pure (*R*)-PZQ or analogues thereof, and in particular a suitable stereoselective route that allows the preparation of enantiomerically enriched or even enantiomerically pure (*R*)-PZQ or analogues thereof without the need for chiral separation. Nonetheless, it was a further object of the present invention to provide an advantageous method for preparing desired enantiomers both in terms of intermediates and final product, including mixtures of enantiomers with a more favorable percentage of the desired enantiomer compared to the starting material. The applicant, Merck Patent GmbH, filed the patent application as a PCT application, *WO 2017/036577*. Before this PCT filing was nationalized in multiple jurisdictions, the European filing was the first, establishing the priority application with a priority date of September 1, 2015. For this case a reduced country list was chosen, and the application was nationalized in the following countries listed by their two-letter codes:

AU; BR; CA; CH; CN; DE; ES; FR; GB; IL; IN; IT; JP; and US [9].

This overview highlights the most relevant patent rights and the associated patent family portfolio for PZQ. The anthelmintic PZQ has been registered, approved, and commercialized as a racemate since the early 1980s. The entire development process is an elegant and not overly complex example, illustrating typical development cycles and the incremental improvements achieved over time. Certainly, the patent portfolio relating to PZQ, while not comprehensive, covers all major and relevant patent rights for the marketed PZQ. It also includes patents for the more active and less bitter-tasting (*R*)-PZQ, the manufacturing processes for an enantioselective synthesis as well as methods for separating both enantiomers [17].

In the following outlook, we will reflect on a few other issues that may come up during development cycles, highlighting all the progress and success but also mentioning possible pitfalls you might encounter.

Outlook

A meaningful patent portfolio around a product is very helpful and often facilitates your business considerably and to exploit it successfully. Owning or controlling certain patent rights does not necessarily mean that you are allowed to practice your invention and is not guaranteeing freedom to operate (FTO) since patent rights are only negative protective rights. In other words, they provide you with the right to stop others from copying, using, or practicing your invention (the "negative right"), but do not automatically allow the owner or the beneficiary to use, practice, and exploit the invention. There might be obstacles like blocking third-party rights or legal restrictions, either territorially or globally. This has to be proofed carefully to be able to respect third-party patent rights, in order to avoid patent infringement and potential costly litigation or violation of law.

If you identify third-party rights or they are quoted in search reports during the examination of your own patent application, it is advisable to carefully analyze them and check the register as well as the file history of the third-party patent rights. Remember that only a granted patent and a patent that is in force provides an enforceable right to the third party. Dependency is best checked by a thorough claim analysis of the relevant third-party patent rights in comparison to the claim set of your own invention. With regard to third-party patent rights, it is very important to know the status of the patent right, i.e., is it granted or pending, the exact expiry date, and if it is still valid and in force? Moreover, in which territory is the patent right effective, in certain countries only or nearly on a worldwide basis. Other important information is: who is the owner or beneficiary, are there already licenses granted or any licensees known already, or is any litigation or opposition pending or even decided? Very relevant is also an assessment to analyze if the pending claims have a realistic chance to become allowed and granted in the light of the legal framework, existing case law, and prior art or references cited. Will the claims hold during a court challenge?

However, there might be a situation where you have to decide how to treat third-party rights. The easiest way of avoiding any uncertainties is if your analysis did not reveal any dependencies, to ignore those rights and not doing anything. However, the owner may see it differently and can take action. Therefore, it is always good to have an independent, external expert opinion from a qualified patent attorney. Sometimes acquiring a license from the patent owner is not too expensive and possible to secure your development. This requires the closing of a contract and negotiations of various aspects subject to the agreement and of course skills and experience to successfully conclude it. Finally, if you have identified weaknesses of the third-party patent right, you may seek options to challenge it. However, such a patent challenge has to happen on a country-by-country basis or at least on a regional level. It requires legal representation in every jurisdiction and is often quite costly, time-consuming, and the outcome is often difficult to predict.

Such a strategy for litigation or opposition proceedings requires a lot of detailed knowledge and cannot be covered here in more detail.

Companies usually have a Code of Conduct stating that they do respect intellectual property and copyrights of other parties. They state often that they will respect the proprietary and confidential information of others, just as it protects its own. This includes written materials, software, patents, trademarks, and other intellectual property.

This should also be mandatory for individuals to follow basic standards and to respect the patent and trademark rights of others and not to use such rights without authorization of the owner or the beneficiary.

Competing Interests A.H. is an employee of Merck Healthcare KGaA.

References

1. Copyright alliance (2024) https://copyrightalliance.org/faqs/difference-copyright-patent-trademark/. Accessed 20 July 2024
2. Details can be found on the websites of the patent offices and may differ slightly from country to country. Here are the requirements for a European patent application: https://www.epo.org/en/legal/epc/2020/a90.html, which are complete and fulfill also the requirements in many other jurisdictions. Accessed 20 July 2024
3. Maillard D, Waechtler A, Maurin J, Wakaresko E, Jasper C (2016) Method for the production of praziquantel and precursors thereof. WO 2016/078765 A1, 26 May 2016
4. World Intellectual Property Organization (2024) https://www.wipo.int/pct/en/pct_contracting_states.html. Accessed 20 July 2024
5. World Health Organization (2024) https://www.who.int/teams/health-product-and-policy-standards/inn/guidance-on-inn. Accessed 20 July 2024
6. World Health Organization (2018) The use of stems in the selection of International Nonproprietary Names (INN) for pharmaceutical substances 2018 (StemBook2018) WHO. Available via DIALOG. https://cdn.who.int/media/docs/default-source/international-nonproprietary-names-(inn)/stembook-2018.pdf. Accessed 20 July 2024
7. It was published as a proposed INN in the Supplement to WHO Chronicle, 1975, Vol. 29, No. 9. Of note, one enantiomer i.e. (R)-praziquantel has also its own INN, Arpraziquantel. "International nonproprietary names for pharmaceutical substances (INN): recommended INN: list 83". WHO Drug Information. 34 (1). hdl:10665/339768
8. Maillard D, (1986) Method for the production of praziquantel and precursors thereof. DE23625390, 7 May 1986
9. European Patent Register (2024) https://register.epo.org/help?lng=en&topic=countrycodes. Accessed 20 July 2024
10. Seubert J, Herbert T, Andrews T (1977) 2-acyl-4-oxo-pyrazino-isoquinoline derivatives and process for the preparation thereof. US Patent 4,001,411, 4 Jan 1977
11. Andrews P, Thomas H, Pohlke R, Seubert J (2024) Praziquantel. Med Res Rev 3(2):147–200. https://doi.org/10.1002/med.2610030204
12. Meyer T, Sekljic H, Fuchs S, Bothe H, Schollmeyer D, Miculka C (2009) Taste, a new incentive to switch to (R)-praziquantel in schistosomiasis treatment. PLoS Negl Trop Dis 3:e357. https://doi.org/10.1371/journal.pntd.0000357
13. Maillard D, Waechtler A, Maurin J, Wakaresko E, Jasper C (2018) Method for the production of praziquantel and precursors thereof. EP 3 221 296 B1, 17 Oct 2018
14. Woelfle M, Seerden J-P, de Gooijer J, Pouwer K, Olliaro P, Todd MH (2011) Resolution of praziquantel. PLoS Negl Trop Dis 5:e1260. https://doi.org/10.1371/journal.pntd.0001260
15. Cedillo-Cruz A, Aguilar MI, Flores-Alamo M, Palomares-Alonso F, Jung-Cook H (2014) A straightforward and efficient synthesis of praziquantel enantiomers and their 4′-hydroxy derivatives. Tetrahedron Asymmetry 25(2):133–140. https://doi.org/10.1016/j.tetasy.2013.11.004
16. Waechtler A, Saleh-Kassim H, Jasper C, Kolb J, Maillard D (2019) Method for the production of praziquantel. EP 3 221 311 B1, 30 Oct 2019
17. Eberhardt L, Waechtler A, Lehmann S (2020) Method for the production of praziquantel and precursors thereof. EP 3 344 605 B1, 13 May 2020

MIX
Papier aus verantwortungsvollen Quellen
Paper from responsible sources
FSC® C105338

If you have any concerns about our products,
you can contact us on
ProductSafety@springernature.com

In case Publisher is established outside the EU,
the EU authorized representative is:
Springer Nature Customer Service Center GmbH
Europaplatz 3, 69115 Heidelberg, Germany

Printed by Libri Plureos GmbH
in Hamburg, Germany